ORGANIC SPINTRONICS

Edited by
Zeev Valy Vardeny

CRC Press
Taylor & Francis Group
Boca Raton London New York

CRC Press is an imprint of the
Taylor & Francis Group, an **informa** business

CRC Press
Taylor & Francis Group
6000 Broken Sound Parkway NW, Suite 300
Boca Raton, FL 33487-2742

First issued in paperback 2017

© 2010 by Taylor and Francis Group, LLC
CRC Press is an imprint of Taylor & Francis Group, an Informa business

No claim to original U.S. Government works

ISBN 13: 978-1-138-11242-1 (pbk)
ISBN 13: 978-1-4398-0656-2 (hbk)

Library of Congress Cataloging-in-Publication Data

Organic spintronics / editor, Zeev Valy Vardeny.
    p. cm.
    "A CRC title."
    Includes bibliographical references and index.
    ISBN 978-1-4398-0656-2 (hard back : alk. paper)
    1. Spintronics. 2. Organic electronics. I. Vardeny, Z. V.

TK7874.887.O74 2010
621.381--dc22   2009049450

Visit the Taylor & Francis Web site at
http://www.taylorandfrancis.com

and the CRC Press Web site at
http://www.crcpress.com

Contents

Preface

The field of organic electronics has progressed enormously in recent years as a result of worldwide activity in many research groups around the world. Advances have been made in the fields of both device science and fabrication, as well as in the underlying chemistry, physics, and materials science. The impact of this field continues to influence many adjacent disciplines, such as nanotechnology, sensors, and photonics. The advances in organic electronics have generated a vital and growing interest in organic materials basic research, and could potentially revolutionize future electronic applications. It is expected that the present worldwide funding in this field would stimulate a major research and development effort in organic materials research for lighting, photovoltaic, and other optoelectronic applications.

The growth of organic electronics has indeed been impressive. The first commercial products, started two decades ago, were based on conducting polymer films, a business now with annual sales in the billion-dollar range. Organic light-emitting diodes (OLEDs) and displays based on OLEDs were introduced to the scientific community about two decades ago, and to the market about ten years; a large expansion in market penetration has been forecasted for the next decade. A bright future awaits organic white light-emitting diodes, which are viable to replace the Edison-type bulb. In addition, thin-film transistor-based circuits, and electronic circuits incorporating several hundred devices on flexible substrates have been recently demonstrated. Organic photodiodes have been fabricated with quantum efficiencies in excess of 50%, and organic solar cells with certified power conversion efficiencies above 6% have been reported. Also, laser action in organics was introduced in 1996. The initial enthusiasm has given way to more realistic expectations on the fabrication of current-injected organic laser action. There continue to be advances made in the synthesis of new compounds and in improved synthetic procedures of important materials.

Until a few years ago, the *electron spin* was ignored in the field of organic electronics. However, in 2002, as substantive magnetoresistance in a two-terminal organic device was obtained at room temperature, a new field was born: *organic spintronics*. The technology of spintronics (or spin-based electronics), where the electron spin is used as the information carrier in addition to the charge, offers opportunities for a new generation of electronic devices that combines standard microelectronics with spin-dependent effects that arise from the interaction between the carrier spin and externally applied magnetic fields. Adding the spin degree of freedom to the more conventional charge-based electronics substantially increases the functionality and performance of electronic devices. The advantages of these new devices may be increased data processing speed, decreased electric power consumption, and increased integration densities. In addition, spintronic devices may also serve as the writing/reading active element in magnetic memory systems.

The discovery of the giant magnetoresistance in 1988 is considered to be the beginning of the new generation of spin-based electronics; this discovery has led to the 2007 Nobel Prize in Physics. As a result, the role of electron spin in solid-state devices and possible technology that specifically exploits spin in addition to charge properties have been studied extensively. The transition from discovery to commercialization has happened very fast for spintronics. For example, the application of giant magnetoresistance and tunneling magnetoresistance to magnetic information storage occurred within a period of six years following their discovery, and a "read-head" device for magnetic hard disk drives based on spintronics was announced in 1997 by IBM.

Until the 2002 discovery of magnetoresistance in organic devices, only slight attention was paid to the use of organic semiconductors, such as small molecules and π-conjugated polymers, as spin transporting materials. The conductive properties of the π-conjugated polymers were discovered in the late seventies, and this gave birth to organic, or plastic, electronics. Organic semiconductors are mainly composed of light atom elements such as carbon and hydrogen, which leads to large spin diffusion length due to very weak spin orbit coupling, and moderate hyperfine interaction. These properties hold promise for the field of organic spintronics, and thus there is no wonder that this field has flourished during the last few years. A series of international meetings on organic spintronics, dubbed "Spins in Organic Semiconductors (SPINOS)," has been established, with ~100 delegates from all over; the third meeting in the series is planned to be held in September 2010 in Amsterdam.

This book is an updated summary of the experimental and theoretical aspects of organic spintronics. A substantial advance has recently occurred in the field of organic magnetotransport, mainly in three directions: spintronic devices, such as *organic spin valves*, where spin injection, transport, and manipulation have been demonstrated; *magnetic field effects* in OLEDs, where both conductivity and electroluminescence have been shown to strongly depend on magnetic field; and optically and *electrically detected magnetic resonance* effects, where coherent spin control has been obtained in organic electronic devices. Therefore, the book is organized to reflect these three avenues. Chapters 1 to 2 are mainly focused on spin injection and manipulation in organic spin valves. Chapter 3 introduces the magnetic field effect in OLEDs, and Chapters 4 and 5 further discuss this spin transport effect in relation with spin manipulation, which is the focus of the first two chapters. Chapter 6 is devoted to the coherent control of spins in organic devices using the technique of electrically detected magnetic resonance, and Chapter 7 summarizes the possibility of using organic spin valves as sensors. In addition, organic magnets as spin injection electrodes in organic spintronics devices are thoroughly discussed in Chapter 4. Both experiment and theory are well represented in this book. Whereas Chapter 2 and some parts of Chapter 3 are devoted to the theoretical aspects of spin injection, transport, and detection in organic spin valves, Chapters 3 to 5 discuss in detail the underlying mechanism of the magnetoresistance and magnetoelecroluminescence in OLEDs.

Upon the completion of this book, I would like thank the various contributors for their diligent and thorough efforts that made it possible to complete this project on time. I am also grateful to my wife, Nira Vardeny, for the support she gave me during the time period that I was busy planning and preparing the book, as well as writing the chapter that I am involved with (Chapter 5)—the relief time that I needed when not working on the book. I am also grateful to the authorities of the Physics Department at the University of Utah for the partial relief they gave me in teaching during the spring 2008–2009 semester—a time period that was crucial for the writing endeavor and advance in the subject matter. I am also grateful to my friend and collaborator over the years, Prof. Eitan Ehrenfreund, for his contribution to our collaborative efforts in the field of organic spintronics, and especially for his friendly hospitality during my sabbatical at the Technion.

Finally, I also thank the DOE and NSF funding agencies for providing the financial support needed to complete this book during the 2009 spring and summer semesters, and the Lady Davis Foundation for the stipend in support of my stay at the Technion in spring 2009, during which a major part of my contribution to this book was formulated.

Salt Lake City, Utah

The Author

Z. Valy Vardeny is distinguished professor of physics at the University of Utah. He received his BS (1969) and PhD (1979) in physics from the Technion, Haifa, Israel. He received the Alon Price in 1982, the University of Utah Research award in 1996, the Willard Award of Art and Science in 1997, the Lady Davis Professorship at the Technion in 2000 and again in 2005 and 2009, the Utah Governor's Medal of Science and Technology in 2005, the 2008 Frank Isakson APS Prize for Optical Effects in Solids, and the University of Utah Rosenblatt Award for Excellence in 2009. He is a fellow of the American Physical Society (1996). He is regional editor of the *Journal of Synthethic Metals* and consultant with two major corporations: Cambridge Display Technologies, and Plextronics. Vardeny has published more than 480 peer-reviewed research articles, is the editor of 4 books, and holds 12 patents and provisional patents.

His research interests include optical, electrical, and magnetic properties of organic semiconductors; fabrication of organic optoelectronic organic light emitting diodes (OLEDs) and solar cells and spintronics (spin valve) devices; optically detected magnetic resonance, laser action, nonlinear optical spectroscopy, and ultrafast transient spectroscopy from the THz to UV of organic semiconductors, amorphous semiconductors, nanotubes, fullerenes, and graphite; fabrication and properties of 3D dielectric photonic crystals; metallodielectric and metallic photonic crystals; 2D plasmonic lattices, quasicrystals, fractals, and other nonperiodic structures of hole arrays in metallic films; magnetoconductivity and magnetoelectroluminescence in OLEDs; and other two-terminal devices.

The Contributors

F. L. Bloom
Department of Applied Physics
Technische Universiteit Eindhoven
Eindhoven, The Netherlands

P. A. Bobbert
Department of Applied Physics
Technische Universiteit Eindhoven
Eindhoven, The Netherlands

Christoph Boehme
Department of Physics and Astronomy
University of Utah
Salt Lake City, Utah

Valentin Dediu
Institute for the Study of Nanostructured
    Materials
National Research Council
Bologna, Italy

E. Ehrenfreund
Physics Department
Technion–Israel Institute of Technology
Haifa, Israel

A. J. Epstein
Departments of Physics and Chemistry
Ohio State University
Columbus, Ohio

B. Koopmans
Department of Applied Physics
Technische Universiteit Eindhoven
Eindhoven, The Netherlands

Dane R. McCamey
Department of Physics and Astronomy
University of Utah
Salt Lake City, Utah

Jagadeesh S. Moodera
Francis Bitter Magnet Laboratory
Massachusetts Institute of Technology
Cambridge, Massachusetts

Mirko Prezioso
Institute for the Study of Nanostructured
    Materials
National Research Council
Bologna, Italy

V. N. Prigodin
Department of Physics
Ohio State University
Columbus, Ohio
and
Ioffe Institute
St. Petersburg, Russia

Karthik V. Raman
Francis Bitter Magnet Laboratory
Massachusetts Institute of Technology
Cambridge, Massachusetts

Alberto Riminucci
Institute for the Study of Nanostructured
    Materials
National Research Council
Bologna, Italy

P. P. Ruden
University of Minnesota
Minneapolis, Minnesota

Tiffany S. Santos
Center for Nanoscale Materials
Argonne National Laboratory
Argonne, Illinois

D. L. Smith
Los Alamos National Laboratory
Los Alamos, New Mexico

Z. V. Vardeny
Physics Department
University of Utah
Salt Lake City, Utah

M. Wohlgenannt
Department of Physics and Astronomy
Optical Science and Technology Center
University of Iowa
Iowa City, Iowa

Jung-Woo Yoo
Departments of Physics and Chemistry
Ohio State University
Columbus, Ohio

1 Spin-Polarized Transport in Organic Semiconductors

Jagadeesh S. Moodera, Tiffany S. Santos, and Karthik V. Raman

CONTENTS

ABSTRACT

Organic spintronics is a hybrid of two hot fields: organic electronics and spintronics. The excitement in this field of spin transport in organics, mainly organic semiconductors, has evolved rapidly in the last five years. Combined with the novelty and the expectation of a large travel length in organic compounds for spins without being perturbed, the field is marching on. With the possibility of creating unique molecular systems from a bottom-up approach, the field has opened up vast opportunities for discovering newer, fundamental phenomena. This is bound to lead toward technological breakthroughs. In this chapter we give the reader an overview of the activities so far in this very young field, mainly from the perspective of spin transport in spin valve structure and tunnel junctions. It is clear that this highly interdisciplinary topic holds much promise, encompassing areas of physics, chemistry, and materials science. We have attempted to highlight the challenges that need to be overcome as well as where activity is needed, hoping to create intense activity in this field to arrive at the anticipated properties.

1.1 INTRODUCTION

The vast area of organic semiconductors (OSs), among literally millions of organic compounds, has seen considerable activity in the past couple of decades aimed toward exploring organic electronics from the fundamental physics point of view as well as with the promise of developing cheaper and flexible devices, such as organic light-emitting diodes (OLEDs) and organic field effect transistors (OFETs).[1-5] One of the great incentives in this field is the ability to chemically tune the properties at the molecular level for a bottom-up approach, which could be expected to lead the future science and technology, as is happening today in biological fields. Commensurate with the vastness of the field, although the parameter space to search and discover is undoubtedly daunting, the prospectus is exciting.

Organic electronics has been emerging in the last few decades as a new field with many unexplored phenomena,[5-7] whereas the even more complicated phenomenon of organic spintronics is already on the horizon. This nascent field of organic spintronics is a hybrid of the two hot fields organic electronics and spintronics.[8,9] From this viewpoint, of growing interest is the potential to transport and manipulate spin information in OS. Spin-orbit and hyperfine interactions are two of the main causes for spins to lose their orientation. Given that OSs are composed mostly of light elements (i.e., C, H, N, O), they have a weaker spin-orbit interaction than inorganic semiconductors. Thus, a relatively longer spin lifetime expected in OS promises a technological breakthrough in organic spintronics. In fact, the weak hyperfine interactions are found to dominate the spin scattering processes in some cases.[10] Certainly, these weak mechanisms suggest longer timescales for spins to relax or lose their phase coherence.[11] However, currently most OSs show a substantially lower mobility in the range of 10^{-8} to 10^{-2} cm^2/Vs, compared to standard, inorganic semiconductors Si and GaAs (hole mobility ≥ 400 cm^2/Vs and electron mobility $\geq 1,500$ cm^2/Vs), limiting coherent transport lengths. Despite this fact, relatively longer spin lifetimes are reported experimentally.[12-14]

While many OS materials are exploited for their tunability of charge carrier transport properties, their spin transport study is the least explored area. This can be attributed to various factors. Even the simplest OSs, oligomers such as pentacene ($C_{22}H_{14}$) and rubrene ($C_{42}H_{28}$), are very large molecules consisting of many atoms (Figure 1.1) compared to Si, Ge, or GaAs. Their inherently complicated structure poses bigger challenges to obtaining "atomically" sharp interfaces, for example, with a metal, not to mention the all-important interfacial chemistry and charge transfer.[15,16] The latter issue in its entire resplendency should inevitably affect the interfacial magnetism—crucial for organic spintronics. As shown schematically in Figure 1.2, this would be

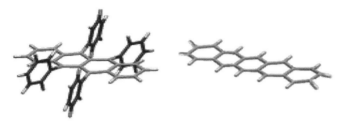

FIGURE 1.1 Molecular structure of organic semiconductors (a) rubrene ($C_{42}H_{28}$) and (b) pentacene ($C_{22}H_{14}$). (Taken from D. Käfer, G. Witte, *Phys. Chem. Chem. Phys.* 7, 2850 [2005].)

strongly dictated by the actual physical structure of the interface, which in itself is becoming a huge branch of research. Along these lines, what may eventually evolve is the ability to create an OS that is also magnetic at higher temperatures (such as V[TCNE]$_x$ discovered by Epstein's group[17,18]) and may seamlessly be grown over some high-mobility OS layer with reduced interfacial defect formation for efficient spin injection in an all-organic structure.

Despite the simplicity of creating the layers of OS, say by either physical vapor deposition or spin coating, handling them becomes nontrivial because of their soft and fragile nature. Standard semiconductor processing to make nanoscale devices is incompatible, as most of the organic solvents dramatically attack the OS layers. Novel techniques such as soft lithography have been developed with some degree of success for nanoscale device fabrication.[19] Maintaining a clean surface thus becomes challenging and complicates the investigation of the intrinsic behavior, especially when using bulk single crystals and *ex situ* processing is involved. Uncontrolled and unknown interfacial conditions can lead to huge contact resistance, irreproducibility, and artifacts in the measurements. These limitations magnify when dealing with spin injection and transport, due to the fact length scales below a nanometer become important. In some cases, what is called the flip method is adopted, wherein the OS crystal is placed on pre-prepared thin-film electrodes, such as Au, for four terminal

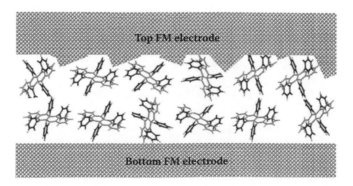

FIGURE 1.2 Growth of a ferromagnetic electrode on top of the organic layer leads to rougher interfaces compared to the bottom one. The interfacial magnetic properties are hence expected to be different in the two cases.

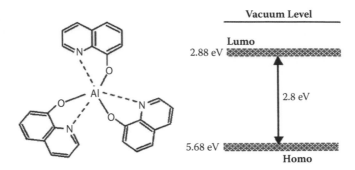

FIGURE 1.3 Energy-level diagram of organic semiconductor Alq_3. The HOMO-LUMO gap is ~2.8 eV.

charge transport measurement. The flip method minimizes the processing of single crystals, thereby reducing the risk of damaging the surface.

Not all is lost. With all of the above limitations in exploring the OS, very interestingly, the field of organic spintronics has taken off in the last few years. Many groups worldwide have begun serious spin transport studies in OS showing considerable success and promise. This is bound to open up hitherto unforeseen phenomena in this open and high-potential field. Among myriad possible OSs, some of the popular ones for spin transport studies are Alq_3, pentacene, rubrene, CuPc, and so on, because they are already being explored for their charge transport potential as well as electroluminescent capability[1,20–22] in organic electronics such as OFETs and OLEDs. For example, the organic π-conjugated molecular semiconductor Alq_3 ($C_{27}H_{18}N_3O_3Al$) is the most widely used electron-transporting and light-emitting material in OLEDs. Alq_3 has been extensively studied for this application since it displayed high electroluminescence (EL) efficiency nearly two decades ago.[1] A band gap of ~2.8 eV separates the highest occupied molecular orbital (HOMO) and the lowest unoccupied molecular orbital (LUMO), shown in Figure 1.3. Typically, the film thickness of the Alq_3 layers in OLED structures is hundreds of nanometers, whereas for spin transport studies it is in the range of tens to a few hundreds of nanometers.

1.2 SPIN TRANSPORT IN OS USING SANDWICH STRUCTURES AND MATRICES

This section presents an overview of work on organic spin valves and magnetic tunnel junctions. Both types of organic spintronic devices comprise an organic spacer layer between two ferromagnetic electrodes, with the main distinction being the *effective* thickness of the organic spacer, which determines the carrier transport mechanism through the OS: by hopping conduction or by tunneling, which can be very different. In general, the probability of electron tunneling from one electrode to the other scales exponentially with thickness and is most effective for spacer layers that are very thin, less than ~5 nm. The organic layer acts as a tunnel barrier, and the tunneling electrons are not injected into the LUMO level. We apply the term *spin valve* to those having a spacer layer in the thick limit, tens of nanometers. Transport of carriers (electrons or holes) in a spin valve begins by injection via tunneling from the magnetic electrode into either the narrow HOMO or LUMO "band" of the organic layer. Conduction across the thick organic layer proceeds by hopping of electrons in the LUMO (or holes in the HOMO), followed by tunneling into the magnetic counterelectrode.[23] At intermediate thicknesses, too thick for single-step tunneling, multistep tunneling can occur, via localized states in the barrier. Although spin information can be lost to some degree during multistep tunneling, it is not lost completely. We describe in Section 1.5 a study of spin polarization as a function of organic layer thickness in magnetic tunnel junctions.

Even though organic electronic devices such as OLEDs and organic thin-film transistors are well on their way toward application,[4] transport in OS is still not well understood. The organic layers in the devices described here are amorphous, or polycrystalline at best. Conduction through this disordered structure occurs by hopping via localized states, and there is a high concentration of charge traps at the grain boundaries, impurities, etc. It is not band conduction via delocalized states as for inorganic semiconductors. For this reason, the mobility in organic semiconductors is much smaller than in inorganic semiconductors by several orders of magnitude. The mobility of OS can improve greatly by having a highly ordered structure. With a more regular arrangement of molecules, there is a greater degree of overlap of the molecular orbitals, so that transport of carriers from one molecule to the next is easier. It is even possible to have band-like conduction in single crystals of OS at low temperature. We are only just beginning to explore how *spin* survives during transport in organic semiconductors.

In several studies of vertical spin valves, a Co top electrode was used, and a common observation was that for organic layers < 100 nm, the device was shorted. This was attributed to penetration of Co atoms to a depth of up to 100 nm into the organic layer, causing a conductive pathway to the bottom electrode.[12] Diffusion of the top electrode or high roughness of the organic layer leads to a much thinner effective spacer layer than the nominal thickness. For this reason, spin transport in spin valves with organic spacer layers having a thickness well above the tunneling limit has been attributed to tunneling through these thinner regions, instead of hopping conduction through the HOMO or LUMO channels. For example, in the study by Xu et al.,[24] in which −15% MR was measured in spin valves with tetraphenyl porphyrin (TPP) spacer layers, the fit to the I-V curve using the Simmons model[25] yielded an effective tunnel barrier thickness of just ~2 nm for TPP that was nominally 15 nm thick. They concluded that the transport was dominated by tunneling. The same conclusion was reached by Jiang et al.[26] in their study of $Fe/Alq_3/Co$ spin valves in the thick limit where no MR was observed. In addition, interdiffusion and chemical reactions at the electrode-organic interface introduce states into the gap of the organic,[27] which are sites for spin scattering. The Moodera group has successfully prepared uniform, ultrathin organic layers as tunnel barriers, with negligible intermixing with the magnetic electrodes (see Figure 1.4), enabling a systematic study of tunneling transport through the OS,[14,28] described in Section 1.4.

The first report of spin transport in a spin valve structure was in 2002 by Dediu et al.,[29] using a lateral structure having two $La_{0.7}Sr_{0.3}MnO_3$ (LSMO) electrodes with

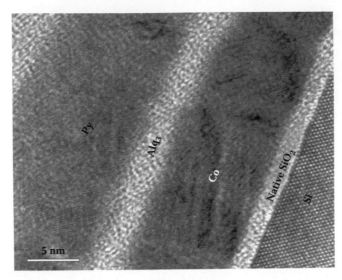

FIGURE 1.4 Cross-sectional high-resolution transmission electron micrograph of a magnetic tunnel junction, showing the continuous Alq$_3$ barrier. (Taken from T. S. Santos, J. S. Lee, P. Migdal, I. C. Lekshmi, B. Satpati, J. S. Moodera, *Phys. Rev. Lett.* 98, 016601 [2007].)

sexithienyl (T$_6$) in between. They observed MR at room temperature for T$_6$ channels 100 to 200 nm thick and estimated a spin diffusion length of 200 nm in the organic layer. The MR was measured not by independently switching the magnetization of the two LSMO electrodes, but by measuring the resistance change between zero field and 3.4 kOe, corresponding to an initial, randomly magnetized state and parallel alignment of the manganite electrodes, respectively. A couple of years later, Xiong et al.[12] were the first to report MR in a vertical LSMO/Alq$_3$/Co spin valve with Alq$_3$ layers 130 to 260 nm thick, measuring MR = −40% at 11 K, which reduced to zero by $T > 200$ K. The same group later performed a detailed study[30] using various organic materials between LSMO and metallic electrodes. They found that at high field, at which the two magnetic electrodes have already reached parallel alignment, the device resistance steadily decreased with increasing field, similar to the earlier MR effect observed by Dediu et al.[29] They attributed this high-field MR to a field-dependent shift in the Fermi level of the LSMO electrode, leading to an enhanced carrier injection across the LSMO-organic interface, an effect that originates from the LSMO and is independent of temperature and the organic spacer material.

Most of the organic spin valve structures investigated use LSMO as the bottom electrode. The main advantages of LSMO are its chemically stable surface and that it is half metallic, with a spin polarization > 90% shown experimentally at low temperature in magnetic tunnel junctions with inorganic barriers.[31,32] Thus, one has a better chance of obtaining at least one good interface and measuring high MR values. The main disadvantage to using LSMO is that even though it has a Curie temperature (T_C) above room temperature (370 K for bulk, but lower for thin films, depending on thickness, composition, and quality), the spin polarization at the LSMO surface (most relevant for spin injection[33]) decreases well below room temperature, so that highly reduced or zero MR effect is observed at room temperature. This is the case for both organic and inorganic spacer layers. Wang et al.[34] performed a study to address the question of whether the organic layer or the LSMO electrode was dominating the temperature dependence of the MR and lack of MR at room temperature. They concluded that the LSMO surface was the culprit and some other high-polarization, high-T_C ferromagnetic electrode must be used in order to obtain room temperature MR. Nevertheless, Dediu et al.,[35] taking care to prevent degradation of the magnetic properties of the LSMO surface, realized a finite MR in a vertical spin valve using LSMO. They reported MR = −0.15% at room temperature in an LSMO/Alq$_3$/Al$_2$O$_3$/ Co spin valve and >10% at 20 K for 100-nm thick Alq$_3$. The temperature dependence of MR scaled with the surface magnetization of the manganite,[33] reaching zero at T_C (= 325 K) and showing that the spin transport through the spin valve is indeed limited by the LSMO electrode.

These authors also improved the Co/Alq$_3$ interface by inserting an ultrathin Al$_2$O$_3$ tunnel barrier,[36] finding that it prevents penetration and chemical reaction[37] of the Co atoms into the Alq$_3$. This was the trick previously shown by Santos et al.[28] in magnetic tunnel junctions, in which an Al$_2$O$_3$ barrier inserted at the interface between Co and Alq$_3$ was found to increase the spin polarization, likely due to the minimization of trap states inherent to the metal/Alq$_3$ interface (see Section 1.4). Usage of an ultrathin, inorganic tunnel barrier, such as Al$_2$O$_3$ or LiF, is common in organic light-emitting diodes (OLEDs), where it is known to reduce the barrier for carrier injection across the cathode-organic interface and to reduce formation of trap states.[38,39] Now, it has been observed to improve spin conservation in organic spintronics devices as well. In the case of rubrene barriers, the Al$_2$O$_3$ barrier is seen to affect the growth morphology of the organic layer,[14] which in turn has a significant effect on the spin transport[40] (see Section 1.4).

There has been a report[13] of spin transport in a Ni/Alq$_3$/Co nanowire spin valve device, where a long spin relaxation time in the range of a few milliseconds to >1 s was estimated. The nanowires synthesized in a porous alumina matrix had a diameter of 50 nm and an amorphous Alq$_3$ thickness of ~30 nm. This estimate of extremely long relaxation time has some uncertainties. For one, the spin polarization (P) values used for the Ni and Co, 33% and 42%, respectively, are for a clean interface, and thus probably too high for their sample. The Ni electrode was deposited inside the porous alumina by chemical electrodeposition and can be expected to have a P value much less than 33% since P_{Ni} is extremely sensitive to any interface contamination.[41,42] Another point of uncertainty is that the estimate of the spin relaxation time depends on the drift mobility of the carriers in the Alq$_3$, which itself depends greatly on the quality of the layer, the deposition rate, and the applied electric field. The spin relaxation time, τ_s, was determined from the relation $\tau_s = e\lambda_s/kT\mu$, where λ_s is the spin diffusion length found using a variation of the Julliere formula (depends on P). The mobility values (μ) used were in the range of 10^{-8} to 10^{-10} cm^2/Vs, which are several orders of magnitude lower than the drift mobility measured in amorphous Alq$_3$ thin films, found in literature[43,44] to be in the range of 10^{-4} to 10^{-6} cm^2/Vs.

Motivated by the first result by Xiong et al.,[12] most studies on organic spin valves or tunnel junctions have utilized Alq$_3$ as the spacer layer. The Alq$_3$ layer is thermally evaporated in vacuum and can be deposited *in situ* with the rest of the layer stack, though this option is not available for all. Film deposition by thermal evaporation is the usual method for small molecules. A few other organic materials have been tried, by thermal evaporation[14,24,30,34] as well as by some other means. In particular, Majumdar et al.[45] reported a small room temperature MR in a spin value using the conjugated polymer regioregular (poly 3-hexylthiophene) (RRP3HT) as the spacer layer between LSMO and Co electrodes. The RRP3HT layer, ~100 nm thick, was spin coated from solution onto the LSMO and then subsequently annealed. Hopping transport of holes through the HOMO is expected for RRP3HT, but it is possible that the MR effect they observed is due to tunneling, as the Co may have penetrated the soft polymer layer. Using the Langmuir-Blodgett technique, Wang et al.[46] prepared a few monolayers of molecular pyrrole derivative 3-hexadecyl pyrrole, made *ex situ* in between sputtered CoFeB electrodes. This structure showed some MR effect along with telegraph noise, indicating that the spacer was highly defective with many localized states. Petta et al.[47] fabricated nanoscale Ni/octanethiol/Ni tunnel junctions, in which a self-assembled monolayer of molecular octanethiol was formed on the Ni when immersed in solution. The strong temperature and bias dependencies of the MR and the presence of telegraph noise again indicated that the barrier was full of defects.

Rubrene has emerged as a promising candidate material for coherent spin transport. Relative to other OSs, single-crystalline rubrene has a very high hole mobility[22] (= 20 cm²/Vs), making it of interest for application in organic field effect transistors and OLEDs. This high mobility implies that rubrene has a low density of trap states, which is favorable when trying to conserve spin. In one of the early works of spin transport through rubrene, Kusai et al.[48] studied a lateral structure, comprised of a rubrene-Co nanocomposite between nonmagnetic Cr/Au electrodes, shown in Figure 1.5. The nanocomposite was prepared by coevaporation of rubrene and Co, to create Co nanoparticles dispersed in the rubrene matrix. They measured MR = 78% at 4.2 K, where MR was defined as $(\rho_{max} - \rho_{min})/\rho_{min}$, where ρ_{max} and ρ_{min} are the maximum and minimum resistivities, respectively. This is significantly larger than typical values in organic magnetoresistance (OMAR) devices, having a thick organic layer between two nonmagnetic electrodes. They concluded that the MR effect

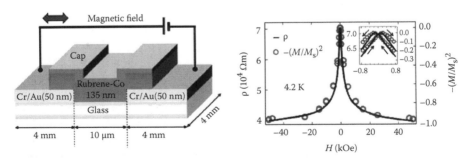

FIGURE 1.5 Lateral device structure of a Co-rubrene nanocomposite between two nonmagnetic electrodes, used in the study of Kusai et al.[48] On the right is the observed MR seen at low temperatures for such a structure. (Taken from H. Kusai, S. Miwa, M. Mizuguchi, T. Shinjo, Y. Suzuki, M. Shiraishi, *Chem. Phys. Lett.* 448, 106 [2007].)

originates from spin-dependent tunneling between the Co nanoparticles through the rubrene molecules. A modified Julliere formula was used to describe the effect: $MR = P^2$, where P is the spin polarization of Co. However, this relation greatly underestimates the observed MR value, and they suggested that a coulomb blockade in the Co nanoparticles or an enhancement of spin polarization at the Co-rubrene interface may play a role in creating this large MR effect. Very recently, Hatanaka et al.[49] reported a large MR in rubrene-Co nanocomposite spin devices. The large MR appeared within the coulomb gap at low temperatures. They point out that the higher-order cotunneling effects induced the enhancement in MR.

One of the not-so-well-understood topics in magnetic tunnel junctions (MTJs) is the observation of inverse magnetoresistance, whether the barrier is $SrTiO_3$ or OS. In the previous section, the work on Co/Alq$_3$/LSMO spin valve structures had showed inverse MR. However, in the work by the Moodera group[14,28] with tunnel junction structures, with and without Al_2O_3 at the Al-organic interface, positive values of P for Co, Fe, and Py electrodes were measured, in agreement with the positive tunneling magnetoresistance (TMR) measured in their MTJs. This contrasting experimental observation has not been fully understood yet. Several possible mechanisms could be at play, keeping in mind the different conduction mechanisms responsible for the GMR and TMR effects.[50] The tunneling behavior shows the expected one, when the positively spin-polarized, itinerant sp electrons dominate spin transport,[51,52] as is the case with amorphous tunnel barriers of Al_2O_3 and $SrTiO_3$.[53] However, inverse TMR seen by de Teresa et al.[54] in epitaxial LSMO/SrTiO$_3$/Co junctions was explained by tunneling of negatively spin-polarized d electrons, brought about by the good matching of the d bands in their epitaxial system. Xiong et al.[12] suggested this explanation as a possibility even in their amorphous Alq$_3$ system. The real answer may lie in the fact that in both experiments LSMO was the counterelectrode. If this is the case, then the d band conduction of LSMO (acting as a spin detector) may define the magnitude and sign of spin polarization. It may be noted here that work done in our group (unpublished) shows positive spin polarization for Co with the $SrTiO_3$ barrier measured with an Al (superconducting) spin detector. Overall, this is one of the fundamental issues in this field that remains unsolved.

1.3 CHARGE INJECTION AND TRANSPORT STUDIES IN ORGANIC SEMICONDUCTORS

Before we deal with spin transport we give the reader a very brief view of the complexity of charge transport in OS. The details of this vast topic can be found in another chapter. Understanding of charge injection and transport in OS has played an important role in the development of organic optoelectronics and recently in organic spintronics. Unlike conventional optoelectronic materials, the physics and chemistry of OS are significantly different. In OS, weak van der Waals intermolecular forces hold the molecules together and sufficiently increase the time spent by the charge carriers within the molecule. This leads to strong coulombic interactions between them that locally distort the potential, leading to the formation of a polaron.[5,7,55] Also, a weak overlap of the molecular levels, viz., the HOMO and LUMO, is expected, which considerably reduces the bandwidth of the HOMO and LUMO bands; typical values are of the order ~0.1 eV.[5,6] Such effects increase the effective mass of the charge carriers (polarons) and greatly reduce their mobility. In addition, the presence of defects and impurities strongly alters the conduction, making transport studies challenging, less reproducible, and difficult to model. Furthermore, the soft nature of the organic surfaces, the formation of an interfacial dipole layer, charge transfer, and influence of molecular morphologies on the interfacial bonding all contribute to the complexity of charge and likely, to a greater extent, the spin transport. In the literature, various charge transport models have been developed, spanning a wide range of organic materials under different growth conditions.[5,7,23,56,57] In most of these studies, transport is found to be dominated by tunneling-like charge injection from the metal electrode into the organic layer, followed by a hopping-like bulk transport in the disordered film. In such cases, the mobility is found to increase with temperature.

For structurally ordered films or in single crystals of OS with negligible defect density, charge transport is expected to show different characteristics, closer to intrinsic behavior. In single crystals with low defect density, strong intermolecular coupling is observed, leading to band-like transport. In this case, the change in mobility with temperature is modeled using a power law given by the following relation[5,58]:

$$\mu = CT^{-n}$$

where n (>1) is empirically varied to fit the observed experimental plots. Such a decrease in mobility with increasing temperature is expected to be due to the electron-phonon coupling. Following Holstein's theory of small polarons,[59] with increasing temperature, vibrational phonons cause bandwidth narrowing, leading to lower mobilities. However, at higher temperature, band-like transport fails as impurities and defects start contributing to conduction. In this case, phonon-assisted hopping dominates such that the mobility increases with temperature. Interplay between band-like and hopping transport may be expected in ordered films. Hence, one can visualize the path toward achieving large spin decay length in OS.

1.3.1 Spin Relaxation Mechanisms in OS

The mechanisms for spin relaxation, injection and transport, are strongly dictated by the conduction processes occurring at the injection interface and within the bulk. For example, in disordered films, the carriers spend sufficient time within the molecule, called trapping time, resulting in a strong interaction of the carrier spins with the local hyperfine fields, mainly coming from the hydrogen atoms of the molecule. This relaxation mechanism is thought to be responsible for the phenomenon of organic magnetoresistance (OMAR), as discussed in another chapter. Spin relaxation processes are necessarily very complex and important to understand for any spin injection and transport to be successfully carried out. This topic, with emphasis on semiconductors, is well discussed in the review article by Žutić, Fabian, and Das,[60] where one can find many relevant, earlier references. The spin relaxation mechanisms at play include Elliot-Yaffet (EY), D'yakonov-Perel' (DP), Bir-Aronov-Pikus (BAP), and also hyperfine interactions. Generally in the case of organics, EY mechanism and hyperfine interactions are thought to play the biggest role. The role of hyperfine interaction is discussed in detail in another chapter. In the EY process, the spin relaxation time, τ_S, is proportional to the momentum scattering time, τ_P. Thus, it depends on the momentum scattering and its sources, which can be phonons, defects, impurities, dopants, boundaries, and molecular vibrations. When dealing with organics, especially amorphous OS layers, these can be present in plenty, leading to substantial spin relaxation even in the presence of negligible spin-orbit interaction. It is clear that the more ordered the OS layer, the higher the expected τ_S and λ_S values, meaning that studies on epitaxial layers or single crystals would be required to reveal the intrinsic nature of spin relaxation in these materials. Thus, spin injection and transport into epitaxial films and single crystals would be exciting. Nearly atomic/molecular level control of the interface becomes critically important for successful spin injection and detection.

1.4 MAGNETIC TUNNEL JUNCTIONS WITH OS BARRIERS

Effective ways to investigate spin conservation of electrons injected across a ferromagnet-OS interface are to create magnetic tunnel junctions (MTJs) and to perform spin-polarized tunneling experiments. In MTJs consisting of two ferromagnetic (FM) electrodes with a tunnel barrier in between, one measures the change in junction resistance, TMR, by switching the relative orientation of magnetization of the two FM layers. Briefly, spin conservation in tunneling and the interfacial density of states for the spin-up and spin-down bands leads to the observation of TMR.[61] Several reviews have been written about this exciting area of MTJs, and the reader may refer to them for details.[62,63] In the second approach, one directly measures the spin polarization of the tunneling electrons through the OS barrier, in a superconductor/OS/FM junction, called the Meservey-Tedrow technique of spin-polarized tunneling (SPT). In this low-temperature method a superconductor such as an ultrathin Al film, with its Zeeman split quasi-particle density of states in a large applied magnetic field, acts as the spin detector for spin-polarized electrons coming from the FM counterelectrode. Excellent coverage of this SPT method can be found in the review by Meservey and Tedrow,[64] as well as briefly in other reviews mentioned above.

1.4.1 OS JUNCTION FORMATION

Tunnel junction preparation with OS is usually done by thermal evaporation or sputtering in a high-vacuum deposition chamber. Certainly the most reliable and best results are obtained when the junctions are prepared *in situ* by thermal evaporation. Fortunately, many of the OSs can be thermally evaporated readily from a powder source at significantly low power using refractory metal crucibles, without any damage to their molecular structure. The deposition rate can be well controlled with the least effort. The MTJs were usually deposited on glass substrates at room temperature (although deposition onto an LN_2-cooled surface can be done) in a chamber with base pressure of 6×10^{-8} Torr. By using shadow masks, the following MTJ structure—1 nm SiO_2/8 nm Co/TB/10 nm $Ni_{80}Fe_{20}$—could be deposited into a cross configuration of area 200×200 μm^2. Here, TB is the tunnel barrier of any of the following types: Alq_3, Al_2O_3/Alq_3, rubrene, Al_2O_3-rubrene, and so on. In the authors' lab, junctions with six different OS thicknesses, ranging from 1 to 10 nm or more, could be prepared in a single run. In some cases an ultrathin layer of Al_2O_3, ~0.6 nm thick, was used as the seed layer at the interface between the Co electrode and the OS, formed by exposing ~2 monolayers of Al metal to a short oxygen plasma. This seed layer influenced the growth of the OS and strongly modified the transport behavior,[40] as will be discussed later. Tunnel junctions for SPT measurements were made similarly and yielded the following structure: 3.8 nm Al/TB/8 nm FM. The Al electrode needed to be deposited onto liquid-nitrogen-cooled substrates to form a stable, continuous layer. The OS layer was deposited on the Al strip either at low temperature or at room temperature, yielding similar tunneling results.

1.4.2 STRUCTURE AND CHARACTERIZATION OF INTERFACES

In order for the OS junctions to qualify as worthy to yield intrinsic behavior, they should meet several criteria: (1) junction resistance (R_J) should be reasonable for its area, expected barrier height, and thickness; R_J should be at least three times larger than lead resistances over the junction area; (2) it should be reproducible and stable with time; (3) R_J should scale exponentially with thickness and increase as T decreases (see such an example in the inset in Figure 1.7 for MTJs with Alq_3 barrier)[28]; and (4) roughness and coverage of the OS layer should be verified by cross-sectional transmission electron microscopy (TEM) (see Figures 1.4 and 1.8), if possible. Additionally, from the bias dependence of tunnel conductance (G), one can ascertain the quality of the barrier and interfaces in a sensitive way. For example, from the shape of $G(V)$ near zero bias, one is able to see if any magnetic inclusions are present at the interfaces or inside the barrier, as these defects have a characteristic signature of a strong dip in $G(V)$ that becomes stronger as T decreases.[65] The low-temperature $G(V)$ curves of our MTJs with Alq_3 barriers show the absence of interfacial diffusion of magnetic atoms, as shown in Figure 1.7. Any leakage due to metallic bridges across the barrier in the junction can be easily noticed in the case of SPT junctions by the bias dependence of the conductance in the superconducting gap region. In our work we were able to go through all these diagnostic checks to ensure that the junctions are of high quality and that tunneling was occurring through the OS barrier.

Growth of good-quality OS, as mentioned earlier, is straightforward. It is encouraging to note that the evaporated films have uniform surface coverage even down to nearly a single monolayer level (keeping in mind that the thickness of a monolayer of Alq_3 or rubrene can be 1 to 2 nm). The cross-sectional high-resolution transmission electron microscope (HRTEM) image of a tunnel junction structure with a rubrene barrier grown either at room temperature or at liquid nitrogen temperature is shown in Figure 1.8. Generally, the OS barriers were amorphous, having complete coverage and smooth, sharp interfaces with the metal electrodes. With a higher sticking coefficient for low-temperature growth, clear differences in barrier thickness are seen not only with substrate temperature but also with the absence or presence of the Al_2O_3 seed layer. The rubrene films grown onto the alumina seed layer at liquid nitrogen temperature were nearly four times thicker than the thickness value given by the *in situ* quartz crystal sensor and the thickness of rubrene grown without the alumina seed layer at room temperature. Strikingly, the thickness of the rubrene films nearly doubled when grown on alumina rather than on the metal, implying different growth modes influenced by the dielectric-OS interface.[40,66,67] A schematic of the possible growth modes of rubrene molecules on metal and oxide seed layer surfaces is shown in Figure 1.9. This becomes even clearer by the tunneling spectroscopy results, discussed below.

A good understanding of the chemical, structural, and electronic nature of the interfaces between the spin-polarized electrode and the transport layer is essential for spin injection into OS. Extensive studies performed in inorganic materials have demonstrated the ubiquitous influence of the interface properties on spin injection. Conductivity mismatch, layer roughness, and intermixing at the interface strongly contribute to spin scattering and poor spin injection efficiencies. The use of a tunnel barrier with clean and sharp interfaces has been shown to remarkably improve the spin injection efficiencies. A similar level of understanding is needed for effective spin transport studies in OS.

Along with the above issues, however, there are several main differences to be considered in dealing with spin injection in organic materials. The soft nature of the organic surfaces, an intrinsic interfacial dipole layer, charge transfer, the 3D nature of electronic coupling in the molecule contributing to conduction anisotropy, interfacial bond chemistry, the different possible morphologies of the organic film, and the different growth modes of the electrode grown on top of the organic film all contribute to the complexity of spin injection studies in OS. Understanding of interfaces is thus nontrivial and involves the application of novel interface characterization tools.

Further insight into the orientation of the deposited molecules and the unchanged chemical structure of molecules during thermal evaporation in vacuum can be ascertained by inelastic tunneling spectroscopy (IETS). The strong electron-phonon coupling in organic materials makes IETS a highly interface-sensitive characterization tool compared to other spectroscopic methods, such as Raman and infrared (IR). Unlike Raman and IR spectroscopy, the wavelength of the probing tool (electrons) in IETS is much smaller. Thus, IETS is a powerful tool to analyze the active vibrational modes of molecules within the barrier, in probing their orientation, structural, electronic, and chemical nature.[68,69] Figure 1.10a shows the IETS for junctions with nonmagnetic electrodes having an OS barrier with and without an alumina seed layer (termed hybrid and rubrene junctions, respectively). The peaks in the spectra match the Raman and IR peaks for bulk rubrene, which indicates that the chemical structure of the rubrene molecules is unchanged during thermal evaporation. The fact that in IETS the tunneling electrons interact preferably with the vibrational modes that involve oscillating bond dipoles parallel to the direction of electron flow helps in identifying the molecular orientation during the growth.[69] With this in mind, for example, the peaks in the spectra observed at 81, 150, and 165 mV show the involvement of oscillations and vibrations along the tetracene backbone axis of a rubrene molecule in hybrid junctions, shown schematically in Figure 1.10b. This strongly suggests azimuthal growth of rubrene molecules (tetracene backbone standing vertically) on the alumina seed layer. However, in rubrene junctions, absence of these modes and presence of other modes at 53, 76, and 163 mV corresponding to oscillations transverse to the tetracene backbone axis implies that the rubrene molecule prefers to grow flat (tetracene backbone lying in the plane of the film), resulting in a disordered growth. Thus, IETS provides information about the growth morphology of the organic molecules at the interface. These different growth modes inferred from IETS well support the schematic shown in Figure 1.9. Implication of this molecular growth is seen in the spin tunneling injection efficiencies into rubrene.[40]

Though we are not aware of any structural characterization of the growth morphology of rubrene at the monolayer level that could corroborate the IETS analysis, the molecular packing in a single layer of pentacene has been recently investigated. Pentacene is a planar molecule with a simpler structure than that of rubrene, having a backbone of five fused benzene rings and no side groups. Using grazing incidence x-ray diffraction (GIXD), Mannsfeld et al.[70] determined the molecular packing motif of the first monolayer of pentacene onto SiO_2. The packing of the molecules dictates the degree of overlap of the orbitals with neighboring molecules, and consequently the carrier mobility. Thus, determining the structure of a pentacene monolayer on SiO_2 is of interest in the study of pentacene thin-film transistors, as the charge transport occurs primarily in the first few monolayers at the interface with the dielectric layer. They found that the molecules adsorb exactly upright on the SiO_2 surface and form a herringbone pattern, enclosing an angle of 52.7°, shown in Figure 1.6. Wetting of the substrate in this vertical, standing arrangement is energetically favorable, as the molecule-substrate interactions are stronger than the molecule-molecule interaction.[66] They found in their calculations that this packing motif results in high mobility, even for a polycrystalline film.

As described above, IETS and GIXD[71] were used to identify the arrangement of the molecules at the bottom interface. The use of x-ray photoemission spectroscopy (XPS), x-ray absorption spectroscopy (XAS), and ultraviolet photoemission spectroscopy (UPS) has also been very influential in understanding the interface chemistry and the energy-level alignments in organic-based devices.[16] Interfacial roughness studies using x-ray reflectometry (XRR) and polarized neutron reflectometry (PNR)

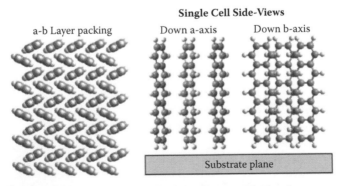

FIGURE 1.6 Herringbone packing motif of the first monolayer of pentacene molecules vacuum deposited onto a SiO_2 substrate. (From S. C. B. Mannsfeld, A. Virkar, C. Reese, M. F. Toney, Z. Bao, *Adv. Mater.* 21, 2294 [2009].)

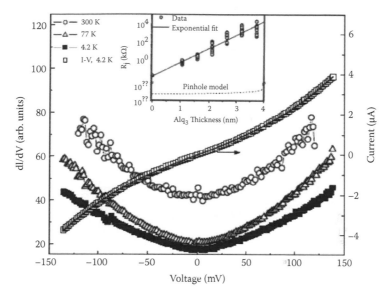

FIGURE 1.7 I-V characteristics for an 8 nm Co/0.6 nm Al_2O_3/1.6 nm Alq_3/10 nm Py junction. The inset shows the exponential dependence of R_J on Alq_3 thickness, for a total of seventy-two junctions made in a single run. (Taken from T. S. Santos, J. S. Lee, P. Migdal, I. C. Lekshmi, B. Satpati, J. S. Moodera, *Phys. Rev. Lett.* 98, 016601 [2007].)

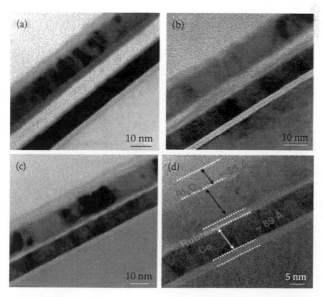

FIGURE 1.8 High-resolution TEM images for 8 nm Co/barrier/10 nm Fe junctions for barriers grown on (100)Si/SiO_2//0.5 nm Al_2O_3/2.2 nm rubrene deposited at 80 K (a) and 295 K (b), and rubrene only deposited at 80 K (c) and 295 K (d). The layers are identified in (d). (From J. H. Shim, K. V. Raman, Y. J. Park, T. S. Santos, G. X. Miao, B. Satpati, J. S. Moodera, *Phys. Rev. Lett.* 100, 226603 [2008].)

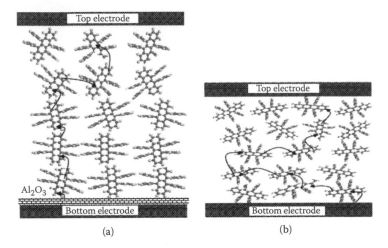

FIGURE 1.9 (a) Stronger intermolecular electronic coupling in azimuthally grown rubrene films leads to efficient spin injection and enhanced spin transport signals in hybrid junctions. (b) Disordered, nonazimuthal growth in rubrene junction leads to hopping-dominated spin transport, resulting in spin scattering and weak magnetotransport signals. (From K. V. Raman, S. M. Watson, J. H. Shim, J. A. Borchers, J. Chang, J. S. Moodera, *Phys. Rev. B* 80, 195212 [2009].)

are some of the other techniques used in characterizing these interfaces.[40,72] PNR, x-ray magnetic circular dichroism (XMCD), and x-ray resonant magnetic scattering (XRMS) are sensitive to magnetization and chemical compositions, making them ideal for investigating magnetic properties at interfaces. These studies have just begun, and there are plenty of open areas for future investigation, which is bound to lead us to exciting interfacial phenomena. One of the limitations in utilizing XAS and XMCD techniques is the beam damage to the OS layers. One has to ascertain that before concluding on the results.

Recent PNR studies have elucidated the FM/OS interface in spin valve structures. For example, recently Liu et al.[72] performed extensive interface-sensitive characterization studies on Alq_3-based spin valve structures and correlated the interface microstructural properties to the observed magnetotransport signals in their devices. A large roughness at the Alq_3/Fe interface was reported, which suppressed the magnetic properties of the first few monolayers of Fe electrode. Preliminary PNR studies by the authors have revealed interesting properties of the top Fe electrode over rubrene for different growth morphologies of the OS layer.[40]

1.4.3 TRANSPORT STUDIES

Charge transport in OS can be quite complex, as mentioned in Section 1.3, and can reveal a lot of information about the properties of the material. The transport behavior is dominated by hopping processes, and in the case of tunnel barriers, the charges can tunnel via localized molecular states or structural defects. For a tunnel junction, traditionally the conductance (G) is measured as a function of temperature and

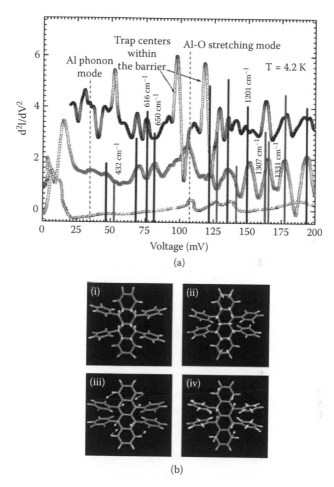

FIGURE 1.10 (a) IETS plots of rubrene junction (8 nm Al/10 nm Rubrene/10 nm Al) (circles) and hybrid junction (8 nm Al/0.6 nm Al₂ O₃/6 nm Rubrene/10 nm Al) (squares) with IETS plot of control junction (8 nm Al/1 nm Al₂O₃/8 nm Al) (triangles), and computed Raman and IR peaks (bars) for single crystal rubrene included for reference. (b) Active molecular vibrational modes observed in hybrid junctions at (i) 1201 cm⁻¹ and (ii) 650.1 cm⁻¹ (arrows represent vibrations in plane along the tetracene axis) and in rubrene junctions at (iii) 432 cm⁻¹ and (iv) 616.4 cm⁻¹ (arrows represent vibrations out of plane, perpendicular to the tetracene axis). (From K. V. Raman, S. M. Watson, J. H. Shim, J. A. Borchers, J. Chang, J. S. Moodera, *Phys. Rev. B* 80, 195212 [2009].)

bias, which gives information about the barrier properties. The low-bias conductance (*G*) for two types of junctions, measured as a function of temperature, is shown in Figure 1.11. For the rubrene junctions *G* decreases dramatically from 295 to 4.2 K, compared to only a small decrease for the hybrid junctions. Mott's variable range hopping (VRH) model was adopted to satisfactorily explain the *G* vs. *T* data.[73,75] This model predicts that δ_{VRH}, the variable range hopping length, increases with

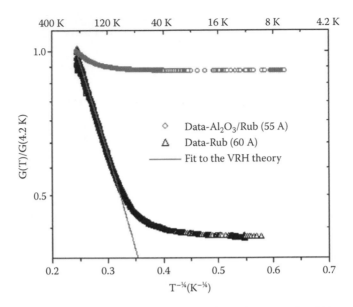

FIGURE 1.11 Temperature dependence of tunnel conductance for 3.8 nm Al/6 nm rubrene/15 nm Co and 3.8 nm Al/0.5 nm Al_2O_3/5.5 nm rubrene/15 nm Co junctions. Mott's VRH theory fit (line) to the data at higher temperatures for rubrene-only barrier is shown. (Taken from J. H. Shim, K. V. Raman, Y. J. Park, T. S. Santos, G. X. Miao, B. Satpati, J. S. Moodera, *Phys. Rev. Lett.* 100, 226603 [2008].)

decreasing temperature as $T^{-\frac{1}{4}}$ for the three-dimensional case. From such a study different localization lengths (δ_{loc}) are obtained for the two types of barriers.[14] The implication is that the electronic states contributing to conduction in OS have different δ_{loc}, consistent with different growth modes for rubrene and hybrid junctions.[40] Inelastic tunneling (via hopping) occurs when the barrier thickness is greater than δ_{loc}. Thus, for hybrid junctions, δ_{loc} is large with conduction occurring mostly by elastic and resonant tunneling, resulting only in a marginal increase of G with T. On the other hand, for the rubrene junctions having a smaller δ_{loc}, the conduction appears to occur via inelastic hopping through localized states, resulting in a stronger $G(T)$ dependence. Thus, with increase in rubrene film thickness, especially in a disordered layer, the probability of direct tunneling diminishes, and the conduction is dominated by inelastic transport in a wider temperature range.

1.4.4 Tunnel Magnetoresistance in OS-Based Magnetic Tunnel Junctions

Charge transport results show that having an Al_2O_3 seed layer should greatly enhance the spin transport in the OS. This turned out to be the case when spin tunneling experiments were performed. Using a planar sandwich structure, spin-conserved tunneling can be done by either SPT or MTJ structure. First, we deal with MTJs for which Julliere's model predicts a large TMR as a result of spin-conserved tunneling.[76] In this model, the resistance of an MTJ depends on the relative orientation of the magnetization of the two ferromagnetic electrodes: lower resistance for parallel alignment (R_P) and higher resistance for antiparallel alignment (R_{AP}). Tunnel magnetoresistance (TMR) is defined as

$$\Delta R/R = (R_{AP} - R_P)/R_P.$$

Such measurements on a $Co/Al_2O_3/rubrene/Fe$ junction are shown in Figure 1.12 at various temperatures. The room temperature TMR of 6% for rubrene increased to 16% upon cooling the sample to 4.2 K, whereas no TMR was observed in junctions without Al_2O_3. Similar behavior was seen in junctions with Alq_3 in the barrier. The relatively lower value of TMR (relative to inorganic Al_2O_3 or MgO barrier junctions) shows significant spin loss in the tunneling process or at the interfaces. Given that these OS barriers are amorphous and have considerable defects in them, as also observed in the charge transport, it is not surprising. The decrease in TMR with increasing temperature is common in MTJs and is discussed in the literature,[33,61] whereas stronger temperature dependence can be expected due to the larger defect density in the barrier.[77] For OS barrier junctions, the temperature dependence is larger for both R_J and TMR. At higher temperatures, tunneling becomes more inelastic in these disordered OS barriers, meaning less likelihood of spin-conserved tunneling, and consequently lower TMR. Relatively speaking, rubrene junctions showed less effect of defects than Alq_3 or pentacene in our studies.

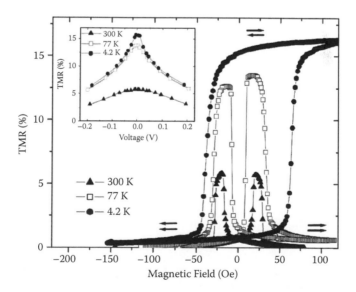

FIGURE 1.12 Resistance variation with applied magnetic fields for the MTJ structure 8 nm Co/0.5 nm Al_2O_3/4.6 nm rubrene/10 nm Fe/1.5 nm CoO junction showing TMR. The data at 4.2 K show the exchange bias due to the antiferromagnetic CoO over Fe. The lines through the data points are a guide to the eye. The inset shows the bias dependence of TMR. The arrows indicate the magnetic configuration of Co and Fe electrodes at various applied fields. (Taken from J. H. Shim, K. V. Raman, Y. J. Park, T. S. Santos, G. X. Miao, B. Satpati, J. S. Moodera, *Phys. Rev. Lett.* 100, 226603 [2008].)

The TMR generally decreases with increasing bias for any MTJ,[62,78] and it is not an exception for junctions with OS barriers. An example of this is shown in Figure 1.12, inset. The TMR decreases symmetrically for $\pm V$, which means that in this case there is significant influence of defects that mask the FM electrode band features or interface effects. Substantial TMR persists beyond ± 200 mV for the rubrene junctions. The magnitude of TMR decreases with increasing bias voltage, depending on the quality of the barrier. Higher-quality barriers have a weaker bias dependence. For junctions with minimal influence of defects in the barrier, the bias dependence has been attributed to the excitation of magnons, phonons, band effects, etc.[78] In addition, for the present junctions with an OS barrier, one can expect chemistry-induced states in the OS band gap[79] that cause increased temperature and bias dependence as well as reduced TMR. Nevertheless, the slow decrease of TMR with applied bias demonstrates that these MTJs are of reasonably high quality and hold promise for reaching full potential for OS-based junctions.

1.5 DIRECT MEASUREMENT OF SPIN-POLARIZED CURRENT

One of the ideal probes to study spin transport is the spin-polarized tunneling, viz., the Meservey-Tedrow method of detecting spin-polarized tunnel current using a superconductor. Here the superconductor in an applied magnetic field, showing Zeeman splitting of the quasi-particle density of states, acts as an excellent spin detector. In essence, this phenomenon ideally allows for the direct determination of spin-polarized current tunneling through an OS. Extensive SPT measurement of this kind has been successfully carried out in the authors' laboratory. With one of the electrodes being a superconductor, it is straightforward to ascertain the leak-free quality of the OS barrier junction. A couple of examples of this are given here for Al/OS/FM junctions. At temperatures far below the superconducting (SC) transition of Al ($T_C \sim 2.7$ K), the tunnel conductance vs. bias in zero field (see Figure 1.13)

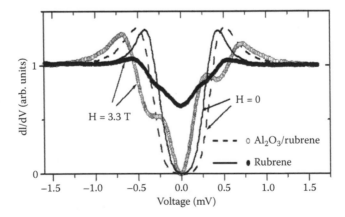

FIGURE 1.13 Tunnel conductance vs. bias at 0.45 K with and without an applied magnetic field for two types of junctions: 3.8 nm Al/barriers/15 nm Co with 0.5 nm Al_2O_3/5.5 nm rubrene or 6.0 nm rubrene as barriers. (Taken from J. H. Shim, K. V. Raman, Y. J. Park, T. S. Santos, G. X. Miao, B. Satpati, J. S. Moodera, *Phys. Rev. Lett.* 100, 226603 [2008].)

shows the characteristic behavior of conduction by tunneling into a superconductor. Here the negligible conductance at zero bias and the observation of a clear SC energy gap in Al with sharp peaks prove leak-free junctions and that the conduction is by tunneling, thus confirming the high quality of the tunnel barriers.[64] However, for the rubrene junctions (without the alumina seed layer), the shape of the conductance with bias in the SC gap region showed a relatively faster increase of conductance with bias than did hybrid junctions (with alumina seed layer) suggesting the existence of inelastic conduction mechanisms at play.[73,80] This is in line with the other charge transport observation discussed in Section 1.4.3.

In an applied magnetic field (H) in the plane of the film, Zeeman splitting of the conductance peaks (quasi-particle density of states of Al) is observed with magnitude $2\mu_B H$. The conductance asymmetry is a result of spin polarization of the tunnel current from which the spin polarization value could be extracted utilizing Maki's theory.[81] For the Al_2O_3/Alq_3 barrier with Co electrode, a P_{Co} of 27% was determined. Similarly, P_{Fe} of 30% and P_{Py} of 38% were determined. These values are smaller than those with a pure Al_2O_3 tunnel barrier prepared in ultra-high vacuum: 42, 44, and 50% obtained for Co, Fe, and Py, respectively.[42,82] For an Al_2O_3-rubrene barrier, a P_{Co} of 30% was determined.[14] However, for nonhybrid barriers (only Alq_3 or rubrene, without Al_2O_3), P values consistently turned out to be a factor of three (for rubrene) or five (for Alq_3) smaller. These results first of all clearly demonstrate spin polarization of the tunnel current from a ferromagnet through an OS. The positive influence of the Al_2O_3 seed layer is evident. In the same token, they demonstrate how the defects in OS suppress spin-conserved tunneling. Furthermore, these observations tell us the direction in which to move to increase the spin transport efficiency.

One of the fundamental issues in the development of organic spintronics is how far the spins can travel in an OS. This is at the heart of the field. Although there has been some estimation of λ_S, as mentioned earlier, one of the direct and unambiguous ways would be by SPT measurement. This was possible in the case of rubrene, by directly measuring P as a function of rubrene thickness. In Figure 1.14, the determined P_{Co} values for increasing thicknesses of rubrene in hybrid junctions are shown. While P decreases with increasing rubrene thickness, the drop is not as detrimental as one would have expected in these amorphous layers. Remarkably, even at 15-nm thick rubrene barrier the polarization is still strong at 12%. The exponential fit to the data yields λ_S of 13.3 nm for amorphous rubrene, giving some optimism for substantially improving λ_S as the quality of the OS layer improves toward higher-mobility organic crystals. It may be noted that as the barrier thickness increases, one would expect to have multistep tunneling, which could be the case in the higher range of rubrene thicknesses. At temperatures closer to RT, inelastic transport or hopping through localized molecular or trap states in disordered films reduce the mobility and λ_S. This result is remarkable considering that spin is not conserved in amorphons Si and Ge tunnel barriers.[73,74] However, in single-crystalline Si, λ_S even up to 100 μm has been reported recently.[83,84] Taking the liberty to compare with amorphous Si, the observation of long λ_S in an amorphous OS is noticeably significant. With improvement in defect density and carrier mobility from $< 10^{-6}$ cm^2/Vs for amorphous state to tens of cm^2/Vs for crystalline OS, very large gains in λ_S can be expected.[85]

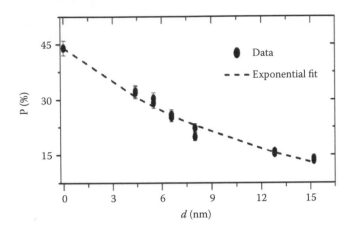

FIGURE 1.14 Measured spin polarization (P) vs. rubrene film thickness d, showing the exponential dependence. Notice the large P even at a d of 15 nm of amorphous rubrene. (Taken from J. H. Shim, K. V. Raman, Y. J. Park, T. S. Santos, G. X. Miao, B. Satpati, J. S. Moodera, *Phys. Rev. Lett.* 100, 226603 [2008].)

1.6 BARRIER PROPERTIES AND EFFECT OF SEED LAYER ON SPIN-CONSERVED TUNNELING

For high-quality tunnel junctions, the resistance depends exponentially on the barrier thickness. For example, as seen in Figure 1.7, R_J scaled exponentially with Alq_3 thickness. This dependence confirms that tunneling is occurring through the Alq_3 layer,[28] and that the OS film coverage is uniform down to a monolayer. This was also the case with rubrene and pentacene. Upon cooling the junctions from room temperature down to 4.2 K, R_J usually increased by a factor of 2 to 3, which is common for junctions with semiconductor tunnel barriers,[73,74] as opposed to alumina barriers, which show an R_J increase of ~20% over the same temperature range. The current-voltage (I-V) characteristics of junctions taken to a few hundred millivolts can give valuable information on the tunnel barrier quality. Hence, performing this test is very important for judging the junction for its suitability to yield reliable results. For the SPT junctions, from the I-V results, the effective barrier height, φ, and thickness, d, could be extracted using the Brinkman, Dynes, and Rowell (BDR) tunneling model.[86] The I-V data at 4.2 K yielded φ (and d) of 0.45 eV (4 nm) for hybrid and 2.2 eV (2 nm) for rubrene junctions, with an asymmetry, $\Delta\varphi$, of 0.2 eV for the hybrid junctions (which was due to the Al_2O_3 present, creating a trapezoidal barrier[86]). Similar values were obtained from the I-V data (see Figure 1.7) for the Alq_3 case as well. The barrier parameters were temperature independent. Insertion of Al_2O_3 seed layer decreased φ in the case of OS with rubrene, as mentioned previously due to molecular orientational change, and in the case of Alq_3 due to the interfacial dipolar modifications,[16,28] whereas the deviation of d from the actual thickness suggests that inelastic processes (more for rubrene junctions) contribute significantly to the conduction process,[73] aside from the growth-related morphological differences.

One of the sensitive areas, particularly for spin tunneling and injection, is the sharpness of the OS/FM interfaces. Given the nature of the OS and their size, it is not uncommon to have significant penetration of the metal electrode atoms into the OS, making the interface ill defined. This will certainly affect the spin transport and tunneling. In the case of tunnel junctions, there are characteristic signatures in the conductance vs. bias near zero bias, particularly at low temperatures: a sharp dip is expected when there are metal atom inclusions in the barrier. We note the absence of any sharp dip at zero bias (known as the zero bias anomaly[65]) for Alq_3 barrier junctions even at 4.2 K (see Figure 1.7). This shows that the barrier and interfaces are free of magnetic inclusions. Presence of such a dip in conductance can be caused by diffusion of magnetic impurities into the barrier, among other possibilities.[64]

The high barrier height for nonhybrid junctions may be attributed to a dipole layer formed at the metal-OS interface. Such a phenomenon can occur due to charge transfer, chemical reactions, and changes of molecular configuration, which introduce states into the OS band gap.[69,77,87,88] The intrinsic trap states, if mainly concentrated at the metal-organic interface, appear to localize the tunneling spins, thereby creating sites for spin loss.[77] Aside from these interface effects, the bulk of the OS film can be a good-quality tunnel barrier, producing a conductance measurement with negligible leakage current. The ultrathin Al_2O_3 seed layer at the electrode-OS interface suppresses the formation of these trap states and effectively lowers the barrier height to electron injection via tunneling across the interface.[89,90] Hence, the seed layer appears to reduce the concentration of trap states, improving the possibility of spin-conserved tunneling. This is manifested in the higher P values for Alq_3 and rubrene barriers with the Al_2O_3 layer. This result demonstrates the degrading effect of interfacial charge states in spin-conserved tunneling and how minimizing the formation of these states greatly improves spin injection efficiency across the ferromagnet-OS interface.

In the amorphous OS barriers, in addition to higher defect concentration at the interface, there may be several other types of defects present that can affect spin-conserved tunneling. Any mechanism that adds or removes the p_z electron from contributing to the π system gives rise to energy states within the energy gap. Such defects include formation of C-H_2 by adding H to the C-H bond, an OH defect by replacing H with a hydroxyl group (–OH), or less likely a dangling bond (formed by losing a H atom).[91] In amorphous Si or Ge, a large number of the dangling bonds ($\approx 10^{19}$ cm^{-3}) exist with unpaired electrons,[92] which is a strong source of spin scattering, as seen experimentally.[73,74] However, in OS, the formation energy for dangling bonds is much higher than for the other defects, and hence least likely to form.[91] This may explain the observation of moderately spin-conserved transport in amorphous OS, but not in amorphous inorganic semiconductors. However, the real explanation needs more research, especially with emphasis on atomic characterization of the ultra-thin OS layers.

1.7 SPIN TRANSPORT IN SINGLE CRYSTALS

Almost all of the spin transport studies have used an organic layer grown by thermal evaporation in vacuum. Such a deposition method often leads to amorphous or at best polycrystalline films that can contain a large density of defects that in turn influence both spin and charge transport properties. Some of the earlier work on spin studies involved the use of Julliere's model in the estimation of the spin-dependent parameters, namely, the spin injection efficiency, the spin diffusion length, and the spin lifetime. However, that model only provides a crude estimate, since its applicability in organic systems may not convey the real picture due to the complex, disordered nature of organic films, both at the interface and within the bulk. Thus, the need for high-quality films with minimal defect density is desired, for understanding the intrinsic spin transport mechanism in OS.

Earlier studies lacked the appropriate characterization tools to obtain the spin properties in a reliable manner. With better understanding in this field, more direct measurements have been attempted. One such technique is spin-polarized tunneling, as discussed previously, which is used to measure the spin diffusion length by the exponential fit to the measured polarization through the organic layer as a function of thickness. This technique is limited to only low temperature, and thus is unable to describe the temperature dependence of the spin-dependent parameters. Using a high-quality MTJ structure, it should be possible to investigate this in the future, although not that straightforward. Alternatively, there have been other methods, such as the two-photon photoemission technique[93] to determine the spin injection efficiency into the organic layer, and low-energy muon spin rotation[94] to measure the spin polarization injection and also determine the temperature dependence of the spin diffusion length. In Section 1.3, we discuss factors contributing to spin relaxation in an OS. Impurities and defects, in addition to directly affecting the mobility, can significantly contribute to spin relaxation and thus reduce the spin diffusion length. With some of those above characterization tools it may now be possible to gain more understanding of the role of defects/impurities to spin relaxation, and thereby the intrinsic mechanisms in OS.

Over the last few years, techniques have been developed to purify organic materials. Vapor phase sublimation method[95] is one of the widely used techniques for purification. This method is effective for removing low vapor phase impurities from the main compound. In the purified form, the quality of the organic films is found to greatly depend on the molecular structure and the preferred crystal morphology of the organic semiconductor. For example, CuPc having a planar structure readily grows ordered and polycrystalline with larger grains, even at substrate temperatures of 300 to 330 K, while it is not easy to grow ordered Alq_3 films. Vapor phase sublimation growth in high-purity inert gas flow has successfully produced high-purity single crystals. Charge transport studies performed on these crystals have shown improved conduction with band-like transport,[5,96] indicating negligible or weak influence of impurities/defects on the transport properties.[97] These are the desired candidates for performing spin transport studies. Thus, along with work on ordered films, spin transport in single crystals is also being actively researched.

Study of single crystals will undoubtedly provide a better insight on spin transport behavior, whereas it is currently faced with device fabrication challenges. The devices are commonly fabricated using the "flip method." In most of these devices, the single crystal is bonded to an inert electrode such as Au that provides a clean interface with the OS.[22,98] However, for spin transport the electrodes are ferromagnetic, and as such, the spin injection into the organic is strongly influenced by the quality of the interface between the FM electrode and the organic crystal. Unfortunately,

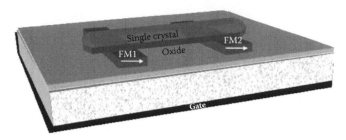

FIGURE 1.15 Schematics of a proposed spin-FET device using single crystals of organic semiconductor stamped using the flip method.

all elemental and alloy FMs have a strong affinity to oxygen, developing an oxide layer that significantly lowers/destroys the interface polarization. In addition to the quality of the interface that can dictate spin scattering, the conductivity mismatch between the FM and the organic crystal adds to the difficulty. These limiting factors are well known in the case of spin injection into inorganic semiconductors such as Si or GaAs.[60] The use of a tunnel barrier to reduce the mismatch problem is also being developed in organic spintronics.[14,47]

Although studies using single crystals are highly challenging and in the nascent state, it is expected that eventual success in this area would open up new directions. One such area is the development of spin-based FET sensors.[99] Single crystals of OS have shown reasonably high mobility in FETs.[22] Figure 1.15 shows the schematic of an organic spin-FET sensor. Unlike conventional FETs, organic-based transistors work in the charge accumulation mode. Under the gated operations, due to the weak intermolecular coupling, charges are accumulated within the first few monolayers of the organic film adjacent to the dielectric layer, dominating the conduction process. This leads to high mobility, analogous to the 2D electron channel observed in inorganic semiconductors. The role of hyperfine interactions, which is expected to be stronger in organic materials than the spin-orbit term,[10] may lead to different effects on the spin dynamics across the channel. This entirely unexplored research field provides ample opportunities for advances in the field.

1.8 FUTURE OUTLOOK

The field of organic spintronics is just coming off the ground. There are already bright signs for an exciting future, in terms of not only plenty of science but also potential for application. What makes it so unique is the vastness of the possibilities, essentially limitless, which might even take us toward biological molecules. Along with the beauty of the whole open area to explore come the one-of-a-kind challenges that require their own unique solutions. For example, in order to measure reliable and reproducible properties, one should very likely deal with epitaxial films and single crystals with controlled interfaces, all *in situ* growth of the structure being preferred. The molecular orientation and conduction along the lengthy chain or orthogonal to it will have a big impact on spin transport, as seen from preliminary studies. This might lead us toward investigating long-chain polymers. On the practical side, the areas that need attention are the types of defects in OS—their role in spin transport and how to minimize them. Important in this direction is the complex nature of interfaces (not to mention defects), whether one is dealing with an OS/FM metal interface with or without a barrier, or perhaps even the development of an organic multilayer system. These challenges will push the characterization techniques beyond what is available at present.

ACKNOWLEDGMENT

We are thankful for funding support from ONR (N00014-09-1-0177), NSF (DMR-0504158), and KIST-MIT for the research reported here. T.S.S. is supported by a CNM Distinguished Postdoctoral Fellowship and a L'Oreal USA Fellowship for Women in Science.

REFERENCES

1. C. W. Tang, S. A. van Slyke, *Appl. Phys. Lett.* 51, 913 (1987).
2. A. J. Heeger, *Semiconducting and Metallic Polymers: The Fourth Generation of Polymeric Materials*, Nobel lecture, 2000.
3. R. H. Friend et al., *Nature* 397, 121 (1999).
4. S. R. Forrest, *Chem. Rev.* 107, 923 (2007).
5. Z. Bao, J. Locklin, *Organic Field-Effect Transistors*, CRC Press, Boca Raton, 2007.
6. V. Coropceanu, J. Cornil, D. A. da Silva, Y. Olivier, R. Silbey, J. L. Bredas, *Chem. Rev.* 107, 926 (2007).
7. M. Pope, C. E. Swenberg, *Electronic Processes in Organic Crystals and Polymers*, 2nd ed., Oxford University Press, New York, 1999.
8. W. J. M. Naber, S. Faez, W. G. van der Wiel, *J. Phys. D* 40, R205 (2007).
9. V. A. Dediu, L. E. Hueso, I. Bergenti, C. Taliani, *Nature Mater.* 8, 707 (2009).
10. P. A. Bobbert, W. Wagemans, F. W. A. van Oost, B. Koopmans, M. Wohlgenannt, *Phys. Rev. Lett.* 102, 156604 (2009).
11. V. I. Krinichnyi, S. D. Chemerisov, Y. S. Lebedev, *Phys. Rev. B* 55, 16233 (1997).
12. Z. H. Xiong, D. Wu, Z. Valy Vardeny, Jing Shi, *Nature* 427, 821 (2004).
13. S. Pramanik, C. G. Stefanita, S. Patibandla, S. Bandyopadhyay, K. Garre, N. Harth, M. Cahay, *Nature Nanotechnol.* 2, 216 (2007).
14. J. H. Shim, K. V. Raman, Y. J. Park, T. S. Santos, G. X. Miao, B. Satpati, J. S. Moodera, *Phys. Rev. Lett.* 100, 226603 (2008).
15. D. Käfer, G. Witte, *Phys. Chem. Chem. Phys.* 7, 2850 (2005).
16. H. Ishii, K. Sugiyama, E. Ito, K. Seki, *Adv. Mater.* 11, 605 (1999).
17. K. I. Pokhodnya, A. J. Epstein, J. S. Miller, *Adv. Mater.* 12, 410 (2000).
18. V. N. Prigodin, N. P. Raju, K. I. Pokhodnya, J. S. Miller, A. J. Epstein, *Adv. Mater.* 14, 1230 (2002).
19. Y. Xia, G. M. Whitesides, *Angew. Chem. Int. Ed. Engl.* 37, 550 (1998).
20. Z. Zhi-lin, J. Xue-yin, X. Shao-hong, T. Nagatomo, O. Omoto, *J. Phys. D Appl. Phys.* 31, 32 (1998).
21. Z. Zhi-lin, J. Xue-yin, X. Shao-hong, T. Nagatomo, O. Omoto, *Chin. Phys. Lett.* 14, 302 (1997).
22. V. Podzorov, E. Menard, A. Borissov, V. Kiryukhin, J. A. Rogers, M. E. Gershenson, *Phys. Rev. Lett.* 93, 086602 (2004).
23. M. A. Baldo, S. R. Forrest, *Phys. Rev. B* 64, 085201 (2001).

24. W. Xu, G. Szulczewski, P. LeClair, I. Navarrete, R. Schad, G. Miao, H. Guo, A. Gupta, *Appl. Phys. Lett.* 90, 072506 (2007).
25. J. G. Simmons, *J. Appl. Phys.* 34, 1793 (1963).
26. J. S. Jiang, J. E. Pearson, S. D. Bader, *Phys. Rev. B* 77, 035303 (2008).
27. A. Rajagopal, A. Kahn, *J. Appl. Phys.* 84, 355 (1998).
28. T. S. Santos, J. S. Lee, P. Migdal, I. C. Lekshmi, B. Satpati, J. S. Moodera, *Phys. Rev. Lett.* 98, 016601 (2007).
29. V. Dediu, M. Murgia, F. C. Matacotta, C. Taliani, S. Barbanera, *Sol. State Commun.* 122, 181 (2002).
30. D. Wu, Z. H. Xiong, X. G. Li, Z. V. Vardeny, Jing Shi, *Phys. Rev. Lett.* 95, 016802 (2005).
31. M. Viret, M. Drouet, J. Nassar, J. P. Contour, C. Fermon, A. Fert, *Europhys. Lett.* 39, 545 (1997).
32. M. Bowen, M. Bibes, A. Barthélémy, J.-P. Contour, A. Anane, Y. Lemaître, A. Fert, *Appl. Phys. Lett.* 82, 233 (2003).
33. C. H. Shang, J. Nowak, R. Jansen, J. S. Moodera, *Phys. Rev. B* 58, R2917 (1998).
34. F. J. Wang, C. G. Yang, Z. Valy Vardeny, X. G. Li, *Phys. Rev. B* 75, 245324 (2007).
35. V. Dediu, L. E. Hueso, I. Bergenti, A. Riminucci, F. Borgatti, P. Graziosi, C. Newby, F. Casoli, M. P. De Jong, C. Taliani, Y. Zhan, *Phys. Rev. B* 78, 115203 (2008).
36. Y. Q. Zhan, X. J. Liu, E. Carlegrim, F. H. Li, I. Bergenti, P. Graziosi, V. Dediu, M. Fahlman, *Appl. Phys. Lett.* 94, 053301 (2009).
37. Y. Q. Zhan, M. P. de Jong, F. H. Li, V. Dediu, M. Fahlman, W. R. Salaneck, *Phys. Rev. B* 78, 045208 (2008).
38. T. Mori, H. Fujikawa, S. Tokito, Y. Taga, *Appl. Phys. Lett.* 73, 2763 (1998).
39. S. T. Zhang, X. M. Ding, J. M. Zhao, H. Z. Shi, J. He, Z. H. Xiong, H. J. Ding, E. G. Obbard, Y. Q. Zhan, W. Huang, X. Y. Hou, *Appl. Phys. Lett.* 84, 425 (2004).
40. K. V. Raman, S. M. Watson, J. H. Shim, J. A. Borchers, J. Chang, J. S. Moodera, *Phys. Rev. B* 80, 195212 (2009).
41. J. S. Moodera, R. Meservey, *Phys. Rev. B* 29, 2943 (1984).
42. T. H. Kim, J. S. Moodera, *Phys. Rev. B* 69, 020403(R) (2004).
43. B. J. Chen, W. Y. Lai, Z. Q. Gao, C. S. Lee, S. T. Lee, W. A. Gambling, *Appl. Phys. Lett.* 75, 4010 (1999).
44. H. Park, D.-S. Shin, H.-S. Yu, H.-B. Chae, *Appl. Phys. Lett.* 90, 202103 (2007).
45. S. Majumdar, R. Laiho, P. Laukkanen, I. J. Väyrynen, H. S. Majumdar, R. Österbacka, *Appl. Phys. Lett.* 89, 122114 (2006).
46. T. X. Wang, H. X. Wei, Z. M. Zeng, X. F. Han, Z. M. Hong, G. Q. Shi, *Appl. Phys. Lett.* 88, 242505 (2006).
47. J. R. Petta, S. K. Slater, D. C. Ralph, *Phys. Rev. Lett.* 93, 136601 (2004).
48. H. Kusai, S. Miwa, M. Mizuguchi, T. Shinjo, Y. Suzuki, M. Shiraishi, *Chem. Phys. Lett.* 448, 106 (2007).
49. D. Hatanaka, S. Tanabe, H. Kusai, R. Nouchi, T. Nozaki, T. Shinjo, Y. Suzuki, H. Wang, K. Takanashi, M. Shiraishi, *Phys. Rev. B* 79, 235402 (2009).
50. P. M. Levy, I. Mertig, in *Spin Dependent Transport in Magnetic Nanostructures*, ed. S. Maekawa, T. Shinjo, Advances in Condensed Matter Science, Taylor & Francis, London, 2002, chap. 2, p. 47.
51. M. B. Stearns, *J. Magn. Magn. Mater.* 5, 167 (1977).
52. M. Münzenberg, J. S. Moodera, *Phys. Rev. B* 70, 060402(R) (2004).
53. A. Thomas, J. S. Moodera, B. Satpati, *J. Appl. Phys.* 97, 10C908 (1999).
54. J. M. De Teresa, A. Barthélémy, A. Fert, J. P. Contour, R. Lyonnet, F. Montaigne, P. Seneor, A. Vaurès, *Phys. Rev. Lett.* 82, 4288 (1999).
55. I. N. Hulea, S. Fratini, H. Xie, C. L. Mulder, N. N. Iossad, G. Rastelli, S. Ciuchi, A. F. Morpurgo, *Nature Mater.* 5, 982 (2006).

56. V. I. Arkhipov, E. V. Emelianova, Y. H. Tak, H. Bässler, *J. Appl. Phys.* 84, 848 (1998).
57. N. F. Mott, E. A. Davis, *Electronic Processes in Non-crystalline Materials*, 2nd ed., Oxford University Press, New York, 1979.
58. E. A. Silinsh, V. Capek, *Organic Molecular Crystals: Interaction, Localization, and Transport Phenomena*, AIP Press, New York, 1994.
59. T. Holstein, *Ann. Phys.* 8, 343 (1959).
60. I. Žutić, J. Fabian, S. Das Sarma, *Rev. Mod. Phys.* 76, 323 (2004).
61. J. S. Moodera, L. R. Kinder, T. M. Wong, R. Meservey, *Phys. Rev. Lett.* 74, 3273 (1995).
62. J. S. Moodera, G. Mathon, Spin polarized tunneling in ferromagnetic junctions, *J. Magn. Magn. Mater.* 200, 248 (1999).
63. E. Y. Tsymbal, O. N. Mryasov, P. R. LeClair, *J. Phys. Condens. Matter* 15, R109 (2003).
64. R. Meservey, P. Tedrow, *Phys. Rep.* 238, 173 (1994).
65. J. Appelbaum, *Phys. Rev. Lett.* 17, 91 (1966).
66. G. E. Thayer, J. T. Sadowski, F. Meyer zu Heringdorf, T. Sakurai, R. M. Tromp, *Phys. Rev. Lett.* 95, 256106 (2005).
67. G. Witte, C. Wöll, *Phase Transitions* 76, 291 (2003); Q. Chen, A. J. McDowell, N. V. Richardson, *Langmuir* 19, 10164 (2003).
68. R. C. Jaklevic, J. Lambe, *Phys. Rev. Lett.* 17, 1139 (1966); J. Lambe, R. C. Jaklevic, *Phys. Rev.* 165, 821 (1968).
69. P. K. Hansma, *Tunneling Spectroscopy-Capabilities, Application and New Techniques*, Plenum Press, New York, 1982.
70. S. C. B. Mannsfeld, A. Virkar, C. Reese, M. F. Toney, Z. Bao, *Adv. Mater.* 21, 2294 (2009).
71. R. J. Kline, M. D. Mcgehee, M. F. Toney, *Nature Mater.* 5, 222 (2006).
72. Y. Liu, S. M. Watson, T. Lee, J. M. Gorham, H. E. Katz, J. A. Borchers, H. D. Fairbrother, D. H. Reich, *Phys. Rev. B* 79, 075312 (2009).
73. R. Meservey, P. M. Tedrow, J. S. Brooks, *J. Appl. Phys.* 53, 1563 (1982); J. S. Moodera, R. Meservey, unpublished tunneling data on amorphous Si barriers.
74. G. A. Gibson, R. Meservey, *J. Appl. Phys.* 58, 1584 (1985).
75. Y. Xu, D. Ephron, M. R. Beasley, *Phys. Rev. B* 52, 2843 (1995).
76. M. Julliere, *Phys. Lett.* 54A, 225 (1975).
77. R. Jansen, J. S. Moodera, *Phys. Rev. B* 61, 9047 (2000).
78. J. S. Moodera, J. Nowak, R. J. M. van de Veerdonk, *Phys. Rev. Lett.* 80, 2941 (1998).
79. C. Shen, A. Kahn, J. Schwartz, *J. Appl. Phys.* 89, 449 (2001).
80. R. C. Dynes, V. Narayanamurti, J. P. Garno, *Phys. Rev. Lett.* 41, 1509 (1978).
81. K. Maki, *Prog. Theor. Phys.* 32, 29 (1964).
82. J. S. Moodera, G. Mathon, *J. Magn. Magn. Mater.* 200, 248 (1999).
83. B. Huang, D. J. Monsma, I. Appelbaum, *Phys. Rev. Lett.* 99, 177209 (2007).
84. B. T. Jonker et al., *Nature Phys.* 3, 542 (2007).
85. S. Seo, B. N. Park, P. G. Evans, *Appl. Phys. Lett.* 88, 232114 (2006).
86. W. F. Brinkman, R. C. Dynes, J. M. Rowell, *J. Appl. Phys.* 41, 1915 (1970).
87. A. N. Caruso, D. L. Schulz, P. A. Dowben, *Chem. Phys. Lett.* 413, 321 (2005).
88. I. G. Hill, A. Rajagopal, A. Kahn, Y. Hu, *Appl. Phys. Lett.* 73, 662 (1998); S. T. Lee, X. Y. Hou, M. G. Mason, C. W. Tang, *Appl. Phys. Lett.* 72, 1593 (1998).
89. K. L. Wang, B. Lai, M. Lu, X. Zhou, L. S. Liao, X. M. Ding, X. Y. Hou, S. T. Lee, *Thin Solid Films* 363, 178 (2000).
90. F. Li, H. Tang, J. Anderegg, J. Shinar, *Appl. Phys. Lett.* 70, 1233 (1997).
91. J. E. Northrup, M. L. Chabinyc, *Phys. Rev. B* 68, 041202(R) (2003).
92. R. A. Street, N. F. Mott, *Phys. Rev. Lett.* 35, 1293 (1975).
93. M. Cinchetti, K. Heimer, J. P. Wüstenberg, O. Andreyev, M. Bauer, S. Lach, C. Ziegler, Y. Gao, M. Aeschlimann, *Nature Mater.* 8, 115 (2009).
94. A. J. Drew et al., *Nature Mater.* 8, 109 (2009).
95. A. R. McGhie, A. F. Garito, A. J. Heeger, *J. Cryst. Growth* 22, 295 (1974).

96. W. Warta, R. Stehle, N. Karl, *Appl. Phys. A Mater. Sci. Process.* 36, 163 (1985).
97. V. Podzorov et al., *Appl. Phys. Lett.* 83, 3504 (2003); V. Podzorov, V. M. Pudalov, M. E. Gershenson, *Appl. Phys. Lett.* 82, 1739 (2003).
98. V. C. Sundar, J. Zaumseil, V. Podzorov, E. Menard, R. L. Willett, M. E. Gershenson, J. A. Rogers, *Science* 303, 1644 (2004).
99. R. W. I. de Boer, M. E. Gershenson, A. F. Morpurgo, V. Podzorov, *Phys. Stat. Solidi A* 201, 1302 (2004).

2 Modeling Spin Injection and Transport in Organic Semiconductor Structures

P. P. Ruden and D. L. Smith

CONTENTS

ABSTRACT

We discuss spin-dependent charge carrier injection and transport in organic semiconductors from a device physics perspective. Structures considered have one or two ferromagnetic contacts to the organic semiconductor, and we discuss the conditions for which injection from the ferromagnetic contact may be strongly spin polarized. Electron tunneling from a ferromagnetic contact can have significant spin dependence because the spatial part of the electron wavefunction is different for majority and minority spin states near the Fermi surface. By contrast, thermionic emission is not very spin dependent. Therefore, if charge injection is dominated by tunneling from a ferromagnetic contact, the injection can be strongly spin polarized, but if it is dominated by thermionic emission or another process that does not depend on spin, the injection will be only weakly spin polarized. We first consider unipolar organic spin valve devices consisting of an organic semiconductor layer sandwiched between two ferromagnetic contacts into which one carrier type (either electrons or holes) is injected. Carrier transport in the organic semi-conductor is modeled by spin-dependent drift diffusion equations. Injected spin currents are related to the charge currents via the transport parameters of the ferromagnetic contacts. We examine the effects of the injected space charge and of spin relaxation in the semiconductor and compare the results of numerical calculations with those of an analytical model. For relatively thick organic semiconductor layers, the injected space charge can have strong effects on both charge injection and spin injection. We subsequently discuss bipolar organic device structures with spin-polarized injection of both electrons and holes. The recombination of injected electrons and holes is modeled as a Langevin process, and the effects of three relevant timescales—the recombination time, the carrier transit time, and the spin relaxation time—are explored. For example, the spatial variation of the spin current depends on the length of the spin relaxation time compared to the carrier transit time. If the applied bias is small, such that the carrier transit time is large compared to the spin relaxation time, the spin current has a minimum at the center of the device when symmetric electron and hole injection occur from both contacts. Last, the effects of spin-polarized electron and hole injection from ferromagnetic contacts on the formation and distribution of singlet and triplet excitons in a conjugated organic semiconductor are discussed for simple organic light-emitting diode structures. The formation of electron-hole pairs at a given site is modeled as a Langevin process, and the subsequent local relaxation into the lowest-energy exciton states is described by rate equations. Once formed, excitons may recombine in the semiconductor or diffuse through the material and recombine at the contact interfaces. The calculations yield steady-state spatial profiles for singlet and triplet excitons. Spin-polarized injection increases the formation of singlet excitons, and the diffusion of excitons has significant effects on the triplet exciton profile.

2.1   INTRODUCTION

In this chapter we discuss spin-polarized charge carrier injection, transport, and detection in organic semiconductor structures from a device physics point of view. This implies that our approach is macroscopic and consequently will not address bulk phenomena that occur on a length scale that is short compared to the mean free path of the charge carriers or their mean hopping distance. Currents are pictured as arising from drift and diffusion of charge (and spin) carriers. However, because the transport coefficients of mobility and diffusivity may be field dependent, the models are not restricted to linear response. Furthermore, the charge and spin carriers may be envisioned as essentially free or as small polarons as appropriate for the materials studied. The specific properties of organic semiconductors enter the models through the transport parameters and the boundary conditions imposed.

The structures considered are based on ferromagnetic metallic contacts, and electrical spin injection/detection occurs predominantly because of spin-dependent tunneling between the ferromagnetic contacts and the organic semiconductor. Electron tunneling from a ferromagnetic metal through a potential barrier into a semiconductor is spin dependent because the spatial part of the electron wave functions near the Fermi surface of a ferromagnetic metal is different for majority and minority spin electrons. Because of this spin dependence, ferromagnetic contacts in which tunneling is a significant electron transport mechanism can be used to both inject and detect spin-polarized currents in semiconductors.[1,2] The process may be described in a device model through contact resistances that are different for majority and minority spin electrons. Thus, the spin type with lower contact resistance will dominate the current flow from a contact biased to inject electrons into the semiconductor, leading to spin-polarized injection. In addition, the voltage drop across a ferromagnetic contact biased to collect electrons from the semiconductor will be different for majority and minority spins in the ferromagnetic contact, and thus this difference in voltage can be used to detect a spin-polarized flow of charge carriers into the contact.

There has been considerable progress in understanding electron spin physics in inorganic semiconductors in recent years. Specifically, electrical spin injection and detection has been clearly established in these materials.[3–9] We will compare electron spin-dependent processes in inorganic and organic semiconductors in order to apply the recent progress in understanding inorganic semiconductor spin physics to organic semiconductor device structures where the present experimental situation is less developed. Magnetoresistance effects in organic spin valve structures have been reported in the literature.[10–14] These devices typically consisted of an organic semiconductor layer sandwiched between two ferromagnetic contacts. If the organic semiconductor layer thicknesses are much larger than tunnel lengths, carrier transport in the semiconductor is expected to be diffusive, and the observed magnetoresistance is not attributed to tunneling from one metal contact to the other. However, not all experimental studies reported magnetoresistance for comparable organic semiconductor spin valves.[15]

Inorganic semiconductors like Si and GaAs are crystalline materials, and their electronic structure can be described by band theory with electronic states labeled by a band index and a wavevector. The materials technology is highly developed. These materials can be doped *n*-type or *p*-type over a wide range of doping levels, and the doping profiles can be precisely controlled. Because of doping, inorganic semiconductors used in devices are fairly conductive. The energies of an inorganic semiconductor's conduction and valence bands relative to the Fermi energy of a metal contact are usually pinned by interface effects that do not depend strongly on the type of metal. Hence, the Schottky barrier between the semiconductor conduction band minimum at the interface and the metal contact Fermi energy, for the common inorganic semiconductors, is largely independent of the type of metal and is not easily manipulated. Because of its favorable spin-dependent optical properties, the spin physics of GaAs has been extensively studied.[5–9] Both the lowest-conduction and highest-valence band states of GaAs occur at the Brillouin zone center. The lowest-energy conduction band states are made up primarily of antibonding *s*-orbitals of the Ga and As atoms, and the highest-energy valence band states are made up primarily of bonding *p*-orbitals of the Ga and As atoms. The spin-orbit interaction ($\vec{L} \cdot \vec{S}$ coupling) strongly mixes spin and orbital angular momentum in the *p*-orbital valence bands so that total angular momentum and its *z*-component (J and J_z) and not spin (S and S_z) are the appropriate quantum numbers. The four $J = 3/2$ states form heavy- and light-hole bands at the top of the valence band, and the two $J = 1/2$ states are shifted to lower energy. A momentum scattering event for a hole in one of the heavy- or light-hole bands changes the value of J_z so the lifetime of the hole angular momentum is very short. By contrast, electron states near the conduction band edge of GaAs are formed from s-orbitals so there is no strong mixing of spin and orbital angular momentum by the spin-orbit interaction. As a result, electron spin is the appropriate quantum number for electrons near the conduction band minimum. Because the spin-orbit interaction acts differently on the electron and hole states, most studies of spin physics in inorganic semiconductors like GaAs concentrate on electron, not hole, spin phenomena.

Organic semiconductors consist of π-conjugated hydrocarbons in which valence states are formed primarily from bonding combinations of π-orbitals centered on carbon atoms, and conduction states are formed primarily from the corresponding antibonding combinations of the π-orbitals. Typically these materials are highly disordered and their electronic states are not labeled by a wavevector. Conduction in most cases occurs by electron or hole hopping rather than by band transport, as described by the Boltzmann transport equation. As a result, carrier mobilities are much smaller in magnitude, and stronger functions of carrier density and electric field strength than in inorganic semiconductors. The materials technology of organic semiconductors is developing rapidly, but it is not yet nearly as advanced as that of the established inorganic semiconductors.

In devices, organic semiconductors are not usually doped, and the materials are essentially insulating. Electrons or holes are introduced into the organic semiconductors by injection from metallic contacts. The energy positions of an organic semiconductor's conduction and valence states relative to the Fermi energy of a metal are usually not strongly pinned by interface effects. The Schottky energy barriers between the organic semiconductor conduction levels and the metal contact Fermi energy are therefore strongly dependent on the type of metal[16] and can often be manipulated using, for example, self-assembled monolayers.[17] Because the principal elements making up most organic semiconductors (hydrogen and carbon) are very light, the spin-orbit interaction in these materials is weak. Unlike the case of inorganic (specifically III-V) semiconductors, one expects the spin physics of electrons and holes to be qualitatively similar in organic semiconductors.

Probably the most studied materials system among the inorganic semiconductors has been n-doped GaAs with iron contacts. Electrical spin injection, electrical spin detection, spin transport, and manipulation of electron spin dynamics using magnetic and strain fields have now all been demonstrated.[18–20] The use of carefully designed doping profiles in GaAs near the iron interface in order to achieve a Schottky contact in which tunneling through the depletion layer dominated current flow was critical for achieving electrical spin injection and detection.[1,2,21,22] Spin-dependent optical probes based on selection rules were important experimental tools in demonstrating electrical spin injection and spin transport. Because of the spin-dependent optical selection rules, arising from the spin-orbit splitting of the valence bands, absorption of circularly polarized light can be used to optically generate nonequilibrium spin-polarized carrier distributions. Likewise, rotation of the polarization angle of linearly polarized light transmitted through or reflected from a sample can be used to detect the presence of such spin-polarized carrier distributions. Circular polarized luminescence can also be used to detect a spin-polarized distribution. Thus, well-understood optical means are available to establish and characterize nonequilibrium spin-polarized carrier distributions in this material. It is very useful to be able to establish separately electrical spin injection and detection using optical means. If spins must be both injected and detected electrically, a null result will occur if either process fails. Then it is not easy to determine experimentally which process has failed and how to correct the problem. Unfortunately, this is the situation regarding spin experiments with organic semiconductors. Simultaneous electrical spin injection and detection has now been demonstrated in GaAs/Fe structures, but only after both processes had been independently demonstrated primarily using optical means.

Electron states in GaAs have spatial s-orbital character at the Brillouin zone center ($\vec{k} = 0$), but some p-orbital character is mixed in for $k > 0$. As a result, the spin-orbit interaction produces a loss of spin coherence for electrons of nonzero wavevector, and the rate with which coherence is lost increases the farther away the electron is from the zone center. Electron states of increasing wavevector are occupied as either the doping level is increased or the temperature is increased. Hence, the electron spin lifetime decreases with both increased doping level and temperature. For many spin physics measurements in n-GaAs, the doping level is therefore below about 10^{17}cm^{-3} and the temperature below about 20 K.

As pointed out above, the spin-orbit interaction is small in most organic semiconductors because they are made up of light elements. Hence, the optical selection rules that have been usefully applied to study spin physics in inorganic (III-V) semiconductors do not apply, at least not in a straightforward way, to most organic semiconductors. However, for the same reason electron *and* hole spin relaxation times are expected to be long in organic semiconductors and not to have the strong dependence on carrier density and temperature that electron spin lifetimes have in n-GaAs. Although the loss of the spin-dependent optical techniques complicates considerably the study of spin physics in organic semiconductors, the increased spin lifetimes, especially at higher carrier densities and temperatures, make organic semiconductors attractive for possible device applications that make use of spin-dependent processes.

The hyperfine interaction provides a significant spin relaxation mechanism for electrons in GaAs.[23] States at the bottom of the conduction band are predominantly derived from atomic s-states, and they therefore have considerable overlap with the gallium and arsenic nuclei. In contrast, the organic semiconductor states of interest are derived from carbon p-orbitals, which have a vanishing amplitude at the carbon nuclei. Furthermore, the natural abundance of ^{12}C is 98.89%, implying that only very few carbon nuclei have nonzero spin. Of course, in most organic semiconductors there are many hydrogen atoms, and even though the π-orbitals tend to have small amplitudes at the hydrogen nuclei, the spin dipole interaction between carriers in the π- or π^*-orbitals and the hydrogen nuclei may play a significant role in some of the magnetoresistance phenomena observed with organic semiconductor structures,[24] and that may also provide the dominant relaxation mechanism for spin-polarized injected carrier distributions. At present, the quantitative situation regarding spin relaxation in organic semiconductors is far from being well understood. There may be considerable variation from material to material, depending on details of the molecular structure. Electron spin resonance experiments with certain organic molecules have yielded line widths that are quite narrow,[25] consistent with the general notion that spin relaxation times in these materials are rather long.[26]

The chapter is organized as follows: In Section 2.2 we describe the models relevant to spin-polarized injection of charge carriers from ferromagnetic contacts into organic semiconductors. Section 2.3 explores unipolar organic semiconductor spin valves. Bipolar spin-polarized injection and recombination are discussed in Section 2.4. Section 2.5 presents a brief summary of the work and some conclusions.

2.2 INJECTION MODEL

To explore spin-polarized charge carrier injection we consider a thin organic semiconductor layer sandwiched between two metallic contacts. The problem is thus essentially one-dimensional. Initially, we will focus on the case of a ferromagnetic metal in direct contact with the organic semiconductor. Later we will consider the presence of thin insulating layers between the metal and the semiconductor.[27]

The (nonmagnetic) organic semiconductor is characterized by an energy gap, E_g, and by bands of conduction and valence states with narrow widths. For the purpose of charge carrier population, these bands are described by equal effective densities of states, n_0, which are of the order of the molecular (or monomer) density of the material.[28] (Neglecting the small spin-orbit coupling, equal effective densities of state $n_0/2$ may be attributed to spin-up and spin-down electrons and holes.) The material is not doped; hence, all mobile charge carriers are injected, and the large effective density of states ensures that nondegenerate statistics apply to essentially all cases of interest.

The charge carrier mobility is independent of spin. It may be taken as field independent (as is reasonable for some organic molecular crystals), or it may be taken to have the Poole-Frenkel form, $\mu_n(F) = \mu_{n0} \exp((|F|/F_0)^{1/2})$, where F is the electric field and μ_{n0} and F_0 are material parameters. The latter form describes well the field dependence of the mobility in many polymers.[29,30] Spin relaxation is described by a time constant, τ_s. For convenience we will formulate the problem in terms of electron injection. Hole injection may be more readily realized in many organic materials, but the similarity of the π- and π^*-states implies that the results derived for electrons will also apply to holes.

The ferromagnetic contacts are characterized by four parameters: conductivities for spin-up and spin-down electrons written in terms of the total conductivity, σ, $\sigma_\uparrow = \alpha\sigma$ and $\sigma_\downarrow = (1 - \alpha)\sigma$, respectively; the spin diffusion length, Λ; and the equilibrium Schottky barrier height, Φ_{B0}.

Figure 2.1 shows a schematic energy-level diagram for the injecting contact under bias. Also shown is the image-charge-induced barrier-lowering effect. The current

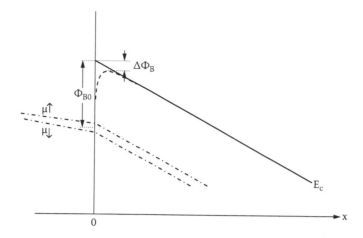

FIGURE 2.1   Schematic energy band diagram of a Schottky contact between a ferromagnetic metal and a semiconductor under bias such as to enable injection of electrons into the semiconductor.

densities for spin-up and spin-down electrons in the ferromagnetic metal contacts can be written as

$$j_\uparrow = (\sigma_\uparrow / e)(d\mu_\uparrow / dx)$$
$$j_\downarrow = (\sigma_\downarrow / e)(d\mu_\downarrow / dx)$$
(2.1)

where $\mu_{\uparrow\downarrow}$ are the electrochemical potentials (or quasi-Fermi levels) for spin-up and spin-down electrons, and e is the magnitude of the electron charge. Evidently, the charge current density is given by $j = j_\uparrow + j_\downarrow$, and the spin current density by $j_s = j_\uparrow - j_\downarrow$. Inside the bulk ferromagnetic contact, $\mu_\uparrow = \mu_\downarrow$ and the different conductivities of spin-up and spin-down electrons give rise to a net spin current that is equal to $(2\alpha - 1)j$. Under steady-state conditions the charge current is constant throughout the entire structure. However, due to spin scattering, the spin current tends to decrease toward the interface to the nonmagnetic semiconductor and the electrochemical potentials split. This may be described by[31]

$$d^2(\mu_\uparrow - \mu_\downarrow) / dx^2 = (\mu_\uparrow - \mu_\downarrow) / \Lambda^2$$
(2.2)

The splitting of the electrochemical potentials results in an increase in the ratio of the majority spin electron density to the minority spin electron density near the interface.

It is readily seen that the spin current at the injecting contact interface $(x = 0^-)$ is

$$j_s(0^-) = (2\alpha - 1)j + 2\alpha(1 - \alpha)(\sigma / \Lambda e)(\mu_\uparrow(0^-) - \mu_\downarrow(0^-))$$
(2.3)

We assume that there is no spin scattering as electrons traverse the interfacial layer between the metal and the semiconductor, so that both charge and spin currents are continuous across this interface.

2.2.1 THERMIONIC EMISSION

In this model the injected current for each spin direction, $J_{inj;\uparrow,\downarrow}$, is the sum of a thermionic emission current and an interface recombination current, which is the time-reversed process of thermionic emission:

$$J_{inj;\uparrow,\downarrow} = -ev_R\left[(n_0/2)\exp(-\Phi_{B;\uparrow,\downarrow}/kT) - n_{\uparrow,\downarrow}(0^+)\right] \qquad (2.4)$$

Here, v_R designates the effective recombination velocity, T the temperature, k Boltzmann's constant, and $\Phi_{B;\uparrow,\downarrow}$ the bias-dependent Schottky barrier height, which depends on spin because the electrochemical potentials in the contact at the interface are different for the two spin directions (see Figure 2.1). Image-charge-induced Schottky barrier lowering is comparatively strong in the organic semiconductors due to their small dielectric constants. The effect may be approximated by lowering the equilibrium Schottky barrier height by $\Delta\Phi_B \propto \sqrt{eF(0^+)/\kappa}$, where $F(0^+)$ is the electric field in the semiconductor in the absence of the image charge field and κ is the static dielectric constant. The electron densities for spin-up and spin-down carriers in the semiconductor at the interface are denoted by $n_{\uparrow,\downarrow}(0^+)$. In equilibrium, the two terms on the right side of Equation 2.4 are related by detailed balance.

For small spin-polarized electron densities in the organic semiconductor, i.e., $|\mu_\uparrow - \mu_\downarrow| \ll kT$, the injected electron charge current is given by

$$J_{inj} = -ev_R\left[n_0\exp(-\Phi_B/kT) - n(0^+)\right] \qquad (2.5)$$

and the spin current is

$$J_{s.inj} = -ev_R\left[n_0\exp(-\Phi_B/kT)(\mu_\uparrow(0^-) - \mu_\downarrow(0^-))/2kT - n_s(0^+)\right] \qquad (2.6)$$

where $\Phi_B = (1/2)(\Phi_{B;\uparrow} + \Phi_{B;\downarrow})$ and $n(x) = n_\uparrow(x) + n_\downarrow(x)$. The spin-polarized electron density is written as $n_s(x) = n_\uparrow(x) - n_\downarrow(x)$. While it presents no fundamental difficulties to relax the limitation to small quasi-Fermi level differences, the resulting problem is more complex numerically, and the limitation to small current polarization represents no severe restriction for the applicability of the results.[32]

2.2.2 INJECTION THROUGH TUNNELING

If the electric field in the organic semiconductor near the injecting contact is large, as is the case under strong bias, then charge carriers may tunnel through the relatively thin potential barrier associated with the Schottky contact. The details of the tunneling process of electrons from a metal into the states associated with diffusive conduction in an organic semiconductor are quite complex. Disorder is likely to be very important, and the standard models that build on translational symmetry parallel to the interface and extended states in both the metal and the semiconductor are not likely to be applicable. However, a reasonably general relationship that expresses the tunneling current in terms of transfer Hamiltonian matrix elements gives some insights[33]:

$$J_{tun} = -\frac{2\pi e}{S\hbar} \sum_{\rho\nu} [f(E_\rho) - f(E_\nu + e\Delta V)] \, |M_{\rho\nu}|^2 \, \delta(E_\rho - E_\nu) \qquad (2.7)$$

Here ρ is a set of quantum numbers (including spin) designating metal states that extend (as evanescent states) into the semiconductor, and ν designates organic semiconductor states. $M_{\rho\nu}$ is the transfer Hamiltonian matrix element, which may be written as[34]

$$M_{\rho\nu} = i\hbar \int dy\,dz < \rho\,|\,j_x(y,z)|\,\nu > \qquad (2.8)$$

where j_x is the x-component of the probability current operator and the integral extends over an area parallel to the contact interface and located at a point x within the tunnel barrier. The last factor in Equation 2.7 ensures energy conservation, but the occupation probabilities of the states involved are different because of the voltage drop, ΔV, across the tunnel barrier. It is evident from Equations 2.7 and 2.8 that the tunneling current can be different for electrons of different spin near the Fermi level in the ferromagnetic metal because their spatial wave functions ($|\rho\rangle$) are different.

We will return to Equation 2.7 in a later section that addresses thin insulating tunnel barriers. Here, for the case of the Schottky junction of interest, we calculate the total tunneling current in a simple first approximation using the WKB (Wentzel, Kramers, Brillonin) approximation and attribute a phenomenological fraction of $(1+f)/2$ to spin-up electrons and a fraction $(1-f)/2$ to spin-down electrons $(0 < f < 1)$. The contribution due to tunneling then enters the charge current simply by adding J_{tun} to the right side of Equation 2.5 and fJ_{tun} to the right side of Equation 2.6.

2.2.3 TRANSPORT IN THE ORGANIC SEMICONDUCTOR AND BOUNDARY CONDITIONS

Expressing the current in the organic semiconductor in drift diffusion approximation, the steady-state spin-up and spin-down carrier densities satisfy continuity equations of the form

$$0 = \frac{d}{dx}\left(\mu_n n_\uparrow F + D_n \frac{dn_\uparrow}{dx}\right) - \frac{n_\uparrow - n_\downarrow}{\tau_s}$$

$$\qquad (2.9)$$

$$0 = \frac{d}{dx}\left(\mu_n n_\downarrow F + D_n \frac{dn_\downarrow}{dx}\right) - \frac{n_\downarrow - n_\uparrow}{\tau_s}$$

Here D_n is the electron diffusivity, which is related to the mobility through the Einstein relation. Evidently, the charge and spin continuity equations become simply

$$0 = \frac{d}{dx}\left(\mu_n n F + D_n \frac{dn}{dx}\right) \tag{2.10}$$

$$0 = \frac{d}{dx}\left(\mu_n n_s F + D_n \frac{dn_s}{dx}\right) - \frac{2n_s}{\tau_s} \tag{2.11}$$

The charge density is coupled to the electric field through Poisson's equation,

$$\frac{dF}{dx} = -4\pi e(n_\uparrow + n_\downarrow)/\kappa \tag{2.12}$$

Assuming negligible spin scattering in the interfacial layer, both the charge and the spin current are continuous at the interface:

$$J_{inj} = j(0^-) \tag{2.13}$$

$$J_{s,inj} = j_s(0^-) \tag{2.14}$$

Thus, boundary conditions for Equations 2.10 and 2.11 are specified. Because the charge current and the spin current problems are decoupled for low spin polarization, the charge transport current continuity equation coupled to Poisson's equation can be solved self-consistently for a given voltage drop in the semiconductor. Subsequently, the spin current and spin density are calculated from the known injected electron density, $n(x) = n_\uparrow(x) + n_\downarrow(x)$, the electric field, $F(x)$, and the charge current density, $j = j_\uparrow(x) + j_\downarrow(x)$.

2.2.4 NUMERICAL RESULTS FOR SPIN-POLARIZED INJECTION

We first consider a structure consisting of an organic semiconductor sandwiched between a ferromagnetic contact ($\sigma_\uparrow > \sigma_\downarrow$, $\sigma = 10^5$S/cm, $\alpha = 0.8$) and a normal non-magnetic metal contact. Both metals form Schottky contacts, and we assume for simplicity that they have equal equilibrium barrier heights, and $f = 1/3$. The organic semiconductor layer thickness is 100 nm, and we assume an electron mobility of Poole-Frenkel form with $\mu_{n0} = 10^{-4}$cm^2/Vs and $F_0 = 7 \times 10^4$V/cm.

Figure 2.2 shows the calculated injected charge and spin current densities near the injecting contact for Schottky barrier heights of 0.3 eV (a) and 0.8 eV (b). The larger Schottky barrier leads to a smaller charge current density for a given applied voltage, but the spin polarization of the current density, for a given charge current density, is greater for the structure with the large Schottky barrier than for that with the small Schottky barrier. The greater spin polarization at high voltages is due to spin-selective tunneling through the Schottky barrier region, and at high

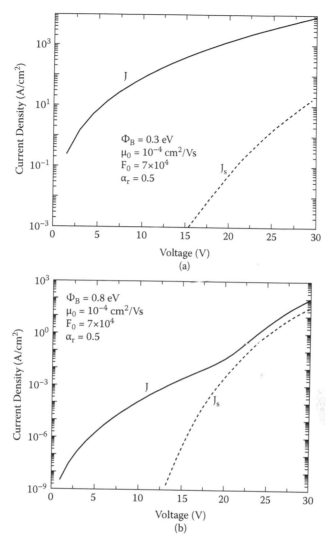

FIGURE 2.2 Charge current density (solid) and spin current density (dashed) for carrier injection from a ferromagnetic metal into an organic semiconductor (polymer) for two different Schottky barrier heights: 0.3 eV (a) and 0.8 eV (b).

voltages j_s/j approaches f, the fractional spin polarization of the tunneling current. The tunneling current increases more rapidly with increasing electric field than does the thermionic emission current. We plot the magnitudes of the individual contributions to the injection current arising from thermionic emission, interface recombination, and tunneling, together with the device current in Figure 2.3 for both Schottky contacts for (a) 0.3 eV and (b) 0.8 eV. For small Schottky barriers and for the large Schottky barriers at low bias, the net injection current is primarily determined by a combination of thermionic emission and its time-reversed process, interface recombination, as is shown in Figure 2.3. When electrical injection is

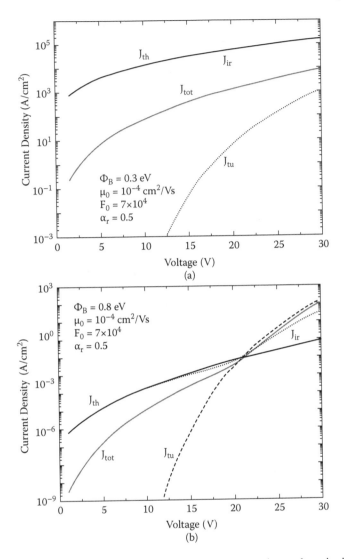

FIGURE 2.3 Magnitudes of the injection current components due to thermionic emission, interface recombination, and tunneling through the Schottky contacts of Figure 2.2. Also shown is the resulting total device current density.

dominated by processes related by time reversal, the electron populations on the two sides of the interface are nearly in equilibrium with each other. Because of the high electron density in the metal contact, the spin population in the contact cannot be driven out of local thermal equilibrium by practical current densities. Therefore, because the organic semiconductor is in quasi-equilibrium with the contact when thermionic emission and its time-reversed process dominate injection, the organic semiconductor-metal interface will be near thermal equilibrium, and there will not be efficient spin injection, as is evident in Figure 2.2. For large Schottky barriers at high bias, however, a combination of tunneling and interface recombination dominates injection, as is seen in Figure 2.3b. Tunneling and interface recombination are not related by time reversal, and the rates for these two processes are not connected by detailed balance. Therefore, when tunneling and interface recombination dominate injection, the electron populations on the two sides of the interface need not be in quasi-equilibrium with each other, and efficient spin injection is possible, as is shown in Figure 2.2b.

We may summarize these results as follows: the rather low mobility of organic semiconductors implies that injection currents are balanced by large interface recombination currents; i.e., injection is essentially limited by charge carrier transport from the injecting contact interface into the bulk of the organic semiconductor. Consequently, in contrast to the case of high-mobility inorganic semiconductors, the discontinuity in the quasi-Fermi level at the interface is very small. Thermionic emission is a rather ineffective process for spin-polarized injection, and effects such as image-charge-induced barrier lowering do not alter that conclusion. Furthermore, unless the ferromagnetic contact is half metallic ($\alpha \to 1$), significant spin polarization of the injected current is achieved only through a spin-selective process, for example, tunneling through the potential barrier associated with the Schottky contact.

Of course, tunneling through a thin insulating barrier layer between the metal contact and the semiconductor may also be suitable for spin-polarized carrier injection. Organic semiconductors may in fact have certain unique advantages over inorganic semiconductors inasmuch as they lend themselves as substrates for self-assembled molecular monolayers of other organic materials. This approach has been shown to be useful in controlling effective Schottky barrier heights.[17]

To model this type of contact we may envision a thin tunnel barrier in the region $-\delta < x < 0$ of Figure 2.4. Again, assuming no spin scattering at the contact interface, i.e., $j_s(-\delta) = j_s(0)$, and describing the spin-selective tunneling process through the

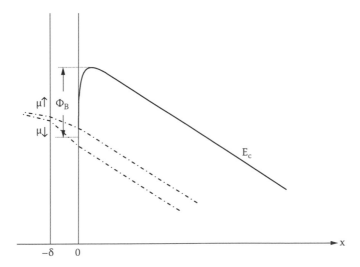

FIGURE 2.4   Schematic energy band diagram for a contact between a ferromagnetic metal and a semiconductor through a thin insulating tunnel layer under bias.

contact barrier by resistances $R_\uparrow$ and $R_\downarrow$, yields the differences of the quasi-Fermi levels at $x = 0^-$.

$$\Delta\mu(0^-) = \Delta\mu(-\delta) + \frac{1}{2}e(R_\uparrow - R_\downarrow)j + \frac{1}{2}e(R_\uparrow + R_\downarrow)j_S(0) \tag{2.15}$$

It is convenient to combine the effects of the ferromagnetic metal and the tunnel contact and to express the polarization effect in the semiconductor in terms of $\Delta\mu(0^-)$:

$$\Delta\mu(0^-) = \frac{1}{2}e\left(R'_\uparrow - R'_\downarrow\right)j + \frac{1}{2}e\left(R'_\uparrow + R'_\downarrow\right)j_S(0) \tag{2.16}$$

where the effective resistances are defined by

$$R'_\uparrow - R'_\downarrow = R_\uparrow - R_\downarrow - \frac{(2\alpha - 1)\Lambda}{\alpha(1 - \alpha)\sigma} \tag{2.17}$$

Because the last term in Equation 2.17 is proportional to Λ/σ, it tends to be very small for conventional ferromagnetic metals on the scale of resistances relevant to semiconductor devices. (This statement is of course incorrect for truly half-metallic materials, where $\alpha \to 1$.)

Evidently, treating the effective contact resistances as linear, i.e., independent of the bias, is only a rather crude approximation. It is unlikely to be a correct description of spin-selective tunneling through an insulating interfacial layer. For that case, a relatively strong decrease of the contact resistance with increasing bias is expected. In the terminology of Equation 2.7, this arises from two effects. First, the lowest energy E_ρ at which tunneling into transport states in the organic semiconductor is possible exceeds the metal Fermi level in equilibrium by an energy Φ_B. However, under bias the voltage drop across the barrier is δ_V, and the lowest-energy tunneling process can occur at an energy of $\Phi_B - \delta_V$. Hence, a simple model that accounts only for the population effects at energies relevant for injection leads to a bias dependence of the contact resistance of the form $R_{\uparrow\downarrow}(\delta V) = R_{\uparrow\downarrow}(0)\exp(-e\delta V/kT)$.[35] Second, the applied bias deforms the potential barrier through which the carriers tunnel. This is expected to have a strong effect on tunneling and will lead to changes in the ratio $R_\uparrow(\delta V)/R_\downarrow(\delta V)$.[36] In the absence of a microscopic model for the tunneling transport, we may explore the effect of the bias dependence of the contact resistance with a relationship of the form

$$R_{\uparrow\downarrow}(\delta V) = R_{\uparrow\downarrow}(0)\exp(-\gamma e\delta V/kT) \tag{2.18}$$

where γ is a phenomenological parameter that describes the strength of the decrease in contact resistance with increasing bias.

2.3   MODELING OF AN ORGANIC SEMICONDUCTOR SPIN VALVE

In the following sections we consider devices that consist of an organic semiconductor layer sandwiched between two ferromagnetic contacts. Following the discussion above we will generally assume that spin-selective tunnel barriers separate the metal contacts from the semiconductor, and that this effect may be adequately described with effective contact resistances. The carrier extracting contact is thus modeled in analogy to the injecting contact, but the relative polarization alignment of the ferromagnetic contacts controls the relative magnitudes of the contact resistances for spin-up and spin-down electrons.

It is instructive first to consider a case in which an essentially analytical solution to the transport problem may be obtained.[35] We assume that the material and bias conditions are such that the electric field dependence of the mobility may be neglected, and make the following approximations: (1) τ_s is long compared to the transit time, i.e., the last terms in Equation 2.9 are neglected; (2) the effect of the injected space charge is neglected, i.e., the electric field is constant throughout the organic semiconductor; and (3) ohmic boundary conditions apply, i.e., $\mu_\uparrow$ and $\mu_\downarrow$ are continuous at $x = 0$ and $x = d$. The resulting charge current and spin current in the organic semiconductor can then be expressed as

$$j = G(V') \left[\exp\left(\frac{eV'}{kT} \right) \times \cosh\left(\frac{\Delta\mu(0)}{2kT} \right) - \cosh\left(\frac{\Delta\mu(d)}{2kT} \right) \right] \qquad (2.19)$$

$$j_s = G(V') \left[\exp\left(\frac{eV'}{kT} \right) \times \sinh\left(\frac{\Delta\mu(0)}{2kT} \right) - \sinh\left(\frac{\Delta\mu(d)}{2kT} \right) \right] \qquad (2.20)$$

V' is the voltage dropped across the semiconductor, and $G(V')$ is given by

$$G(V') = \frac{-e\mu_n \frac{V'-V_0}{d} n_0 \exp\left(-\frac{\Phi_{BR}}{kT} \right)}{\exp\left(\frac{eV'-eV_0}{kT} \right) - 1} \qquad (2.21)$$

Here eV_0 is the built-in potential ($\Phi_{BR} - \Phi_{BL}$) and $\Phi_{BR(L)}$ are the barrier heights of the right (left) contacts. We assume $\tau_s \gg d^2/(\mu_n kT/e)$; hence, the injected spin current density is constant throughout the semiconductor. From Equations 2.19 and 2.20 we see that if $eV' \gg kT$, the charge and spin currents depend only on the difference of the quasi-Fermi levels for spin-up and spin-down electrons at the left (injecting) electrode. However, in the low-bias regime, both contacts control the currents in the semiconductor. Equation 2.16, an analogous equation for the right (carrier-extracting) contact, and Equations 2.19 and 2.20 can be solved self-consistently for a particular bias, V'. The total voltage applied to the device is subsequently obtained from

$$V = V' + \Delta V_L + \Delta V_R \qquad (2.22)$$

$$\Delta V_{L,R} = -(1/4) \left[\left(R'_{\uparrow L,R} + R'_{\downarrow L,R} \right) j + \left(R'_{\uparrow L,R} - R'_{\downarrow L,R} \right) j_s \right] \qquad (2.23)$$

The difference between parallel (P) and antiparallel (AP) alignments of the contact magnetizations is expressed through the terms involving $R'_{\uparrow R} - R'_{\downarrow R}$, which have the same sign as $R'_{\uparrow L} - R'_{\downarrow L}$ in P configuration, but the opposite sign in AP configuration. Thus, the currents and total applied voltages for the two contact alignments are obtained as a function of V'. Finally, the magnetoresistance ratio is defined as

$$\text{MR} = \frac{V_{AP}}{V_P} \frac{J_P}{J_{AP}} - 1 \qquad (2.24)$$

Here V_{AP} and V_P denote the applied biases for the P and AP configurations, and J_P and J_{AP} are the corresponding charge current densities.

We use the analytical calculation outlined for relatively thin organic semiconductor layers where space charge effects due to the injected carriers are small under low bias. The thickness, d, is taken to be 10^{-5}cm, and the mobility is 10^{-2} cm^2/Vs. The spin relaxation time thus needs to be much greater than 200 ns to be consistent with our approximations above. For simplicity, we take $V_0 = 0$, and choose a barrier height of 0.2 eV. All calculations are for room temperature conditions, and ferromagnetic metal transport parameters of $\sigma = 10^5$S/cm, $\Lambda = 10^{-5}$cm, and $\alpha = \frac{3}{4}$ are assumed. Figure 2.5 shows the calculated charge and spin currents vs. total applied voltage for P and AP alignments. The ratios $R'_{\uparrow}/R'_{\downarrow}$ for both left and right contacts are 1/3 and $R'_{\uparrow L} = 10^{-2}\,\Omega\text{cm}^2 = R'_{\uparrow R}$. We also consider cases in which the contact resistance values of the injecting and extracting contacts are different. The current polarizations for three cases are depicted in Figure 2.6. Here the results for effective contact resistances are compared with those obtained for reduced effective resistances on

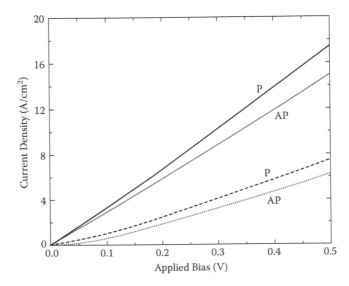

FIGURE 2.5 Charge current density (solid curves) and spin current density (dashed curves) for P (thick lines) and AP (thin lines) contact magnetization as a function of applied bias. The ratios of $R'_{\uparrow}/R'_{\downarrow}$ for both left and right contacts are 1/3 and $R'_{\uparrow L} = 10^{-2}\,\Omega\text{cm}^2 = R'_{\uparrow R}$.

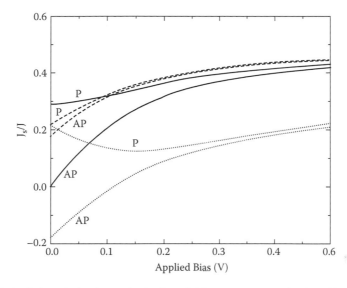

FIGURE 2.6   Spin polarization, J_s/J, for P and AP contact magnetization alignments and different effective contact resistances. The ratios of $R'_\uparrow/R'_\downarrow$ for both left and right contacts are 1/3 in all cases. The magnitudes of the contact resistances are $R'_{\uparrow L} = 10^{-2}\Omega\text{cm}^2 = R'_{\uparrow R}$ (solid lines), $R'_{\uparrow L} = 10^{-2}\Omega\text{cm}^2 = 10R'_{\uparrow R}$ (dashed lines), and $10R'_{\uparrow L} = 10^{-2}\Omega\text{cm}^2 = R'_{\uparrow R}$ (dotted lines).

the injecting or collecting side. It can be seen that in the nonsymmetric situations, even in the *AP* alignment case the current polarization extrapolates to a nonzero value at vanishing bias. If the effective contact resistance of the extracting contact is larger than that of the injecting contact, it tends to determine the spin polarization at low bias, which in the sign convention adopted here is negative at low voltage and becomes positive only when the injecting contact begins to dominate (here near $V = 0.1$ V).

Figure 2.7 shows the calculated MR values for three sets of effective contact resistances. It is evident that the MR may decrease with increasing bias (at small voltages) if the effective contact resistances of the injecting contact are smaller than those of the extracting contact.

Next, we explore the effects associated with the injected space charge, as is particularly relevant under high applied bias, and we do not presuppose the continuity of the quasi-Fermi levels at $x = 0$ and $x = d$. (The discontinuity, however, is very small for all cases examined due to the rather low carrier mobility of organic semiconductors, as discussed in Section 2.2.)[37] We also extend the investigation to thicker semiconductor layers and allow for finite spin relaxation times. Relaxing the three approximations described above implies that we have to solve the continuity equations (Equation 2.9) numerically, together with the Poisson equation (Equation 2.12).

Results for the current polarization in a device with $d = 10^{-5}$ cm are shown in Figure 2.8a. Additional results for a $d = 2 \times 10^{-5}$ cm semiconductor layer are shown in Figure 2.8b. In these calculations the current polarization is determined at $x = 0^+$.

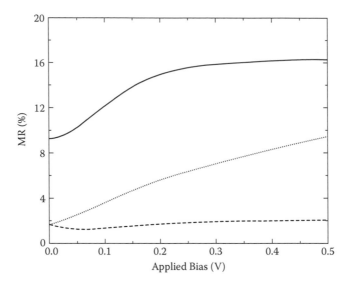

FIGURE 2.7 MR plotted as a function of the applied bias for three cases of different effective contact resistances shown in Figure 2.6.

In both figures we compare the analytical results (without space charge effects) to the numerical results that include the self-consistent treatment of the Poisson equation. Clearly, the buildup of the injected space charge reduces both the charge and the spin current. It has a greater effect in thicker devices, and evidently it suppresses the spin current to a greater extent than the charge current, thus diminishing the relative current polarization. The results also show that at high voltage the current polarization tends to saturate. The saturation value is equal to $(1 - R'_{\uparrow L}/R'_{\downarrow L})/$ $(1 + R'_{\uparrow L}/R'_{\downarrow L})$; i.e., it depends only on the ratio $R'_{\uparrow}/R'_{\downarrow}$ of the injecting contact and approaches 1/2 for the cases shown. However, the voltage at which saturation is reached clearly depends on the magnitude of the effective contact resistances.

The finite spin relaxation time implies that the current polarization decreases between $x = 0^+$ and $x = d^-$, as is demonstrated by the numerical results depicted in Figure 2.9. In this calculation we also include the effect of image-charge-induced barrier lowering, and to avoid unrealistically high current densities, we choose the equilibrium barrier heights to be $\Phi_B = 0.4$ eV. Since the spin current at the extracting contact is critical for the magnetoresistance of the spin valve, the decreasing spin current with decreasing spin relaxation time inevitably reduces the magnetoresistance, as is shown in Figure 2.10. Clearly, the carrier transit time decreases with increasing applied bias, and consequently the current polarization between the injecting and extracting contacts is maintained to a greater extent. This also increases the magnetoresistance, which therefore increases with increasing bias.

The model calculations above yield MRs that increase or decrease moderately with increasing voltage at low bias, and that increase monotonically with increasing

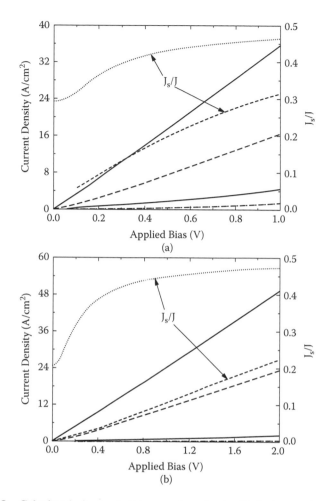

FIGURE 2.8 Calculated charge (solid curves) and spin (dashed curves) current densities for numerical (thick lines) and analytical (thin lines) models (left scale). The current polarizations for the two cases are also shown (right scale). The values of the contact resistances are $R'_{\uparrow R} = R'_{\uparrow L} = 10^{-2} \Omega cm^2, (R'_{\uparrow}/R'_{\downarrow} = 1/3)$; (a) is for device length 100 nm and (b) is for device length 200 nm.

voltage at high bias. Most experiments to date appear to show decreasing magnetoresistance with increasing bias.[38] Thus far, we have treated the effective contact resistances as linear, i.e., independent of the bias. We now explore the effect of a contact resistance that decreases with increasing bias as outlined above. In doing so, we return to the simple analytical model formulated through Equations 2.19 to 2.24. The resulting current polarization and MR are plotted in Figures 2.11 and 2.12. It is evident that current polarizations decrease relative to the case of constant effective contact resistances, but most striking is the decrease of the MR with increasing bias.

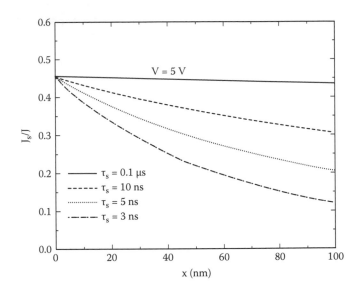

FIGURE 2.9 Spin polarization inside the organic semiconductors for different spin relaxation times. The applied bias is 5 V. The values of the effective contact resistances are $R_{\uparrow R} = R_{\uparrow L} = 10^{-2}\ \Omega\mathrm{cm}^2$, $(R_\uparrow / R_\downarrow = 1/3)$.

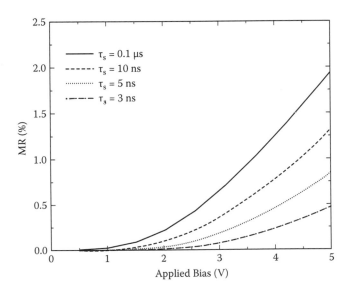

FIGURE 2.10 Calculated MR plotted as a function of applied bias for different spin relaxation times ($d = 100$ nm). The values of the effective contact resistances are $R_{\uparrow R} = R_{\uparrow L} = 10^{-2}\ \Omega\mathrm{cm}^2$, $(R_\uparrow / R_\downarrow = 1/3)$.

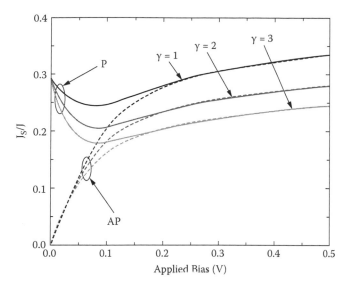

FIGURE 2.11 Spin polarization, J_s/J, for P and AP contact magnetization for three cases of different γ parameters. Here the contact resistances are allowed to vary with bias across the layer as described in the text. The $\delta V = 0$ values are $R'_{\uparrow L0} = 10^{-2}$ $\Omega cm^2 = R'_{\uparrow R0}$, $(R'_{\uparrow L0}/R'_{\downarrow L0} = 1/3)$.

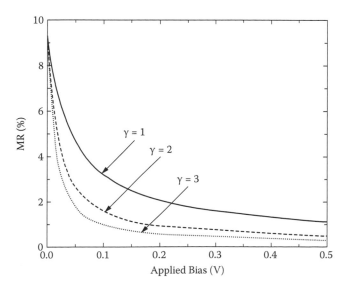

FIGURE 2.12 MR plotted as a function of the applied bias for the three cases as shown in Figure 2.11.

2.4 BIPOLAR SPIN-POLARIZED INJECTION, TRANSPORT, AND RECOMBINATION

Unlike the case of inorganic materials where the Schottky energy barriers weakly depend on the metal work function, the Schottky energy barriers between the metal and organic semiconductors strongly depend on the metal work functions. Hence, low-work-function metals can be used to inject electrons and high-work-function metals can be used to inject holes in devices fabricated from organic semiconductors. The most widely studied devices fabricated using conjugated organic semiconductors are organic light-emitting diodes (OLEDs), which are already seeing commercial use.[39] Recent experiments suggest that spin-polarized injection enhances the electroluminescence in OLEDs.[40]

In this section, we consider ambipolar carrier injection in structures consisting of a relatively thin organic semiconductor film, sandwiched between two ferromagnetic contacts. The work functions of the two contacts are different so that one contact can inject electrons and the other contact can inject holes. A schematic device structure is shown in Figure 2.13.

Numerical results are used to investigate the effect of contact polarization, contact conductivity, and additional spin-dependent interfacial contact resistances on spin injection.

2.4.1 BIPOLAR INJECTION AND TRANSPORT MODEL

We considered the structure shown in Figure 2.13, but in general allow for the possible existence of thin insulating layers between the metal contacts and the organic semiconductor, which will be modeled with simple linear contact resistances that are different for the two spin directions. Figure 2.14 shows the schematic energy band diagrams of the device structure under bias for the cases of parallel (a) and antiparallel (b) magnetization of the contacts.[32] Also indicated are the tunnel barrier layers of thickness δ. In the case of parallel magnetization, the quasi-Fermi levels of spin-up and spin-down carriers (both electrons and holes) have to cross inside the organic semiconductor. For antiparallel contact magnetization the quasi-Fermi levels do not cross.

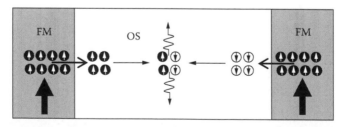

FIGURE 2.13 Schematic device structure of organic semiconductor light-emitting diode with spin-polarized injection of charge carriers.

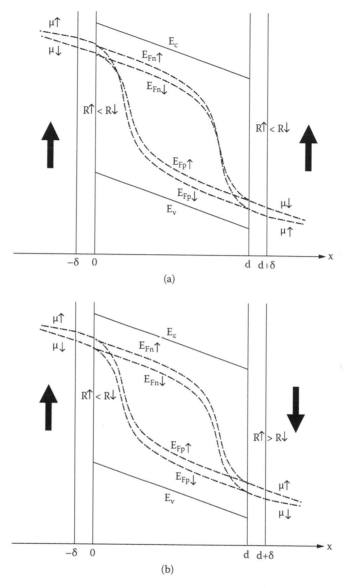

FIGURE 2.14 Schematic band diagrams of the device structure, including spin-selective tunneling contact resistances between the ferromagnetic contacts and the organic semiconductors: (a) parallel polarization and (b) antiparallel polarization of the contact magnetization. (From M. Yunus, P. P. Ruden, D. L. Smith, *J. Appl. Phys.* 103, 103714 [2008].)

Once the carriers are injected into the organic semiconductor, their transport is governed by spin-dependent continuity equations coupled with Poisson's equation:

$$0 = +\frac{1}{e}\frac{\partial j_{n\uparrow}}{\partial x} - \gamma_{AP}n_{\uparrow}p_{\downarrow} - \gamma_{P}n_{\uparrow}p_{\uparrow} - \frac{n_{\uparrow} - n_{\downarrow}}{\tau_s} \tag{2.25a}$$

$$0 = +\frac{1}{e}\frac{\partial j_{n\downarrow}}{\partial x} - \gamma_{AP}n_{\downarrow}p_{\uparrow} - \gamma_{P}n_{\downarrow}p_{\downarrow} - \frac{n_{\downarrow} - n_{\uparrow}}{\tau_s} \tag{2.25b}$$

$$0 = -\frac{1}{e}\frac{\partial j_{p\uparrow}}{\partial x} - \gamma_{AP}p_{\uparrow}n_{\downarrow} - \gamma_{P}p_{\uparrow}n_{\uparrow} - \frac{p_{\uparrow} - p_{\downarrow}}{\tau_s} \tag{2.25c}$$

$$0 = -\frac{1}{e}\frac{\partial j_{p\downarrow}}{\partial x} - \gamma_{AP}p_{\downarrow}n_{\uparrow} - \gamma_{P}p_{\downarrow}n_{\downarrow} - \frac{p_{\downarrow} - p_{\uparrow}}{\tau_s} \tag{2.25d}$$

$$\frac{\partial F}{\partial x} = \frac{4\pi e}{\kappa}(p_{\uparrow} + p_{\downarrow} - n_{\uparrow} - n_{\downarrow}) \tag{2.26}$$

Here $n_{\uparrow,\downarrow}$ and $p_{\uparrow,\downarrow}$ are the spin dependent electron and hole concentrations, γ_{AP} and γ_P are the recombination coefficients for spin parallel pairs ($\uparrow\uparrow$, $\downarrow\downarrow$) and spin antiparallel pairs ($\uparrow\downarrow$, $\downarrow\uparrow$), and τ_s is the spin relaxation time. Electrons and holes approach each other due to their mutual coulomb attraction. Because the carrier mobilities in organic semiconductors are relatively low, this diffusive transport process is the dominant limitation for recombination of the electron-hole pairs, and it does not depend on the relative orientation of the electron and hole spins. Hence, we assume that γ_{AP} and γ_P are equal in the organic semiconductor and that they are given by the Langevin recombination coefficient, γ_L:

$$\gamma_L = \frac{4\pi e(\mu_n + \mu_p)}{\kappa} \tag{2.27}$$

where μ_n and μ_p are the mobilities of electrons and holes, respectively. The electron and hole current densities can be written as

$$j_{n\uparrow,\downarrow} = e\mu_n\left(n_{\uparrow,\downarrow}F + \frac{kT}{e}\frac{\partial n_{\uparrow,\downarrow}}{\partial x}\right) \tag{2.28a}$$

$$j_{p\uparrow,\downarrow} = e\mu_p\left(p_{\uparrow,\downarrow}F - \frac{kT}{e}\frac{\partial p_{\uparrow,\downarrow}}{\partial x}\right) \tag{2.28b}$$

where the Einstein relation has been used.

The boundary conditions are given by specifying the particle currents for each spin type at the boundaries, i.e., at $x = 0$ and $x = d$. For each spin type, the injected particle current, $J_{n(p);\,\uparrow,\downarrow}$, is the sum of a thermionic emission current and the interface

TABLE 2.1

Transport Parameters

	σ (S/cm)	α	Λ (cm)	
FM metal contacts	10^5	0.9	10^{-5}	
Organic semiconductor	E_g (eV)	μ_n, μ_p (cm²/Vs)	n_0 (cm⁻³)	τ_s (s)
	2.4	10^{-3}	10^{21}	10^{-7}

recombination current. For the present model, we assume that tunneling through the depletion regions at the Schottky contacts is negligible. We assume that the left electrode ($x = 0$) injects electrons and extract holes, and vice versa, for the right electrode ($x = d$).

We assume that the Schottky barriers are such as to enable electron injection for one contact and hole injection for the other. To minimize parameters, we take the Schottky barrier heights for electron and hole injection to be the same and equal to 0.4 eV in all calculations. For the same reason, we take the electron and hole mobilities to be equal and independent of the electric field. All relevant material parameters are listed in Table 2.1. The organic semiconductor layer thickness is 100 nm, and the polarizations of the contacts are parallel in all examples.

For the thin insulating interface layers we assume that the tunneling probability of spin-up electrons is greater than that of spin-down electrons, i.e., $R_\uparrow < R_\downarrow$. The resulting injected spin and charge current densities are plotted in Figure 2.15. The values of the spin-selective contact resistances are $R_\uparrow = 5 \times 10^{-3}\ \Omega\ cm^2$ and $R_\downarrow = 10^{-2}\ \Omega\ cm^2$, such that the ratio of $R_\uparrow/R_\downarrow$ is 1/2. In the plot, we exclude the voltage drop across the

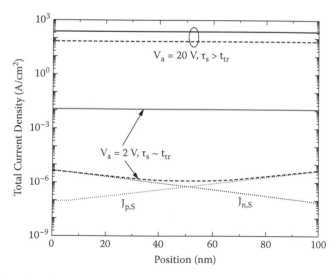

FIGURE 2.15   Spatial dependence of the total charge (solid) and spin (dashed) current densities inside the organic semiconductor. (From M. Yunus, P. P. Ruden, D. L. Smith, *J. Appl. Phys.* 103, 103714 [2008].)

tunneling contact resistance, as it is small compared to the overall applied bias (but it is critical for the magnetoresistance).

We investigate the effects of the three relevant timescales—the recombination time, the spin relaxation time, and the carrier transit time—on the spin and charge carrier transport through the organic semiconductor. The carrier transit time is defined as $t_{tr} = d^2/\mu_{n,p}V_a$, where V_a is the applied bias. The average recombination time for the electrons and holes can be defined as $d/[\gamma_L \int_0^d p\,dx]$ and $d/[\gamma_L \int_0^d n\,dx]$, respectively, where γ_L is the Langevin recombination coefficient. The carrier spin relaxation time, τ_s, is held fixed at 0.1 μs.

First, we explore the effect of the transit time on the charge and spin current inside the organic semiconductor. We plot the total charge and spin current densities inside the organic semiconductor in Figure 2.15.[32] Results are shown for two cases; applied bias 2 V (lower curves), and applied bias 20 V (upper curves). At low applied bias, $t_{tr} = 0.05$ μs, which is comparable to τ_s, but at large bias, $t_{tr} < \tau_s$ ($t_{tr} = 5$ ns). Due to the conservation of charge, the total charge current must be constant throughout the device (solid lines). However, total spin current depends on position. The spatial dependence of the spin current arises from the finite spin relaxation time. When the carrier transit time is small compared to the spin relaxation time, carrier transport from the injecting contact to the extracting contact occurs without spin flip and the spin current is constant, as shown by the upper dashed curve. On the other hand, when the carrier transit time is comparable to, or larger than, the spin relaxation time, some carriers flip their spin during their transit. The spin current decreases from the injecting contact toward the extracting contact. Thus, electron spin current decreases from left to right, and vice versa, for hole spin current, as shown in the figure by $J_{n,s}$ and $J_{p,s}$. Due to the electron-hole symmetry in our example (same mobilities, same Schottky barrier heights), the global minimum of the total spin current occurs at the center of the device.

Next we consider the effects of the average recombination time on the spin and charge carrier transport. Figures 2.16 to 2.18 present charge carrier density, spin density, and current density profiles inside the organic semiconductor, respectively.[32] Results are again shown for the two applied biases: 2 V and 20 V. Here, we focus on the recombination time between electrons and holes and the transit time. The upper curves show the results when the recombination time is small compared to the transit time, and the lower curves show the results when the recombination time is large compared to the transit time. The recombination time depends on the level of injection and therefore on the bias. For 2 V, the average carrier concentration is 8.5×10^{14}cm^{-3}, which corresponds to a recombination time of 1 μs. However, the transit time is 0.05 μs, and thus the carrier recombination time is long compared to the carrier transit time at 2 V.

On the other hand, at 20 V the average carrier concentration is 4×10^{17} cm^{-3}, which corresponds to a recombination time of 2 ns, but the carrier transit time is 5 ns. Thus, in this instance the carrier recombination time is short compared to the carrier transit time. When the recombination time is long (at 2 V), carriers travel without recombination and their densities are almost constant throughout the device. However, at the contact, carrier concentrations are determined by the boundary conditions. The Schottky barrier height for electrons is very large at the right (hole-injecting) contact,

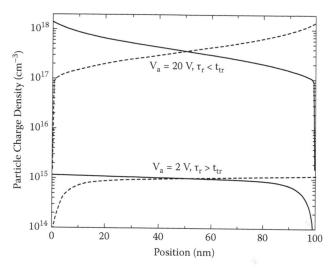

FIGURE 2.16 Charge carrier—electrons (solid) and holes (dashed)—density profiles inside the organic semiconductor for different bias voltages. (From M. Yunus, P. P. Ruden, D. L. Smith, *J. Appl. Phys.* 103, 103714 [2008].)

forcing a sudden decrease in the electron concentration near the contact. Similarly, the hole concentration decreases rapidly at the left (electron-injecting) electrode. By contrast, when the recombination time is short compared to the transit time (at 20 V), electrons and holes recombine everywhere and carrier concentrations vary appreciably across the semiconductor.

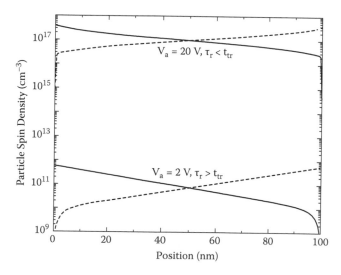

FIGURE 2.17 Spin—electrons (solid) and holes (dashed)—density profiles inside the organic semiconductor for different bias voltages. (From M. Yunus, P. P. Ruden, D. L. Smith, *J. Appl. Phys.* 103, 103714 [2008].)

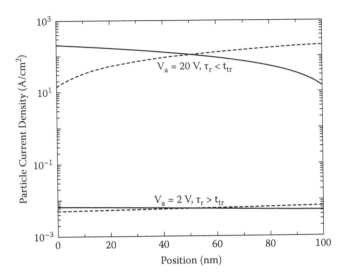

FIGURE 2.18 Particle current—electrons (solid) and holes (dashed)—density profiles inside the organic semiconductor for different bias voltages. (From M. Yunus, P. P. Ruden, D. L. Smith, *J. Appl. Phys.* 103, 103714 [2008].)

The spin density profiles (Figure 2.17) almost follow the carrier density profiles except for one difference. At the extracting contact (either carrier), the spin density is negative. As explained in Figure 2.14, the quasi-Fermi levels for spin-up and spin-down carriers must cross inside the semiconductor when the contact polarizations are parallel. In the cases examined here, this crossover always occurs near the extracting contact.

The recombination of electrons and holes also affects the particle current profile inside the semiconductor. Figure 2.18 shows the electron and hole charge current densities. When the bias voltage is 2 V, recombination in the bulk of the organic semiconductor is negligible. Thus, the electron and hole currents are nearly constant throughout the material. However, when the recombination time is short compared to the transit time, the electron and hole current densities decrease from the injecting contact toward the extracting contact, as evident from the upper curve of Figure 2.18. Although the individual components of electron and hole current vary across the semiconductor, the total charge current is constant, as was shown in Figure 2.15.

2.4.2 EFFECTS OF SPIN-POLARIZED BIPOLAR INJECTION ON EXCITONS

Electrons and holes in organic semiconductors tend to form strongly bound excitons. Excitons can have spin $S = 0$ (singlet) or $S = 1$ (triplet), depending on the total spin of the electron-hole pair. Only singlet excitons can recombine radiatively (and quickly), giving rise to light output. Triplet excitons eventually recombine via nonradiative processes. Thus, the electroluminescence quantum efficiency, η_{EL}, of organic semiconductors is limited by the fraction of excitons that form as singlets, χ_S. Due to ratio of triplet to singlet states (3:1), χ_S is only ¼, in the simplest picture, and the maximum quantum efficiency should be less than 25%. (Experiments have yielded a wide range of χ_S values[41] and different approaches[42] to explain these results.)

Here, we discuss a device model for spin-polarized electron and hole injection, exciton formation, and recombination.[43] The model can be divided into three stages: Injected spin-polarized electrons and holes propagate through the material and form local electron-hole pairs due to their mutual coulomb attraction. The charge carriers may undergo spin flips due to interaction with nuclear spins or due to other scattering processes that involve the spin-orbit interaction. Once electron-hole pairs share a particular molecule, or a monomer in the case of a polymer material, they relax rapidly into tightly bound exciton states. The excitons then recombine or diffuse, but they are not expected to dissociate.

The coupled spin-dependent device equations are solved for the steady state with boundary conditions expressed through the (spin-polarized) injection and extraction current. We again take $\gamma_P = \gamma_{AP}$, and given by the Langevin recombination coefficient, γ_L. Solution of the transport problem yields densities of spin parallel and antiparallel, electron-hole pairs. The electron-hole pairs can be in the singlet state $\{1/\sqrt{2}[|\uparrow\downarrow> - |\downarrow\uparrow>]\}$ or in one of the three triplet states $\{(1/\sqrt{2}[|\uparrow\downarrow> + |\downarrow\uparrow>], |\uparrow\uparrow>, |\downarrow\downarrow>\}$. Evidently, an electron-hole pair with antiparallel spins may be in either a singlet or a triplet state, but all electron-hole pairs with parallel spins are in triplet states. The densities of electron-hole pairs in these states are denoted by $N_{S,AP}$, $N_{T,AP}$, and $N_{T,P}$. The electron and hole forming a local pair are correlated; hence, the relaxation into the lowest-energy exciton states can be spin dependent. The governing local rate equations for $N_{T,P}$, $N_{T,AP}$, and $N_{S,AP}$ may be written as[44]

$$\frac{\partial N_{T,P}}{\partial t} = \gamma_L(n_\uparrow p_\uparrow + n_\downarrow p_\downarrow) - k_T N_{T,P} - \frac{(N_{T,P} - N_{T,AP} - N_{S,AP})}{\tau_s} \quad (2.29)$$

$$\frac{\partial N_{T,AP}}{\partial t} = \frac{\gamma_L(n_\uparrow p_\downarrow + n_\downarrow p_\uparrow)}{2} - k_T N_{T,AP} - \frac{N_{T,AP} - N_{T,P}/2}{\tau_s} \quad (2.30)$$

$$\frac{\partial N_{S,AP}}{\partial t} = \frac{\gamma_L(n_\uparrow p_\downarrow + n_\downarrow p_\uparrow)}{2} - k_S N_{S,AP} - \frac{N_{S,AP} - N_{T,P}/2}{\tau_s} \quad (2.31)$$

Here k_S and k_T are the spin-dependent formation rates of singlet and triplet excitons, and τ_s is the spin relaxation time constant for localized electron-hole pairs, which may be different from the single-carrier spin relaxation time that appears in the device equations. However, to minimize the introduction of parameters, we neglect this distinction. The excitons may recombine at the site where they form, or they may diffuse through the semiconductor and recombine at other sites or at the contact interfaces. Diffusion of singlet and triplet excitons is governed by

$$\frac{\partial N_S}{\partial t} = D_{ex} \frac{\partial^2 N_S}{\partial x^2} + k_S N_{S,AP} - \frac{N_S}{\tau_{S,rec}} - \frac{N_S - N_T}{\tau_{isc}} \quad (2.32)$$

$$\frac{\partial N_T}{\partial t} = D_{ex} \frac{\partial^2 N_T}{\partial x^2} + k_T(N_{T,P} + N_{T,AP}) - \frac{N_T}{\tau_{T,rec}} - \frac{N_T - N_S}{\tau_{isc}} \quad (2.33)$$

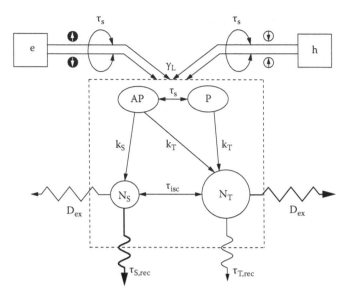

FIGURE 2.19 Schematic representation of singlet and triplet exciton formation, recombination, and diffusion. (From M. Yunus, P. P. Ruden, D. L. Smith, *Appl. Phys. Lett.* 93, 123312 [2008].)

Here N_S and N_T are the densities of singlet and triplet excitons, D_{ex} is the diffusivity of the excitons, $\tau_{S,rec}$ and $\tau_{T,rec}$ are the recombination lifetime of the singlet and triplet excitons, and τ_{isc} is the intersystem crossing time. A schematic representation of the whole process is shown in Figure 2.19.[43] The definition of the fraction of singlet excitons generated is given by $\chi_S = k_S N_{S,AP}/(k_S N_{S,AP} + k_T N_{T,AP} + k_T N_{T,P})$.

As an example, we consider an organic semiconductor layer of 100 nm thickness sandwiched between two ferromagnetic contacts. We assume that the Schottky barriers are such as to enable electron injection for one contact and hole injection for the other, and equal to 0.6 eV in all of the following calculations. We assume that tunnel barriers at the contacts enable spin-polarized carrier injection, and we express the level of spin injection through the current polarization (CP), defined as the ratio of the spin current averaged over the semiconductor to the charge current. Charge conservation ensures that the charge current is constant throughout the device, but the spin current depends on position due to the finite spin relaxation time. The applied bias is 20 V for all examples.

First, we investigate the effect of spin injection on the P and AP electron-hole pair generation rates (Figure 2.20).[43] When injection is unpolarized, the P and AP pair generation rates are equal. However, spin-polarized injection with parallel contact polarizations enhances the AP generation rate relative to the P generation rate. The effect is strongly nonlinear due to the exponential dependence of the injected current density on the (spin-dependent) quasi-Fermi levels at the semiconductor interfaces. For strongly polarized injection (CP = 0.99) and $\mu_n = \mu_p$, the AP generation rate is two orders of magnitude larger than the P generation rate, and the current density in the polarized case exceeds that of the unpolarized case by about one order of magnitude. The profiles of the pair generation rates depend on the relative carrier

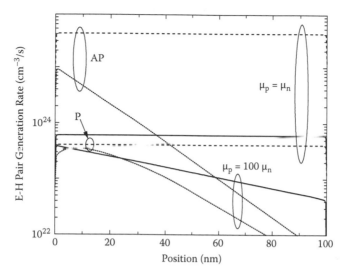

FIGURE 2.20   AP and P pair generation rates as a function of position inside the organic semiconductor for different carrier mobilities. Solid lines are for unpolarized cases and dashed lines are for current polarizations of 0.99 ($\mu_p = \mu_n$) and 0.964 ($\mu_p = 100\,\mu_n$). (From M. Yunus, P. P. Ruden, D. L. Smith, *Appl. Phys. Lett.* 93, 123312 [2008])

mobilities. When the mobilities of electrons and holes are equal, generation occurs uniformly throughout the device. For many organic semiconductors the hole mobility is larger than the electron mobility. Pair generation then occurs primarily near the electron-injecting contact (left) and decreases toward the hole-injecting contact (right). One interesting point is that for $\mu_n \ll \mu_p$, the P generation rate and $n_\downarrow$ have maxima near the electron-injecting contact. The reason for this is that the spin flip term (the last term in Equation 2.25b), which is positive, counteracts the recombination terms. However, the AP generation rate does not depend strongly on $n_\downarrow$, because $n_\downarrow p_\uparrow$ is negligible compared to $n_\uparrow p_\downarrow$. Hence, the AP generation rate decreases monotonically from the electron-injecting to the hole-injecting contact.

The enhancement of the AP pair generation through spin injection increases the singlet fraction, χ_S. Figure 2.21 plots χ_S as a function of position for the different cases shown in Figure 2.20.[43] When injection is strongly spin polarized, $\chi_S \sim 1/2$. The profiles depend on the relative carrier mobilities. For equal electron and hole mobilities, AP and P pair generation rates are uniform throughout the semiconductor, as shown in Figure 2.20 by dashed bold lines, and hence χ_S is uniform throughout the semiconductor. When the mobility of one carrier type is reduced (here μ_n), its injection rate becomes smaller and so does the current polarization. The reduction of current polarization also reduces χ_S, as evident in Figure 2.21.[43] However, the profile varies spatially. When the electron mobility is reduced, AP and P pair generation occur primarily at the electron-injecting contact and decrease toward the hole-injecting contact. But the relative decrease of the AP pair generation rate is more than that of the P pair generation rate due to spin flips. Only the AP pair generation contributes strongly to χ_S. Thus, χ_S decreases slowly toward the right electrode. The

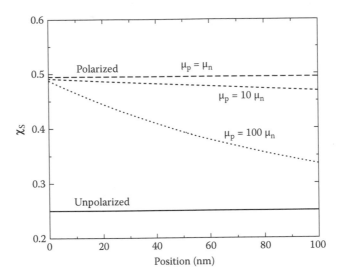

FIGURE 2.21 Profiles of the singlet exciton fraction, χ_S, inside the organic semiconductor for the different cases shown in Figure 2.20. (From M. Yunus, P. P. Ruden, D. L. Smith, *Appl. Phys. Lett.* 93, 123312 [2008].)

average values of χ_S for the three polarized cases shown in Figure 2.20 are 49.5, 48, and 46%; the corresponding CP values are 0.99, 0.975, and 0.964, respectively. The average value of singlet fraction increases approximately parabolically with CP and tends toward ½ when CP = 1. On the other hand, when injection is unpolarized, χ_S is ¼, independent of the carrier mobilities.

The tightly bound excitons recombine or diffuse through the device. Singlet excitons recombine radiatively and thus have a short recombination lifetime. Triplet excitons recombine nonradiatively with a long recombination lifetime. We assume that the exciton formation rates (k_S, k_T) are much larger than the spin relaxation rate ($1/\tau_s$). The intersystem crossing time, τ_{isc}, is assumed to be long (1 ms); hence, it does not play any role in the examples discussed here. The diffusivity is the same for singlet and triplet excitons and is calculated from the lower of the charge carrier mobilities by Einstein's relation. Figure 2.22 represents the singlet and triplet exciton profiles inside the organic semiconductor.[43] The diffusion length of singlet excitons is short, and their steady-state density profile is similar to the AP generation rate. On the other hand, diffusion is important for triplet excitons. When the carrier mobilites are equal to 10^{-2}cm²/Vs, the resulting diffusivity of 2.5×10^{-4}cm²/s gives a diffusion length of 500 nm, which is larger than the device size. Due to this long diffusion length, most of the triplet excitons recombine at the contact interfaces. The diffusion lengths for the other two cases are 158 nm ($\mu_p = 10\,\mu_n$) and 50 nm ($\mu_p = 100\,\mu_n$). As seen from the dashed lines in Figure 2.22, the triplet exciton profiles differ appreciably from the generated P and AP electron-hole pair profiles.

Exploring the effect of spin injection on the formation and distribution of excitons leads to the following conclusions. The relative carrier mobilities strongly affect the spatial distributions of exciton formation. The fraction of singlet excitons created

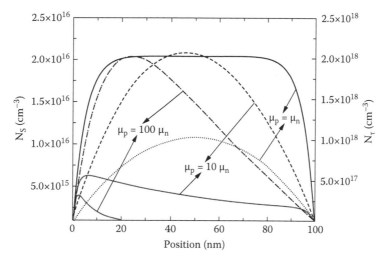

FIGURE 2.22   Steady-state profiles of singlet (solid) and triplet (dashed) exciton densities inside the organic semiconductor for different carrier mobilities. (From M. Yunus, P. P. Ruden, D. L. Smith, *Appl. Phys. Lett.* 93, 123312 [2008].)

also varies spatially, assuming values between ¼ and 1/2. Diffusion is expected to be important for triplet excitons.

2.5   SUMMARY AND CONCLUSIONS

Device modeling of spin injection and transport in organic semiconductor structures can be helpful in understanding the underlying spin physics relevant for these materials. Although many of the basic relationships that control the phenomena of interest are also encountered in inorganic materials, quantitative differences are important. This is, for example, the case for injection, where the low mobility of organic semiconductors ensures a much greater backflow of injected charge (and spin) carriers than is typical for inorganic semiconductors. Quite generally, device modeling points to the relevance of space charge effects in organic semiconductors, both near contacts and throughout the rest of the structure. As is the case for inorganic semiconductors, efficient spin-polarized carrier injection requires a tunneling process. However, organic semiconductors may offer new opportunities to create well-controlled tunnel barrier layers through self-assembly and through their generally greater compatibility with foreign substrates.

Bipolar spin-polarized injection is of particular interest in organic semiconductors, as it can impact the relative proportion of singlet excitons created in the device structure. Device modeling allows one to explore the interplay of the multiple time-scales that control the exciton formation and recombination: carrier transit times, recombination times, and spin relaxation times.

To some extent, one may look at the modeling presented in this chapter as an exploration of a range of scenarios that may be realized for different materials and device structures. The approach taken is such as to be essentially insensitive to microscopic details of the transport mechanism, i.e., the distinction of hopping transport and so-called band transport. It also is an approach that does not attempt to explore the microscopic origin of spin relaxation, which may occur due to residual spin-orbit interactions or hyperfine coupling. Material specific effects enter the device equations through the transport parameters (and their dependences on external parameters), including spin relaxation and carrier recombination coefficients, and through boundary conditions. Device modeling can point out the relative importance of ranges of parameter values that are currently not yet determined, such as spin relaxation and tunneling contact resistances. These parameters will impact the experimental observation of measurable spin effects, such as, for example, magnetoresistance or electroluminescence efficiency.

ACKNOWLEDGMENTS

The authors thank M. Yunus for his help in the calculations and the preparation of figures. Work at the University of Minnesota was supported by NSF-ECCS. Work at Los Alamos National Laboratory was supported by DOE Office of Basic Energy Sciences Work Proposal 2010LANLE412.

REFERENCES

1. E.I. Rashba, *Phys. Rev. B* 62, R16267 (2000).
2. D. L. Smith, R. N. Silver, *Phys. Rev. B* 66, 113303 (2001).
3. G. A. Prinz, *Science* 282, 1660 (1998).
4. I. Zutic, J. Fabian, S. DasSarma, *Rev. Mod. Phys.* 76, 323 (2004).
5. A. T. Hanbicki, B. T. Jonker, G. Itskos, G. Kioseoglou, A. Petrou, *Appl. Phys. Lett.* 80, 1240 (2002).
6. A. T. Hanbicki, O. M. J. van't Erve, R. Magno, G. Kioseoglou, C. H. Li, B. T. Jonker, G. Itskos, R. Mallory, M. Yasar, A. Petrou, *Appl. Phys. Lett.* 82, 4092 (2003).
7. C. Adelmann, X. Lou, J. Strand, C. J. Palmstrom, P. A. Crowell, *Phys. Rev. B* 71, 121301 (2005).
8. S. A. Crooker, M. Furis, X. Lou, C. Adelman, D. L. Smith, C. J. Palmstrom, P. A. Crowell, *Science* 309, 5744 (2005).
9. X. Lou, C. Adelman, S. A. Crooker, E. S. Garlid, J. Zhang, K. S. M. Reddy, S. D. Flexner, C. J. Palmstrom, P. A. Crowell, *Nature Phys.* 3, 197 (2007).
10. V. Dediu, M. Murgia, F. C. Matacotta, C. Taliani, S. Barbanera, *Solid State Commun.* 122, 181 (2002).
11. Z. H. Xiong, Di Wu, Z. Valy Vardeny, J. Shi, *Nature* 427, 821 (2004).
12. F. J. Wang, C. G. Yang, Z. Valy Vardeny, X. G. Li, *Phys. Rev. B* 75, 245324 (2007).
13. S. Majumdar, H. Huhtinen, H. S. Majumdar, R. Laiho, R. Österbacka, *J. Appl. Phys.* 104, 033910 (2008).
14. Y. Liu, S. M. Watson, T. Lee, J. M. Gorham, H. E. Katz, J. A. Borchers, H. D. Fairbrother, D. H. Reich, *Phys Rev. B* 79, 075312 (2009).
15. J. S. Jiang, J. E. Pearson, S. D. Bader, *Phys. Rev. B* 77, 035303 (2008).
16. I. H. Campbell, D. L. Smith, *Solid State Physics*, ed. H. Ehrenreich, F. Spaepen, Vol. 55, Academic, New York, 2001.
17. I. H. Campbell, S. Rubin, T. A. Zawodzinski, J. D. Kress, R. L. Martin, D. L. Smith, N. N. Barashkov, J. P. Ferraris, *Phys. Rev. B* 54, 14321 (1996); I. H. Campbell, J. D. Kress, R. L. Martin, D. L. Smith, N. N. Barashkov, J. P. Ferraris, *Appl. Phys. Lett.* 71, 3528 (1997).

18. Y. Kato, R. C. Myers, A. C. Gossard, D. D. Awschalom, *Nature* 427, 50 (2003).
19. S. A. Crooker, D. L. Smith, *Phys. Rev. Lett.* 94, 236601 (2005).
20. M. Hruska, S. Kos, S. A. Crooker, A. Saxena, D. L. Smith, *Phys. Rev. B* 73, 075306 (2006).
21. G. Schmidt, D. Ferrand, L. W. Molenkamp, A. T. Filip, B. J. van Wees, *Phys. Rev. B* 62, R4790 (2000).
22. J. D. Albrecht, D. L. Smith, *Phys. Rev. B* 66, 113303 (2002).
23. J. Strand, X. Lou, C. Adelmann, B. D. Schultz, A. F. Isakovic, C. J. Palmstrøm, P.A. Crowell, *Phys. Rev. B* 72, 155308 (2005).
24. P. A. Bobbert, T. D. Nguyen, F. W. A. van Oost, B. Koopmans, M. Wohlgenannt, *Phys. Rev. Lett.* 99, 216801 (2007).
25. S. Funaoka, I. Imae, N. Noma, Y. Shirota, *Synth. Metals* 101, 600 (1999); T. Wangwijit, H. Sato, S. Tantayanon, *Polym. Adv. Technol.* 13, 25 (2002).
26. C. G. Yang, E. Ehrenfreund, Z. V. Vardeny, *Phys. Rev. Lett.* 99, 157401 (2007).
27. P. P. Ruden, D. L. Smith, *J. Appl. Phys.* 95, 4898 (2004).
28. P. S. Davis, I. H. Campbell, D. L. Smith, *J. Appl. Phys.* 82, 6319 (1997).
29. P. W. Blom, M. J. M. Dejong, M. G. VanMunster, *Phys. Rev. B* 55, R656 (1997).
30. M. Abkowitz, H. Bässler, M. Stolka, *Philos. Mag. B* 63, 201 (1991).
31. P. C. van Son, H. van Kempen, P. Wyder, *Phys. Rev. Lett.* 58, 2271 (1987).
32. M. Yunus, P. P. Ruden, D. L. Smith, *J. Appl. Phys.* 103, 103714 (2008).
33. M. H. Cohen, L. M. Falicov, J. C. Phillips, *Phys. Rev. Lett.* 8, 316 (1962).
34. J. Bardeen, *Phys. Rev. Lett.* 6, 57 (1961).
35. M. Yunus, P. P. Ruden, D. L. Smith, *Synth. Metals* (2009) (in press).
36. D. L. Smith, P. P. Ruden, *Phys. Rev. B* 78, 125202 (2008).
37. P. P. Ruden, J. D. Albrecht, D. L. Smith, *Organic Thin-Film Electronics,* ed. A. C. Arias, N. Tessler, L. Burgi, J. A. Emerson, Materials Research Society Proceedings. 871E, Warrendale, PA, 2005, p. I1.6.1.
38. A. Dediu, presented at SPINOS 2, Salt Lake City, UT, February 2009, and private communication.
39. R. H. Friend, R. W. Gymer, A. B. Holmes. J. H. Burroughes, R. N. Marks, C. Taliani, D. D. C. Bradley, D. A. D. Santos, J. L. Brédas, M. Löglund, W. R. Salaneck, *Nature* 397, 121 (1999).
40. Y. Wu, B. Hu, J. Howe, An-Ping Li, J. Shen, *Phys. Rev. B* 75, 075413 (2007).
41. M. Segal, M. A. Baldo, R. J. Holmes, S. R. Forrest, Z. G. Soos, *Phys. Rev. B* 68, 075211 (2003); M. Wohlgenannt, K. Tandon, S. Mazumdar, S. Ramasesha, Z. V. Vardeny, *Nature* 409, 494 (2001); J. S. Wilson, A. S. Dhoot, A. J. A. B. Seeley, M. S. Khan, A. Köhler, R. H. Friend, *Nature* 413, 828 (2001).
42. Z. Shuai, D. Beljonne, R. J. Silbey, J. L. Brédas, *Phys. Rev. Lett.* 84, 131 (2000); T. Hong, H. Meng, *Phys. Rev. B* 63, 075206 (2001).
43. M. Yunus, P. P. Ruden, D. L. Smith, *Appl. Phys. Lett.* 93, 123312 (2008).
44. M. Wohlgenannt, Z. V. Vardeny, *J. Phys. Condens. Matter* 15, 83 (2003).

3 Magnetoresistance and Spin Transport in Organic Semiconductor Devices

M. Wohlgenannt, P. A. Bobbert,
B. Koopmans, and F. L. Bloom

CONTENTS

3.1   INTRODUCTION

3.1.1   THE ELECTRON SPIN IN ORGANIC MATERIALS

The study of electron-spin transport through nonmagnetic spacer materials sandwiched in between ferromagnetic electrodes is an extremely active field, because of the rich physics involved and the important applications in the area of magnetic sensors.[1] If the spin diffusion length is larger than or comparable to the distance between the electrodes, the current through such sandwich structures can depend strongly on the mutual orientation of the magnetizations of the electrodes, which is called the spin valve effect. Switching of this orientation by an external magnetic field, B, can then lead to a strong dependence of the current on B, an effect called giant magnetoresistance (GMR).[2,3] This effect can be used in magnetic sensors, e.g., in reading magnetic information in hard disks.

Traditionally, nonmagnetic metals are used as the spacer layer material in these structures. Spintronic devices utilizing spin injection and transport through a *semiconducting* spacer layer offer additional functionalities, such as spin transistors and the possibility to realize quantum computation logic. Consequently, a lot of effort is put into finding suitable materials. Spin randomization in the inorganic crystalline materials traditionally used in these structures, containing relatively heavy atoms, is mediated by spin-orbit coupling.[4] Organic (semiconducting) materials are a very interesting alternative because of the enormous versatility of organic chemistry and because the light atoms from which they are composed cause very little spin-orbit coupling.[5,6] Recent years have seen the first demonstrations of GMR devices[7–10] as well as magnetic tunnel junctions[11] using organic semiconducting materials as the spacer layer. In the former devices the GMR effect disappears on a typical length scale of the order of 100 nm as the thickness of the organic spacer layer increases.[7,8] Therefore, questions such as "What is the cause of the remaining spin scattering?" and "What factors determine the spin diffusion length?" arise.

We believe that the most direct clues to answering these questions arose from studies of magnetic field effects in organic devices, which were conducted in parallel, and even prior to the emergence of interest in organic spintronics. In 1996, Frankevich and coworkers[12] detected a significant (some percent) effect of small magnetic fields (usually less than 100 mT) on the photoconductivity in poly-phenylene-vinylene. More recently, Kalinowski and coworkers[13,14] showed that the electroluminescence intensity can be modulated in organic light-emitting diodes (OLEDs) made from the small molecule tris(8-hydroxyquinoline aluminum) (Alq_3) by the application of a magnetic field. These photoconductivity and electroluminescence experiments are spin selective. Photoconductivity is spin selective since the photoexcitation of organic semiconductors leads primarily to singlet excited states, and because exciton dissociation is more probable in the less tightly bound singlet exciton, and electroluminescence is spin selective because emissive recombination usually occurs only from the singlet exciton. Accordingly, these effects were interpreted using a model where the singlet and triplet states were mixed by the hyperfine field of the hydrogen atoms within the organic compounds. Building upon this work, we studied the magnetic field effect on the conductivity: Francis et al.[15] demonstrated magnetoresistive effects of up to 10% at room temperature for magnetic fields of $B \approx 10$ mT. The effect was dubbed organic magnetoresistance (OMAR). OMAR is among the largest magnetoresistive effects of any bulk material, and is therefore an interesting spintronics effect in its own right. Later, Mermer et al.[16] extended these results to also include small-molecule organic semiconductors and many different polymers, where equally large effects were found. In distinction to magnetic field effects in photoconductivity and electroluminescence, the model interpretation of OMAR is far less obvious, since the conductivity measurement is not spin selective: carriers are injected and collected in the device in a spin-independent way since OMAR devices use nonmagnetic electrodes. The correct model for OMAR is still debated.

3.1.2 ORGANIZATION OF THIS CHAPTER

We believe that the observation of OMAR most clearly demonstrates that the carrier transport in organics is sensitive to the hyperfine field, and that indeed an understanding of spin diffusion in organic spin valves must be based on hyperfine coupling as the dominant spin randomization mechanism. The experimental and theoretical study of OMAR occupies a large part of the first half of this chapter (Section 3.2), and we will use the insights gained to develop a theory of spin diffusion in the second part (Section 3.3). A general introduction precedes these two parts.

The section on OMAR (for which a definite theory is not yet available) begins with a discussion of possible mechanisms. The section focuses on the bipolaron model, which we recently developed. This model is first formulated within a highly simplified essentially two-site model, and proceeds with a discussion of detailed Monte Carlo simulations of single-carrier (i.e., either hole or electron conduction-dominated) devices. Several experiments have shown that the OMAR effect can be distinctly different in bipolar devices (including a sign change), and therefore modeling of bipolar devices occupies a significant part of this section.

The section dealing with spin valve effects in organic semiconductors begins with a discussion of the experimental results obtained thus far. In our opinion, a clear and complete picture has not yet emerged from the experimental studies. Nevertheless, the central importance of hyperfine coupling regarding spintransport has become evident. We therefore develop a theory of spin diffusion based on this insight, and finally attempt a comparison with the currently available experimental results.

3.1.3   π-Conjugated Organic Semiconductors, a General Introduction

Interest in organic electronics stems, in part, from the ability to deposit organic films on a variety of low-cost substrates such as glass or plastic, and the relative ease of processing of the organic compounds. High-efficiency, very bright and colorful thin displays based on organic light-emitting diodes[17,18] (OLEDs) are already in commercial production. Significant progress is also being made in the realization of thin-film transistors[19–21] (TFTs) for use in low-cost electronics and thin-film organic photovoltaic cells[22–24] for low-cost solar energy generation. Organic electronic materials are soft and flexible due to their weak bonding through van der Waals forces, whereas inorganic semiconductors are hard and brittle. The softness of organic materials has also opened the door to a multitude of innovative fabrication methods that are simpler and cheaper to implement on a large scale than those for inorganic semiconductors. Processes for organic semiconductor device fabrication involve direct printing through use of stamps, or alternatively via ink jet techniques and other solution-based methods. However, these simple fabrication techniques generally lead to disordered films.

Based on their molecular weight, organic semiconductors are classified into two groups: π-conjugated polymers and small molecules. Both types have in common a π-conjugated chemical structure that results in the delocalization of their highest energy electrons (π-electrons) over the entire extent of the π-conjugation. Simple organic semiconductor molecules are planar, with three of the four carbon valence electrons forming sp^2 hybrid orbitals (σ-bonds), while the fourth valence electron is in a π-orbital perpendicular to the molecular plain. The σ-bonds are the building blocks of the molecular skeleton and are thus responsible for the molecule's integrity. The π-orbitals form the highest occupied molecular orbitals (HOMOs) and the lowest unoccupied molecular orbitals (LUMOs), roughly equivalent to the inorganic semiconductor's valence and conduction band edges, respectively.

3.1.3.1   π-Conjugated Polymers

A π-conjugated polymer is a carbon-based macromolecule over which the valence π-electrons are delocalized. Research into the electronic and optical properties of conjugated polymers began in the 1970s after a number of seminal experimental achievements. First, the synthesis of polyacetylene thin films[25] and the subsequent success in doping these polymers to create conducting polymers[26] established the field of synthetic metals. Second, the synthesis of phenyl-based polymers, e.g., poly(para-phenylene-vinylene) (PPV) (see Figure 3.1b), and the discovery of electroluminescence under low voltages in these systems[27] established the field of polymer optoelectronics. The discovery and development of conductive polymers was recognized by the award of the Nobel Prize for chemistry in 2000 to Heeger, MacDiarmid, and Shirakawa.

π-Conjugated polymers exhibit electronic properties that are quite different from those observed in inorganic metals or semiconductors. These unusual electronic properties may essentially be attributed to the fact that conjugated polymers are quasi-one-dimensional systems owing to their strong intramolecular but relatively weak intermolecular interactions. The quantum mechanical wavefunction is therefore usually confined to a single chain. This quasi-one-dimensionality results in weakly screened electron-electron interactions. Thus, electronic correlations are important in determining the character of the electronic states, and the neutral excited states are dominated by excitons. Another important factor in determining the character of the electronic states is that the electrons and lattice are strongly coupled. Just like the effects of electron-electron interactions, the effects of electron-phonon coupling are enhanced in low dimensions. Therefore, the charge carriers in these materials are positive and negative polarons, rather than holes and electrons.

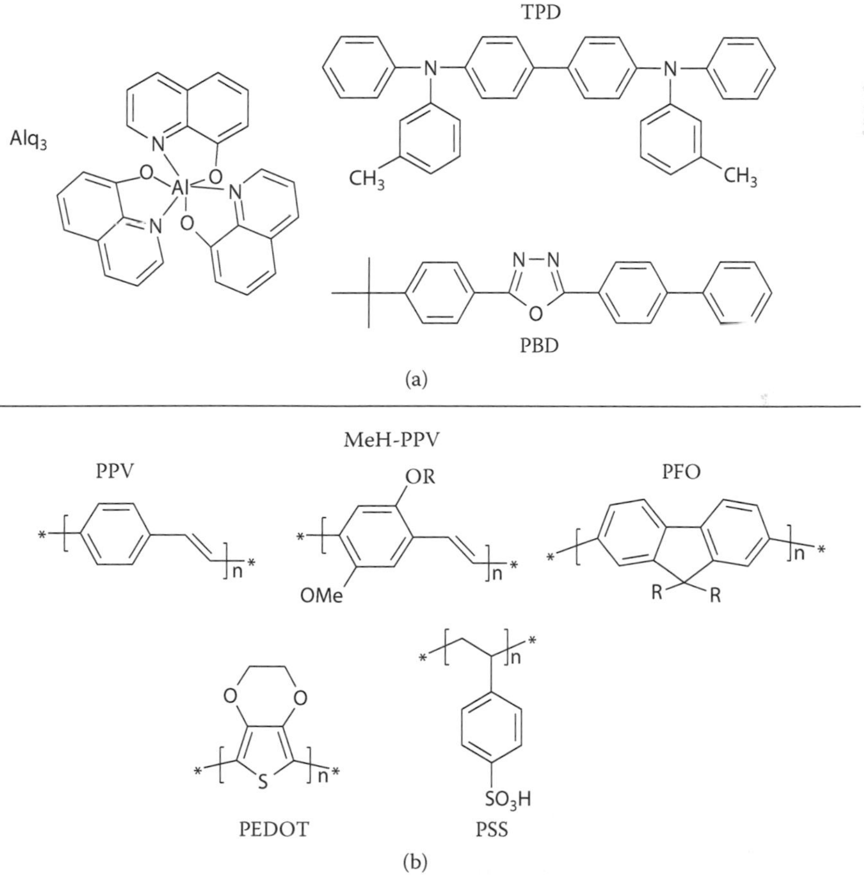

FIGURE 3.1 (a) Structures of some molecular semiconductors that have been used in thin-film electroluminescent devices. Alq$_3$ is used as an electron transport and emissive layer, TPD is used as a hole transport layer, and PBD is used as an electron transport layer. (b) Polymers used in electroluminescent diodes. The prototypical (green) fluorescent polymer is poly(p-phenylene-vinylene) (PPV). One of the best-known (orange-red) solution-processable conjugated polymers is MEH-PPV. Poly(dialkylfluorene)s (PFO) are blue-emitting, high-purity polymers, which show high luminescence efficiencies. Polymers such as poly(3,4-ethylenedioxythiophene) (PEDOT) doped with polystyrenesulfonic acid (PSS) are widely used as hole injection layers.

3.1.3.2   π-Conjugated Small Molecules

In addition to π-conjugated polymers, low-molecular-weight organic compounds (see Figure 3.1a) have also been extensively investigated. Electroluminescence from OLEDs made from small molecules was first observed and extensively studied in the 1960s.[28] In 1987, a team at Kodak introduced a double-layer OLED, which combined modern thin-film deposition techniques with suitable materials and structure to give moderately low bias voltages and attractive electroluminescence efficiency.[29] Intense research in both academia and industry has yielded small-molecule OLEDs with remarkable color fidelity, device efficiency, and operational stability.[30–32]

3.1.4   ORGANIC LIGHT-EMITTING DIODES

The early studies on organic electroluminescence established that it requires injection of electrons from one electrode and holes from the other, the capture of oppositely charged carriers (so-called recombination), and the radiative decay of the excited electron-hole state (exciton) produced by this recombination process. The carrier injection is necessary since the organic layer is (nominally) undoped and therefore possesses only an insignificant charge carrier density. The simplest OLED device structure consists of a thin film of indium tin oxide (ITO) coated onto the substrate, as the bottom electrode, upon which the organic semiconductor thin film is deposited. Finally, the top electrode is conveniently formed by thermal evaporation. ITO is a transparent metal that allows the light generated within the diode to leave the device. Diodes of this type can be readily fabricated by spin coating of the semiconducting polymer, or evaporating the small molecular film, onto the ITO-coated glass, where the film thickness is typically on the order of 100 nm. Injection of charge usually requires surmounting or tunneling through a barrier at the interface. The electrodes are chosen to facilitate charge injection; ITO has a relatively high work function and is therefore suitable for use as a hole-injecting electrode; low-work-function metals such as Ca, Mg, or Al are suitable for injection of electrons. Organic semiconductors have an energy gap between the bonding and antibonding π- and π∗-states, respectively, of between about 2 and 3 eV, covering the whole visible spectrum.

Since organic solids are typically bonded by weak van der Waals forces, they are more prone to disorder than inorganic solids. Therefore, charge carrier transport usually occurs via hopping between molecular sites, or from chain to chain. In this case, the carrier mobilities are quite low compared with inorganic semiconductors, where the room temperature mobilities typically range from 100 to 10^4 cm^2 V^{-1} s^{-1}. In contrast, in disordered molecular systems and polymers, the mobilities are typically between 10^{-5} and 10^{-3} cm^2 V^{-1} s^{-1}.

3.2   MAGNETORESISTANCE IN NONMAGNETIC DEVICES

3.2.1   EXPERIMENTAL RESULTS

3.2.1.1   Organic Magnetoresistance in Polymer Devices

Figure 3.2 shows a schematic OMAR device configuration and the measurement process. OMAR devices consist of an organic layer sandwiched between anode and cathode electrodes, a structure typical for organic light-emitting diodes. Figure 3.3

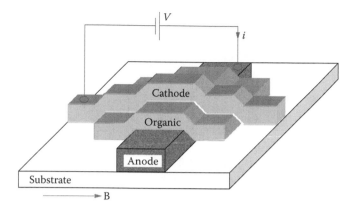

FIGURE 3.2   A schematic drawing of an OMAR device and the magnetoresistance experiment. (From Mermer et al., *Phys. Rev. B* 72, 205202 [2005]. With permission.)

shows measured OMAR traces in a polyfluorene sandwich device (details are given in the caption) at room temperature at different voltages.[15] The magnetoresistance ratio, $\Delta R/R$, in polyfluorene reaches up to 10% at 10 mT at room temperature. It was found[15] that the measured OMAR traces in polyfluorene are independent of the angle between film plane and applied magnetic field. Figure 3.4 shows OMAR traces[15] in

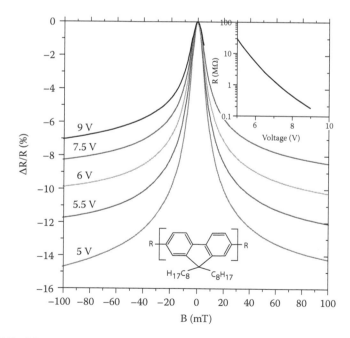

FIGURE 3.3   Magnetoresistance, $\Delta R/R$ curves, measured at room temperature in an ITO (30 nm)/PEDOT ($\approx$ 100 nm)/PFO ($\approx$ 100 nm)/Ca ($\approx$ 50 nm, including capping layer) device at different voltages. The inset shows the device resistance as a function of the applied voltage. (From Francis et al., *New J. Phys.* 6, 185 [2004]. With permission.)

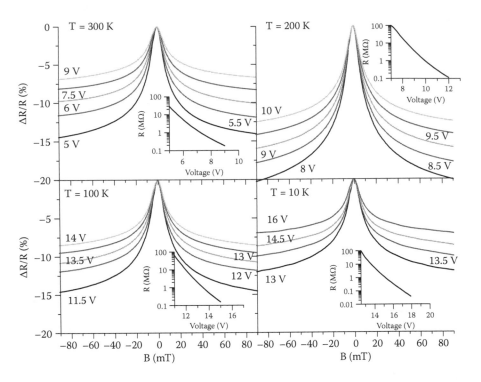

FIGURE 3.4 Magnetoresistance, $\Delta R/R$ curves, in the device of Figure 3.2 measured at different temperatures, namely, 10, 100, 200, and 300 K. The applied voltages are assigned. The insets show the device resistance as a function of the applied voltage. (From Francis et al., *New J. Phys.* 6, 185 [2004]. With permission.)

a PEDOT/polyfluorene/Ca device for four different temperatures between 300 K and 10 K (PEDOT stands for the conducting polymer poly(3,4-ethylenedioxythiophene)-poly(styrenesulfonate) and was purchased from H. C. Starck). The magnitude of the OMAR cones is only weakly temperature dependent, and the width of the cones is temperature independent. The lack of a strong temperature dependence, although found in many materials, is not a universal feature of OMAR: Figure 3.5 shows measured OMAR traces[33] as a function of temperature in an ITO/PEDOT/RR-P3HT/Ca device (RR-P3HT stands for regioregular poly(3-hexylthiophene)). The magnitude of the observed OMAR effect is small (less than 1.5%). The data show that the OMAR effect can be both positive and negative in RR-P3HT, depending mostly on the temperature. At room temperature, the OMAR effect is completely positive, whereas at 100 K the effect is negative. At 200 K, a transition from positive to negative magnetoresistance occurs as the voltage increases.

3.2.1.2 Organic Magnetoresistance in Small-Molecule Devices

Figure 3.6 shows measured OMAR traces[16] in an Alq$_3$ sandwich device (details are given in the caption) at room temperature at different voltages, reaching values for the magnetoresistance ratio of up to 10% at 10 mT. In agreement with the previous

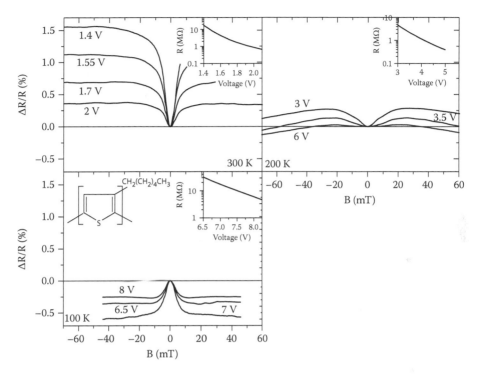

FIGURE 3.5 Magnetoresistance, $\Delta R/R$ curves, in an ITO/PEDOT/RR-P3HT ($\approx$ 100 nm)/Ca device measured at different temperatures (100, 200, and 300 K). The insets show the device resistance as a function of the applied voltage. (From Mermer et al., *Phys. Rev. B* 72, 205202 [2005]. With permission.

results in polyfluorene,[15] the measured OMAR traces in Alq$_3$ devices are independent of the angle between film plane and applied magnetic field. All measurements shown here are for an in-plane magnetic field.

3.2.1.3 Universality of OMAR Traces

Figure 3.7 shows a summary[33] of *normalized* OMAR traces in devices made from different organic semiconductors at room temperature. It is evident that there are two groups of OMAR traces in organic semiconductor materials: one group (pentacene, RR-P3HT, and RRa-P3OT [regiorandom poly(3-octyl-thiophene)]) where the OMAR effect has saturated at 50 mT and the other (polyfluorene, Alq$_3$, Pt-PPE and PPE [platinum-containing and platinum-free poly(phenylene-ethynelene)]) where the effect is still unsaturated. We will refer to these two kinds of traces as weakly saturated and fully saturated traces, respectively. The width of the OMAR traces appears to be approximately universal within all the materials studied, all of an order of magnitude typical of hyperfine interaction in organic semiconductors (several mT). The functional dependence of OMAR on B, i.e., $(\Delta I/I)(B)$, could also be determined. For this purpose a typical room temperature OMAR trace in a PEDOT/polyfluorene/Ca

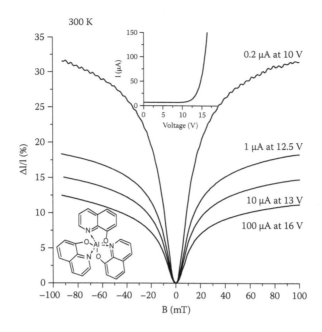

FIGURE 3.6 Magnetoconductance, $\Delta I/I$ curves, measured at room temperature in an ITO (30 nm)/PEDOT ($\approx$ 100 nm)/Alq$_3$ ($\approx$ 100 nm)/Ca ($\approx$ 50 nm, including capping layer) device at different voltages. The inset shows the device current-voltage characteristics. (From Mermer et al., *Phys. Rev. B* 72, 205202 [2005]. With permission.)

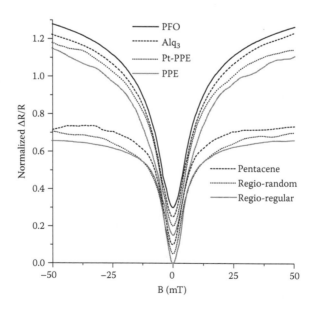

FIGURE 3.7 Normalized magnetoresistance, $\Delta R/R$ curve, of PFO, RRa-P3OT, RR-P3HT, Pt-PPE, PPE, Alq$_3$, pentacene devices measured at room temperature. The data have been offset for clarity. (From Mermer et al., *Phys. Rev. B* 72, 205202 [2005]. With permission.)

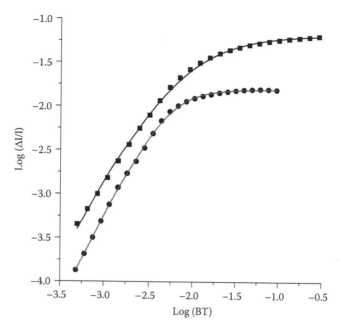

FIGURE 3.8 Room temperature OMAR traces in a PEDOT/polyfluorene/Ca device (squares) and a PEDOT/RR-P3HT/Ca device (circles) plotted on a log-log scale. The solid curves are fits using an empirical law of the form $\Delta I/I = (\Delta I/I)_{\max} [B/(B+B_0)]^2$ and $\Delta I/I = (\Delta I/I)_{\max} B^2/(B^2 + B_0^2)$, respectively. (From Mermer et al., *Phys. Rev. B* 72, 205202 [2005]. With permission.)

device (squares) and a PEDOT/RR-P3HT/Ca device (circles), representatives for the weakly saturated and fully saturated traces, respectively, is plotted on a log-log scale in Figure 3.8. The solid line through the circles (fully saturated trace) in Figure 3.8 is a fit using the function $\Delta I/I = (\Delta I/I)_{\max} B^2/(B^2 + B_0^2)$ (Lorentzian), which gives a very accurate agreement, where B_0 is the half-saturation field. The value $B_0 = 5.15$ mT results in the best fit. This Lorentzian function fits all the traces for materials with fully saturated behavior shown in Figure 3.7, but does not fit the weakly saturated behavior shown by some materials in Figure 3.7. Instead, the solid curve through the polyfluorene data is a fit using an empirical law of the form $\Delta I/I = (\Delta I/I)_{\max} [B/(B + B_0)]^2$, where $(\Delta I/I)_{\max}$ denotes the fractional change in current extrapolated to infinite field, and B_0 denotes the quarter-saturation field. The fit is excellent over almost three orders of magnitude in B. The value $B_0 = 5.89$ mT results in the best fit.

3.2.1.4 Single- and Double-Carrier Injection

In devices with undoped organic semiconductor layers the carriers that result in electrical current must be injected from the electrodes. If both the anode and cathode are chosen suitably, both form (near-) ohmic contacts and the device is bipolar and shows efficient electroluminescence. If one of the electrodes is chosen to enforce a large barrier to injection of this carrier type, then the device is (almost) unipolar, and therefore ideally shows no electroluminescence. To achieve a variety of injection conditions, we fabricated devices with different cathode materials. The fabrication started with glass substrates coated with ITO, or the conducting polymer PEDOT spin coated onto ITO, as the anode (PEDOT is known to give very good hole injection). The Alq_3 (sublimed, H.W. Sands Corp.) layer was thermally evaporated in high vacuum (10^{-6} mbar) onto the bottom electrode, yielding an organic semiconductor layer thickness of about 100 nm, without breaking the vacuum. The cathode, either Ca (with an Al capping layer), Al, or Au, was then deposited by thermal (Ca) or electron beam evaporation (Al, Au) on top of the organic thin film. Ca is known to be a very good electron injector, whereas Au is largely unsuitable for electron injection and should result in largely unipolar devices.

Figure 3.9a shows the measured OMAR ratio, $\Delta I/I$, at $B = 100$ mT and several constant applied voltages resulting in a current flow between 1 μA and 100 μA in devices ranging from bipolar to almost hole only. The degree of unipolar/bipolar behavior was assessed by measuring the electroluminescence efficiency η, where higher η corresponds to better balanced bipolar injection. It is seen that the magnitude of the maximum $\Delta I/I$ in Figure 3.9 remains constant within a factor of 2, whereas η changes by more than two orders of magnitude. For completeness, the

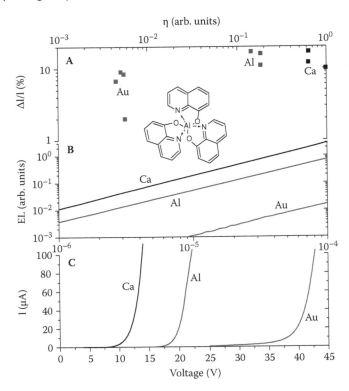

FIGURE 3.9 (a) Magnetoconductance ratio, $\Delta I/I$ at $B = 100$ mT, in several PEDOT/Alq$_3$ ($\approx$ 100 nm)/cathode devices with Ca, Al, or Au as the cathode as a function of the exciton/carrier ratio, η. (b) Electroluminescence as a function of current. (c) Current-voltage (I-V) characteristics. (From Nguyen et al., *Phys. Rev. B* 77, 235209 [2008]. With permission.)

electroluminescence-current characteristics and current-voltage characteristics of the devices are shown in Figure 3.9b and c. We also note that there exists a clear relation between onset voltage and η for the data of Figure 3.9c. The increase in driving voltage when going from bipolar to unipolar devices is usually attributed to a reduced cancelation of the space charge fields of the electrons and holes.[34] These data therefore strongly suggest that OMAR exists also in unipolar devices, which, as we will see, puts a significant constraint on possible theoretical models of OMAR. We note, however, that in most materials OMAR decreases significantly when going from bipolar to unipolar devices,[35] indicating that either bipolar conditions are favorable for OMAR, or bipolar mechanisms make a significant contribution to OMAR. In Section 3.2.3 a detailed modeling study of bipolar injection-limited devices is presented, and in Section 3.2.4 the various aspects of sign changes occurring in OMAR are studied and discussed.

3.2.1.5 Hyperfine Interaction and Magnetoresistance

In this section we show that occurrence of hydrogen in the organic semiconductor's molecular structure is essential for the observation of OMAR. For this purpose we fabricated devices entirely from hydrogen-free compounds, specifically ITO/C_{60}/Ca/Al. First, we report on the nonobservation of any OMAR effect in ITO/C_{60}/Ca/Al devices (an example trace for a null result is shown in Figure 3.10, black line).

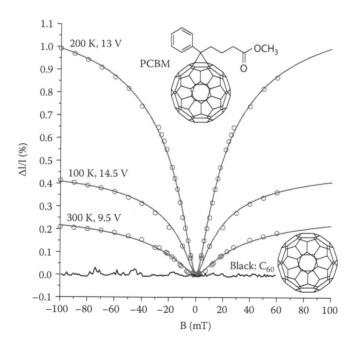

FIGURE 3.10 Magnetoconductance ratio, $\Delta I/I$, in an ITO/PCBM (≈ 160 nm)/Ca device (red) measured at three different temperatures and a constant voltage that results in a current of 100 μA (scatter plot). Black line: Example of null-result trace for OMAR in an ITO/C_{60} (≈ 160 nm)/Ca device. The molecular structure of PCBM and the current-voltage characteristics of the device are shown as insets.

Next, we deliberately reintroduce hyperfine coupling by using a substituted fullerene derivative with a hydrocarbon side group (1-(3-methoxycarbonyl) propyl-1-phenyl-[6,6]-methanofullerene, PCBM). Figure 3.10 shows the magnetoconductance ratio, $\Delta I/I$, in an ITO/PCBM ($\approx$ 160 nm)/Ca device measured at different temperatures and at a constant voltage, which results in a current of about 100 μA. This demonstration is strong evidence for the notion that hyperfine coupling causes OMAR. In the next section we discuss the various models that have been put forward to explain OMAR.

3.2.2 Models of Organic Magnetoresistance

3.2.2.1 General Considerations

To the best of our knowledge, the mechanism causing OMAR is currently not known with certainty, although it is generally agreed upon that it is related to spin-dependent interactions and mixing of different spin states by the hyperfine interaction. Before following this line, we first want to give the arguments by which more conventional magnetoresistance mechanisms can be excluded as possible explanations of OMAR. Most magnetoresistance mechanisms rely on the presence of ferromagnetic materials and are therefore not applicable to OMAR devices. The most widely known types of magnetoresistance in nonmagnetic materials are: (1) classical magnetoresistance, (2) hopping magnetoresistance,[36] (3) magnetoresistance caused by electron-electron interactions,[37] and (4) magnetoresistance caused by weak localization.[38]

Classical magnetoresistance (mechanism 1) has a positive sign because the applied magnetic field causes the electrons to follow periodic cyclotron orbits, which increases the resistance. Its magnitude is on the order of $\mu^2 B^2$, with μ the electron mobility. Using a typical value for the mobility, $\mu \approx 10^{-4}$ cm^2/Vs, we estimate a magnetoresistance ratio, $\Delta R/R \approx 10^{-20}$, at $B \approx 10$ mT for classical magnetoresistance. Classical magnetoresistance is therefore much too small to explain OMAR at these fields.

In hopping magnetoresistance (mechanism 2), an applied magnetic field shrinks the electron wavefunction, and this reduces the overlap between hopping sites, leading to an increase in resistance of the system, which results in a positive magnetoresistance effect. The size of this effect is only appreciable if the magnetic length, λ, is comparable to the hopping distance. In our case, $\lambda (= \sqrt{\hbar/eB}$, where $\hbar$ and e have their usual meanings) is around 200 nm at 10 mT, which is much bigger than the hopping distance, which we estimate to be about 1 nm.

In disordered systems at low temperatures (mechanism 3), corrections to the transport due to electron-electron interactions become important. This is mainly due to the fact that carriers interact often when they diffuse slowly. It can be shown that the electron-electron interaction is modified in the presence of a magnetic field. However, this occurs only if the thermal energy, kT, is less than or comparable to the Zeeman energy, $\Delta E = g\mu_B B$, where g is the g-factor and μ_B is the Bohr magneton. In the case of organics, $g \approx 2$,[39] and therefore $\Delta E \approx 1$ μeV at a field of 10 mT. Hence, ΔE is much smaller than the thermal energy at room temperature (≈ 25 meV). Therefore, this mechanism also fails to explain OMAR. Mechanisms 1 to 3 lead to only positive magnetoresistance, whereas OMAR can also be negative. This further excludes mechanisms 1, 2, and 3.

Weak localization (WL) due to quantum corrections to Drude-like transport is another mechanism (4) for magnetoresistance. This mechanism is well known from the study of diffusive transport in metals and semiconductors.[38,40,41] It is based on back-scattering processes due to constructive quantum interference. When a magnetic field is applied to the system, the quantum interference is destroyed by the magnetic field if the phase delay due to the enclosed magnetic flux exceeds the coherence length. Therefore, the resistivity is decreased (negative magnetoresistance effect). In WL theory, it is assumed that the spin-orbit coupling is weak. A strong spin-orbit coupling would cause weak antilocalization (WAL), leading to a positive magnetoresistance effect due to destructive quantum interference. Indeed, the observed OMAR traces resemble magnetoresistance traces due to weak localization (WL, negative magnetoresistance) and weak antilocalization (WAL, positive magnetoresistance). However, the weak temperature dependence of OMAR is contrary to most, if not all, of the literature on WL and WAL in inorganic conductors. Furthermore, in WL theory the width of the magnetoresistance cones is inversely proportional to the phase-breaking length, which in turn is intimately related to the mobility and temperature. In OMAR, however, equal magnetoresistance cone widths are observed in materials with largely different mobilities, and at largely different temperatures (see Figure 3.7). So, WL and WAL can also be ruled out as possible mechanisms behind OMAR.

3.2.2.2 Survey of Possible Models

Three kinds of models based on spin mixing between spin states induced by hyperfine interaction have recently been put forward to explain OMAR. All three of them are based on spin-dependent recombination/scattering events between different paramagnetic species. The following possibilities have been considered: (1) electron-hole pair models,[42–46] based on spin-dependent singlet/triplet exciton formation from oppositely charged radical pairs[47,48]; (2) the triplet exciton-polaron quenching model,[49] which is based on the quenching reaction between a triplet exciton and a polaron; and (3) the bipolaron model,[50] which treats the spin-dependent formation of doubly occupied sites (bipolarons) during the hopping transport through the organic film. Other possibilities of spin-dependent recombination/scattering events also exist in principle, but have not been considered to the best of our knowledge.

Although conclusive evidence regarding the three above models is still missing, some remarks can be made. In the electron-hole pair models the magnetic field lifts the degeneracy between the singlet (S) and two of the triplet electron-hole pairs (T_1 and T_{-1}, the triplets with spin along or opposite to the magnetic field; the triplet T_0 with zero spin along the magnetic field remains degenerate with the singlet). Since this reduces the mixing between triplets and singlets and therefore deprives the triplets from the efficient recombination channel via the singlets, the efficiency of recombination is reduced. Since the device current depends on this recombination efficiency,[42] this leads to a dependence of the current on the magnetic field. The sign change often observed in OMAR is explained by a nonmonotonic dependence of the current on the recombination efficiency.[46] However, the electron-hole pair models are difficult to reconcile with the measurement of OMAR in (close-to) unipolar devices.

In the triplet exciton-polaron quenching model[49] the idea is that the presence of the magnetic field increases the intersystem crossing rate from triplet to singlet excitons (no mechanism for this increase of the intersystem crossing rate is provided in Desai et al.[49]). Since this reduces the amount of triplet excitons and therefore the triplet exciton-polaron quenching, the polaron mobility will increase. However, it is known that the number of triplets in OLEDs is very strongly dependent on temperature,[51] and therefore this model is difficult to reconcile with the measured weak temperature dependence of OMAR. An experimental result that is difficult to reconcile both with the electron-hole pair models and with the triplet exciton-polaron quenching model is the measurement of equal magnetic field dependencies of singlet and triplet exciton densities in an OLED of a ladder-type polymer.[52]

In the bipolaron model[50] the idea is that the magnetic field reduces the rate of bipolaron formation from polarons with equal charge sign. Depending on parameters, this can lead to an increase or decrease of the current.[50] The model is obviously consistent with the measurement of OMAR in (close-to) unipolar devices and with the experiments on a ladder-type polymer-OLED.[52] The model predicts a weak temperature dependence, in agreement with the experiment. Moreover, depending on a "branching ratio" for bipolaron formation, the Lorentzian as well as the non-Lorentzian line shapes discussed in Section 3.2.1.3 are obtained. In the next section we will elaborate further on the bipolaron model.

3.2.2.3   The Bipolaron Model

Charge transport in disordered organic materials occurs by hopping of charge carriers between localized sites. The density of states (DOS) of these sites is often assumed to be Gaussian.[53] The standard deviation of this Gaussian, σ, is typically 0.1 to 0.2 eV. In addition to a random energy drawn from this Gaussian we introduce an energy penalty, U, for a doubly occupied site, i.e., a bipolaron. Due to the shared lattice deformation of the two charges forming the bipolaron, the energy penalty, U, is not very large: experimental indications are that $U \approx \sigma$.[54] The large on-site exchange effects will lead to energies of triplet bipolarons that are much higher than those of singlet bipolarons.[55,56] Hence, we will assume that bipolarons can only occur as spin singlets.

The hydrogen nuclei in hydrocarbon molecules couple to the spin of polarons. This occurs (1) via the exchange interaction of the carbon π-electrons with the hydrogen s-electrons,[57] which couple with the proton nuclear spin, and (2) by through-space dipole-dipole coupling between the proton nuclear spin and the π-electron spin. In the solid state both couplings are of comparable magnitude. Since typically many (order of ten or more) hydrogen nuclei couple to a single π-electron spin, the coupling at a site i can be described by a classical hyperfine field $\boldsymbol{B}_{hf,i}$, which has a three-dimensional Gaussian distribution.[58] This hyperfine field can to a good approximation also be treated as static, since the flip of the spin of a π-electron is accompanied by the flip of only one of the many proton nuclear spins. We will call the standard deviation of the Gaussian distribution of hyperfine fields B_{hf}. Typically, $B_{hf} \approx 5$ mT. In the presence of an external magnetic field $\boldsymbol{B}$, the total field at a site i is then $\boldsymbol{B}_i = \boldsymbol{B} + \boldsymbol{B}_{hf,i}$.

For the polaron hopping rate from site $i = \alpha$ to site $i = \beta$ we assume the Miller-Abrahams form,[59] valid for phonon-assisted tunneling in the case of coupling of a charge to a bath of acoustical phonons,

$$\omega_{\alpha\beta} = \omega_{hop} \min\left[\exp\left(\frac{\varepsilon_\alpha - \varepsilon_\beta}{kT}\right), 1\right] \tag{3.1}$$

with

$$\omega_{hop} \equiv \omega_{ph} \exp(-2\alpha/\xi) \tag{3.2}$$

where ω_{ph} is a typical phonon frequency, kT the thermal energy, α the distance between the sites α and β, and ξ the decay length of the involved localized wavefunctions. Usually, ξ is short enough ($\approx 0.1\alpha$) to allow consideration of only nearest-neighbor hopping at not too low temperatures.[60] The energies ε_α and ε_β are those of the localized wavefunctions at α and β, respectively. When an electric field E is applied, the corresponding electrostatic energy is added to these energies. In some of the Monte Carlo simulations we performed we took into account the coulomb repulsion between charges. For computational convenience we used a finite range variant of Coulomb's law, specifically $V(R) = (a/R - 1/5)V$ for $R < 5a$, $R \neq 0$, and zero otherwise, where R is the distance between two charges. The corresponding coulomb energy is also added to the energies. Inserting the typical values $\varepsilon \approx 3$ for the dielectric constant and $a \approx 1$ nm for the nearest-neighbor distance leads to $V \approx U$. We will treat V as a parameter in some of the calculations below.

In the case that a polaron is already present at β, bipolaron formation at this site is possible by hopping of a polaron at α to β. The mutual spin orientation of the polarons now becomes a crucial ingredient. If the spins of the polarons are perfectly parallel, no bipolaron formation can occur, since the singlet component of such a configuration is zero. In this case spin blocking will occur. However, the random hyperfine fields at the two sites can change the mutual orientation of the spins with time, leading to a finite singlet component and the possibility of bipolaron formation. In Appendix B this is worked out in detail. In order for the hyperfine field to have any influence on the probability for bipolaron formation, the hopping frequency, $\omega_{\alpha\beta}$, should be low compared to the hyperfine frequency, $\omega_{hf} = \gamma B_{hf} \approx 10^8$ s^{-1} (γ is the gyromagnetic ratio). This is basically a consequence of the energy-time uncertainty relation, $\Delta t \Delta E \geq \hbar/2$, where Δt is the time during which the configuration with a polaron at α and a polaron at β exists and ΔE is the hyperfine energy splitting. If this condition is not fulfilled, the hyperfine field is quenched. If this condition is fulfilled (we will call this the case of slow hopping), it is shown in Appendix B that within a density matrix formulation one can assume that hopping occurs between eigenstates of the spin Hamiltonian (see Figure 3.11). One can then distinguish a parallel configuration, P, with the spins at α and β both along or both opposite to the total local magnetic fields, $\mathbf{B}_\alpha$ and $\mathbf{B}_\beta$, and an antiparallel' configuration, AP, with one spin along and the other opposite to the local magnetic fields. The singlet probabilities of these configurations are

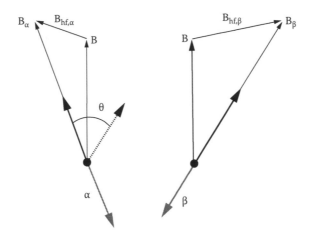

FIGURE 3.11 Two hopping sites, α and β, for polarons in a disordered organic semiconductor, with random hyperfine fields $\mathbf{B}_{hf,\alpha}$ and $\mathbf{B}_{hf,\beta}$. In the presence of an external magnetic field $\mathbf{B}$ this gives rise to effective magnetic fields $\mathbf{B}_\alpha$ and $\mathbf{B}_\beta$, with an angle θ in between. The eigenspinors at the two sites are symbolically indicated with the black ($\sigma = 1$) and gray ($\sigma = -1$) arrows.

$P_p = \frac{1}{2}\sin^2(\theta/2)$ and $P_{Ap} = \frac{1}{2}\cos^2(\theta/2)$, respectively, where θ is the angle between the total magnetic fields at α and β (see Figure 3.11). One should now multiply the hopping frequency $\omega_{\alpha\beta}$ by P_P or P_{AP} in order to obtain the transition rate for bipolaron formation from a parallel or antiparallel configuration. In Appendix A it is shown that under the condition of slow hopping one can also assume that hopping of polarons occurs between spin eigenstates. Starting with a polaron at α with spin σ along $\mathbf{B}_\alpha$ that hops to β with a spin σ' along $\mathbf{B}_\beta$ ($\sigma, \sigma' = \pm 1$), the transition rate, $\omega_{\alpha\beta}$, should now be multiplied with a factor $\pi_{\sigma\sigma'}^2$, with $\pi_{11} = -\pi_{-1-1} = \cos(\theta/2)$ and $\pi_{1-1} = \pi_{-11} = \sin(\theta/2)$.

We will first consider the simplest nontrivial model that leads to a magnetic field-dependent bipolaron probability (see Figure 3.12). We again consider the two sites, α and β. We assume that the energy of site β is so low that this site permanently holds at least one polaron. A bipolaron can be formed by hopping of a polaron to β from the site α, which we call the branching site. We will assume that the electric field is large enough to prevent the occurrence of the backward process. Instead, we assume that dissociation of the bipolaron occurs by hopping of one of the polarons with a rate $\omega_{\beta e}$ to other sites, which we consider to be part of the environment. We assume that polarons with random spin enter α with a rate $\omega_{e\alpha}$ by hopping from sites in the environment, leading to an influx, $\omega_{e\alpha}p/2$, into both parallel and antiparallel spin channels, where p is a measure for the average number of polarons in the environment. We also consider the possibility that a polaron at α directly hops back to an empty site in the environment, with a rate $\omega_{\alpha e}$. Neglecting double occupancy of α and single occupancy of α simultaneously with double occupancy of β, the corresponding steady-state rate

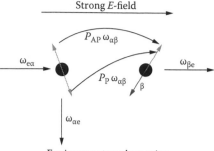

FIGURE 3.12 Two hopping sites, α and β, with a polaron present at β (red arrow). A polaron present at α can be in either a parallel or antiparallel configuration (pink arrows at α). This polaron can hop to β to form a bipolaron (pink arrow at β). A strong electric field prevents the backward process. The bipolaron can unbind by hopping of one of the polarons to the environment (gray area). The polaron at α can also hop to the environment, where its spin is randomized. The site α is filled with random spins from the environment.

equations become

$$\frac{1}{2}\omega_{e\alpha}p - (\omega_{\alpha e} + P_P\omega_{\alpha\beta})p_P = 0$$

$$\frac{1}{2}\omega_{e\alpha}p - (\omega_{\alpha e} + P_{AP}\omega_{\alpha\beta})p_{AP} = 0 \tag{3.3}$$

$$P_P\omega_{\alpha\beta}p_P + P_{AP}\omega_{AP} - \omega_{\beta\to e}p_\beta = 0$$

with $p_{P,AP}$ the probabilities for a parallel and antiparallel spin configuration, and p_β the probability to have a bipolaron at β. These equations can be straightforwardly solved for p_β, resulting in

$$p_\beta = \frac{\omega_{e\alpha}}{\omega_{\beta e}} f(B)p \tag{3.4}$$

where the dependence on the external magnetic field B is absorbed into the function

$$f(B) \equiv \frac{P_P P_{AP} + 1/(4b)}{P_P P_{AP} + 1/(2b) + 1/b^2} \tag{3.5}$$

and where

$$b \equiv \frac{\omega_{\alpha\beta}}{\omega_{\alpha e}} \tag{3.6}$$

is the branching ratio for bipolaron formation.

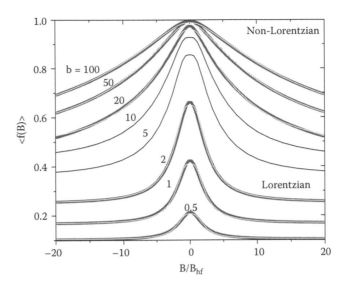

FIGURE 3.13 Average $\langle f(B) \rangle$ of the function $f(B)$, defined by Equation 3.5, over the Gaussian distribution of hyperfine fields, with standard deviation B_{hf}, for different values of the bipolaron branching ratio b, defined by Equation 3.6. The upper thick lines are fits to a Lorentzian function, whereas the lower thick lines are fits to the non-Lorentzian function discussed in Section 3.2.1.3.

For an ensemble of sites the bipolaron probability is obtained by averaging the function $f(B)$ in Equation 3.5 over the Gaussian distribution of hyperfine fields. This average, $\langle f(B) \rangle$, can be calculated numerically without much effort. For different values of the branching ratio, b, the results for $\langle f(B) \rangle$ are shown in Figure 3.13. For increasing branching ratio, the line shape changes from Lorentzian to non-Lorentzian, as discussed in Section 3.2.1.3, with a rather sharp transition in between. This transition from one line shape to the other can be understood immediately from Equation 3.5. For small b, $P_P P_{AP}$ is the small parameter in a Taylor expansion of $f(B)$ and the average, $\langle P_P P_{AP} \rangle$, has a Lorentzian line shape. For large b, on the other hand, $(P_P P_{AP})^{-1}$ is the small parameter and the average, $\langle (P_P P_{AP})^{-1} \rangle$, has the non-Lorentzian line shape.

It is possible to extend the two-site model discussed above to a model that not only predicts a bipolaron probability that depends on the magnetic field, but also predicts a magnetic-field-dependent current.[61] Moreover, this extended model predicts positive as well as negative magnetoresistance, depending on the parameters. In order to go beyond the calculation of only the bipolaron probability, it is crucial to work out the model in terms of many-electron states. Within the aforementioned restrictions and excluding time-reversed states, we have five of them: $|01\rangle$, $|11_P\rangle$, $|11_{AP}\rangle$, $|02\rangle$, and $|12\rangle$, where $|nm\rangle$ denotes an $n(m)$-fold occupation at site $\alpha(\beta)$. Again, we consider only downstream (along the electric field) flow of polarons. We define occupation probabilities, A_{nm}, for the respective many-electron states, $|nm\rangle$, with $\sum_{nm} A_{nm} = 1$, and construct a set of corresponding rate equations for these states, similar to Equation 3.3. As an example, putting $dA_{11p}/dt = 0$ yields:

$$0 = A_{01}p\omega_{e\alpha} + A_{12}\omega_{\alpha e} - A_{11P}(P_P\omega_{\alpha\beta} + \omega_{\alpha\beta}/b) \tag{3.7}$$

while the other equations can be constructed in a similar way. An additional spin-orbit-induced spin relaxation on site α can be included as well,[61] but is neglected in the present review. Solving the set of equations results in analytical expressions for A_{nm}. The current through the system equals the total rate from the upstream environment to α:

$$I/e = p < (A_{01} + A_{02}) > \omega_{e\alpha} \tag{3.8}$$

where $<\cdots>$ denotes again the ensemble average over hyperfine fields. The explicit expression for I is lengthy, but can be rewritten in a generic form:

$$I = I_\infty + I_B < 1 - \frac{1}{1 + \Gamma P_P P_{AP}} > = I_\infty + I_B g\left(\Gamma, \frac{B}{B_{hf}}\right) \tag{3.9}$$

where I_∞, I_B, and Γ are straightforward analytical expressions in terms of the model parameters. We note that the field dependencies are identical to the ones found for the prefactor of the bipolaron probability, $<f(B)>$ (Equation 3.5), and are described here by the model function $g(\Gamma, B/B_{hf})$. Thus, the line shape is fully described by a single parameter, Γ, whereas the line width is monotonously increasing with increasing Γ. For a more detailed treatment of the $MC(B)$ behavior of the two-site model, refer to Wagemans et al.[61] Here we focus on a few elucidating results.

In the limit of low carrier density, i.e., in lowest order of p, the line width parameter can be written as

$$\Gamma = \frac{2}{1/b + 2/b^2} \tag{3.10}$$

which demonstrates the quantitative resemblance to the prefactor of the bipolaron probability, $<f(B)>$. Equation 3.10 shows that in the limit of small p the line width monotonously increases as a function of increasing branching ratio (so decreasing b^{-1}). Stated differently, a larger applied magnetic field is required to saturate the OMAR effect when the current is more forced to flow *through* the bipolaron (β) site. We note that this basic version of the bipolaron model only predicts a negative MC,*

---

* We stress that this conclusion is in contrast with that of Wagemans et al.[61] Unlike claimed in that paper, a sign change to a positive MC is only possible in a modified version of the two-site model, in which rates depend on occupation of the α and/or β-site, or both. More specifically, the results in Wagemans et al.[61] represent the case when the rate from environment to α is reduced by a factor of two in case of a doubly occupied β-site, and the rate from β to the environment is doubled for a singly occupied α-site. Such modifications could be qualitatively thought of as being the result of coulomb repulsion between the charge carriers, but have no strict quantitative justification.

which corresponds to a blocking of the current because of the inability of triplet states to form a bipolaron. One can show analytically that this blocking becomes complete (MC = −1) in the limit $b^{-1} = 0$, i.e., when all carriers are forced through site β. As will be discussed in more detail below, for certain parameters the bipolaron model predicts a positive MC. In the two-site version of it, this situation can occur once some of the rates (ω_{ij}) or the density of polarons in the environment (p) are assumed to be a function of the occupation of the β site. Doing so, for small branching ratio (large b^{-1}), a sign change to a positive MC can be witnessed. Within the two-site model, positive MC is a consequence of the reduction of the current in case of a doubly occupied β site. Such a situation can occur in organic devices when the polaron density increases at the expense of the bipolaron population at increasing B—a mechanism that will be explored in more detail while treating the extended Monte Carlo implementation of the bipolaron model. Note that *the magnitude of the positive MC is generally smaller than the negative MC*, making the negative MC the more "dominant" result of the bipolaron mechanism. It was found that the behavior of the line width derived for small p is more generally valid, i.e., independent of carrier density, Γ increases for increasing b. Furthermore, although a one-to-one correspondence does not hold, there is a trend that *the negative MC is accompanied by a larger line width*.

We note that the essence of the two-site model is the competition that occurs between the possibility that in a certain configuration of two neighboring polarons a bipolaron is formed and the possibility that this configuration is broken up by hopping of one of the polarons away from the other, with a subsequent randomization of the spins. The branching parameter, b, is a measure of the amount of competition between these two processes. We also note that the effect is essentially a nonequilibrium one, which can only occur in the presence of an electric field, which drives the system out of equilibrium. In equilibrium, i.e., in the absence of an electric field, there cannot be a change in the bipolaron probability as a function of magnetic field, since the Zeeman splitting corresponding to a field of 5 mT, which is of the order of 1 μeV, is much smaller than the thermal energy, kT. The essence of the effect is that the dynamic reaction of two polarons is influenced by the magnetic field. In chemistry, the influence of a (small) magnetic field on the reaction between two paramagnetic species is known as chemically induced dynamic electron polarization (CIDEP).[62] The bipolaron mechanism is of this type.

Although the two-site approach provides some interesting trends, it is questionable how reliably the predictions can be transferred to realistic devices. The inevitable approximations involved can be bypassed by performing Monte Carlo simulations. The simulations we performed proceed as follows. Randomly picked sites are occupied according to a Fermi-Dirac distribution with Fermi energy E_F ($E_F = 0$ corresponds to the center of the DOS), taking into account the coulomb field of all sites already occupied. The system is then propagated in time, keeping track of site occupations, spins, and coulomb field, until it reaches a steady state. Subsequent hops are then used to calculate the current, which is evaluated as the difference in number between down-field and up-field hops divided by the simulation time. This procedure is repeated for different disorder configurations until the average difference in the current, ΔI, with and without applied magnetic field has converged to within 1%

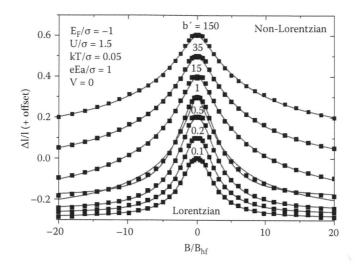

FIGURE 3.14 Symbols: Magnetoconductance, $\Delta I/I$, obtained from Monte Carlo simulations for the indicated parameters and for different multiplication factors, b', for the bipolaron formation rate. Lower lines: Fits to a Lorentzian function. Upper lines: Fits to the non-Lorentzian function discussed in Section 3.2.1.3. From bottom to top, consecutive offsets of 0.1 have been added to the curves.

absolute. The simulations are performed on cubic lattices with typically 100^3 sites. Periodic boundary conditions are imposed in all three directions.

To study the effect of the branching ratio we multiplied the rate for each bipolaron formation event by an extra factor, b'. For a combination of parameters yielding a particularly large effect we show the results for the relative magnetoconductance $\Delta I/I$ in Figure 3.14. Again, now with increasing b', the line shape changes from Lorentzian to non-Lorentzian, fully in agreement with the two-site model. Our conjecture is that in organic materials showing a Lorentzian OMAR (pentacene, RR-P3HT, RRa-P3OT), the microscopic mechanism of bipolaron formation is such that a small branching ratio results, whereas in organic materials showing the non-Lorentzian OMAR (polyfluorene, Alq_3, Pt-PPE, PPE), this mechanism is such that a large branching ratio results. Quantum chemical calculations taking into account the specific microscopic details of the various materials could verify this conjecture.

The results of the Monte Carlo simulations shown in Figure 3.14 demonstrate that the magnetic field dependence of the bipolaron probability found in the two-site model translates to a magnetic field dependence of the current. We note that we find a negative MC (positive MR). Analysis of our simulations shows that this negative MC is caused by blocking of the current by sites at which bipolaron formation can take place. The increase of the external magnetic field leads to a decrease of bipolaron formation at these sites, leading to a reduced current through these sites. Further analysis shows that percolating paths of current occur, which considerably enhances the effect. With only a small amount of bipolarons in the system, the blocking effect can be large: a single site where bipolaron formation occurs can block a complete current path.

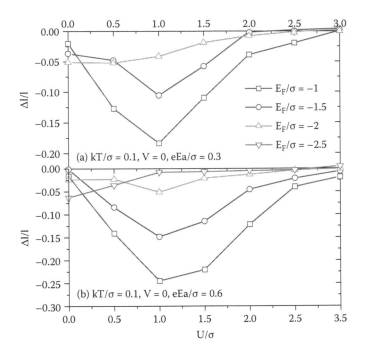

FIGURE 3.15 Simulated magnetoconductance for $B/B_{hf} = 100$ and $kT/\sigma = 0.1$ as a function of the bipolaron energy U for (a) $eE\alpha/\sigma = 0.3$ and (b) $eE\alpha/\sigma = 0.6$ for several choices of the Fermi level, E_F. The model parameters are indicated. (From Bobbert et al., *Phys. Rev. Lett.* 99, 216801 [2007]. With permission.)

For the case $b' = 1$ (Lorentzian line shape), we have studied the dependence on various model parameters. First, we ignore the coulomb interaction and put $V = 0$. Figure 3.15a and b shows a negative MC for $B/B_{hf} = 100$ as a function of the energy penalty, U, for bipolaron formation, for two typical choices of parameters, and several choices for the Fermi energy, E_F. Although the effect decreases with decreasing carrier densities, it still persists at rather low carrier densities, i.e., large $|E_F/\sigma|$. We attribute this to the percolation effect mentioned above. Remarkably, the effect is largest for $U/\sigma \approx 1$, which can be explained as follows. For large U the effect is small because of the high energy penalty for creating bipolarons. However, for small U the effect is also small because of the small number of single polarons: too many polarons have combined to form bipolarons. Figure 3.16 shows that the negative MC effect increases with decreasing temperature (For room temperature and below we typically have $\sigma/kT > 4$. For smaller electric fields the effect becomes less temperature dependent, which is in agreement with experimental findings.)[33]

Switching on the coulomb interaction by taking a nonzero V leads to a dramatic change of the situation. We can now obtain positive MC (see Figure 3.16). Analysis shows that there are now two competing effects contributing to the MC: (1) blocking of transport through bipolaron states (negative MC), and (2) an increase in polaron population at the expense of the bipolaron population with increasing B (positive MC). Apparently, effect 1 dominates for the parameters used in Figure 3.15, where

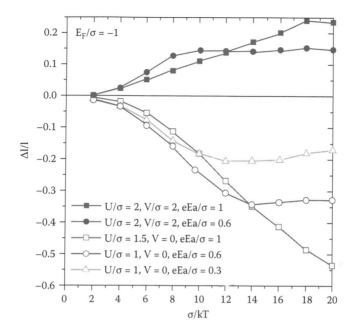

FIGURE 3.16 Simulated magnetoconductance for $B/B_{hf} = 100$ as a function of σ/kT for several choices of the model parameters. The model parameters are indicated.

$V = 0$, while effect 2 dominates for large V. Figure 3.17a shows that the sign of the MC changes with increasing V. The inset in Figure 3.17b shows that the number of polarons decreases with increasing V, which means that the number of bipolarons increases. At first sight this increase of the number of bipolarons is contradictory, but it can be explained as follows. For $V = 0$ bipolaron formation requires a hop to sites with a particularly low energy to compensate for the bipolaron formation energy, U, and such sites are rare. However, upon inclusion of the coulomb interaction the energy that has to be compensated for is reduced to $U - V$, leading to more sites at which bipolaron formation is possible. Therefore, inclusion of V leads to enhanced bipolaron formation. This leads to an enhanced sensitivity of the bipolaron population on the magnetic field and an enhancement of effect 2. Figure 3.17b shows that the effect of the magnetic field on the polaron population increases with V, reflecting this enhancement.

We end our discussion of the bipolaron model by making a few remarks. The first is that the model does not strictly require the two polarons to be at the same site. It is sufficient that the overlap of the wavefunctions of the polarons is large enough to yield a considerable exchange energy, and hence splitting between the singlet and triplet spin configurations of the pair. In particular, so-called π-dimers can exist in π-conjugated polymers and molecules, where the polarons are located on neighboring polymer chains/molecules.[63] The bipolaron mechanism would work in just the same way for this case. The second remark is that the recombination of an electron and a hole polaron to form a singlet exciton obeys the same spin rules as the

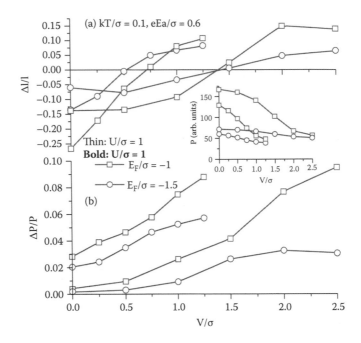

FIGURE 3.17 (a) Simulated magnetoconductance for $B/B_{hf} = 100$ as a function of the strength V of the coulomb repulsion for several choices of the model parameters. (b) Magnetic-field-induced change, $\Delta P/P$, in polaron population, P, as a function of V. The inset shows P vs. V. The model parameters are indicated.

formation of a singlet bipolaron. Because the triplet exciton in organic semiconductors has a much lower energy than the singlet exciton, with an energy difference of ≈ 0.5 eV, the rate for triplet exciton formation will be lower than that for singlet exciton formation.[64] As a result, the analysis in this section of the line shapes for bipolaron formation can be repeated for exciton formation, with the same conclusions. In particular, a branching ratio for singlet exciton formation can be introduced, which will determine the line shape in the same way as the branching ratio for bipolaron formation does. The last remark is that various small magnetoresistive effects (<1%) measured in inorganic amorphous semiconductors in the seventies were attributed to a similar spin-blocking mechanism.[65] The difference between the mechanisms is the cause of the spin randomization. In the present mechanism spins are randomized by the hyperfine fields, whereas in inorganic semiconductors the spin randomization predominantly takes place by the spin-orbit coupling.

3.2.3 DEVICE MODELING OF MAGNETORESISTANCE FOR BIPOLAR INJECTION-LIMITED DEVICES

Within the bipolaron model, OMAR is described as a bulk property, and can be translated to a magnetic field effect on the mobility. Within the present section, we address the nontrivial behavior that arises in space-charge-limited (SCL) devices in the bipolar regime, when carriers of opposite sign are simultaneously present in the device. We do so for the case where the magnetic field effect on the mobility is described by a unipolar mechanism, i.e., induced by spin correlations between quasi-particles of equal charge (e-e or h-h, like in the bipolaron model).

We start our analysis by defining the magnetomobility as the relative change in mobility

$$\Delta\mu_i/\mu_i \equiv \frac{\mu_i(B) - \mu_i(0)}{\mu_i(0)} \tag{3.11}$$

at applied field B and for a specific carrier type; $i = e$ or h, for electrons and holes, respectively. Furthermore, we introduce the concept of *normalized magnetoconductivity* (NMC), defined as the ratio of the magnetoconductance of the whole device and the magnetomobility for that carrier type:

$$\text{NMC}_i \equiv \frac{\Delta I/I}{\Delta\mu_i/\mu_i} \approx \left(\frac{\partial I}{\partial\mu_i}\right) \bigg/ \left(\frac{I}{\mu_i}\right) \tag{3.12}$$

Using this notation, it is easily seen that the MC of an SCL device can be written as a sum of the electron and hole NMC:

$$\text{MC} = \frac{\Delta I}{I} = \frac{1}{I}\left(\frac{\partial I}{\partial\mu_e}\Delta\mu_e + \frac{\partial I}{\partial\mu_h}\Delta\mu_h\right)$$

$$= \frac{\Delta\mu_e}{\mu_e}\text{NMC}_e + \frac{\Delta\mu_h}{\mu_h}\text{NMC}\, h \tag{3.13}$$

in the limit of relatively small magnetomobilities, $|\Delta\mu_i/\mu_i| = 1$. For strictly single-carrier (unipolar) devices one obtains as a trivial relation for that carrier:

$$\text{NMC}_i = 1 \tag{3.14}$$

meaning that an increase/decrease of the electron (hole) mobility by a certain percentage leads to a change of the current by the same percentage. In the present section, we address the case of bipolar devices, in which nontrivial and quite surprising deviations from this trivial behavior occur—even leading to sign changes and negative values of NMC in specific cases. Although only very recently such behavior has been addressed for the first time,[66] the behavior is thought to be very generic, and intrinsic for bipolar SCL devices.

Anomalous behavior can be anticipated based on intuitive arguments.[66] Let us assume a case where the injection of the majority carrier (denoted as M) is ohmic and the injection of the minority carrier (m) is injection limited. In this case, at low bias the device acts as a unipolar one, with $\text{NMC}_M = 1$ and $\text{NMC}_m = 0$. Increasing the bias, the transition from unipolar to bipolar behavior occurs once the electric field at the minority carrier contact becomes large enough, so that minority charges

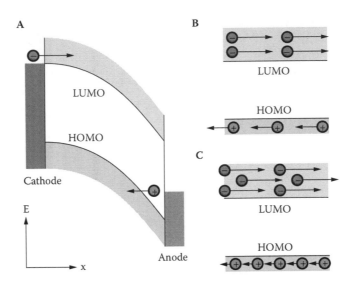

FIGURE 3.18 (a) Schematic band diagram of the modeled device, showing an ohmic electron (majority carrier) contact and injection-limited hole (minority carrier) contact. (b, c) The effect on the charge concentrations of the hole and electron channel when the hole mobility, μ_h, is decreased from panel (b) to (c) (μ is represented by the length of the arrows).

start to be injected, and the device becomes slightly bipolar. At this point the electric field throughout the device is still entirely dominated by the majority carriers. As a consequence, the electrical field at the contact is still insensitive to the density and mobility of minority carriers, causing this contact to act like a constant current source. The consequences of this effect are schematically shown in Figure 3.18. If the mobility (represented by the arrows) in the minority channel decreases in an applied magnetic field, i.e., $\Delta\mu_m/\mu_m < 0$, the density of minority charges increases (Figure 3.18b → Figure 3.18c) because the injected current in the minority channel is constant. The increase in the minority carrier density further compensates the coulomb repulsion in the majority channel, allowing the density of the majority charges to increase—a behavior that is characteristic for bipolar SCL devices[67] and that will be discussed more quantitatively later on. Since the majority carriers carry the bulk of the current, an increase in their density increases the device current. Thus, we obtain a negative NMC for the minority carriers, $\mathrm{NMC}_m < 1$, or in simple words, a *decrease* in the minority carrier mobility results in an *increase* of the current. This counterintuitive behavior has not been signaled in either the original work by Parmenter and Ruppel[67] or in more recent treatments[68] of SCL current in the bipolar regime.

In order to put this effect on a more quantitative footing, we follow the analytical approach as introduced by Parmenter and Ruppel,[67] but extend it to derive the behavior in the intermediate regime, right at the onset of minority carrier injection, not treated explicitly in Parmenter and Ruppel.[67] Following Parmenter and Ruppel,[67] we make three approximations: (1) the diffusion current is neglected, (2) thermal equilibrium carrier densities are neglected, and (3) traps are neglected.

We consider a planar device of thickness d and dielectric constant ε, ranging from $x = 0$ (cathode) to $x = d$ (anode), with local densities of electrons and holes (n and p, respectively), and with mobilities (μ_e and μ_h, respectively) independent of the densities and (local) electrical field, E. Based on the foregoing approximations, the task is to find the current density:

$$J = (n\mu_e + p\mu_h)eE \tag{3.15}$$

subject to Poisson's law,

$$\frac{dE}{dx} = \frac{e}{\varepsilon\varepsilon_0}(n - p) \tag{3.16}$$

while fulfilling the continuity equation

$$-\mu_e \frac{d(nE)}{dx} = \mu_h \frac{d(pE)}{dx} = \frac{2enp\mu_0}{\varepsilon\varepsilon_0} \tag{3.17}$$

where μ_0 is the so-called recombination mobility. For the simple case of a single-carrier device (e.g., $p = 0$), this set of equations leads to the well-known SCL behavior, with the current density proportional to the square of the applied voltage:

$$J = \frac{9}{8}\varepsilon\varepsilon_0\mu_M \frac{V^2}{d^3} \tag{3.18}$$

In order to derive a corresponding relation for a bipolar device, one may introduce a dimensionless parameter, B^{67} (not to be confused with the magnetic field strength):

$$B \equiv J_e/J = en\mu_e E/J \tag{3.19}$$

representing at any point x the fraction of the total current carried by electrons. Integrating Equations 3.15 through 3.17 then leads to an expression for the total voltage over the device,[67] which is easily rewritten to get an expression for the current density as a function of applied bias:

$$J = \left(\frac{v_e v_h \mu_0 \varepsilon\varepsilon_0 V^2}{2d^3}\right) \frac{\left[\int_{B_a}^{B_c} B^{v_e-1}(1-B)^{v_h-1}dB\right]^3}{\left[\int_{B_a}^{B_c} B^{\frac{3}{2}v_e-1}(1-B)^{\frac{3}{2}v_h-1}dB\right]^2} \tag{3.20}$$

where $B_a = B(0)$, $B_c = B(d)$. Furthermore, for the sake of a compact notation, mobilities relative to the recombination mobility have been defined, $v_e \equiv \mu_e/\mu_0$ and $v_h \equiv \mu_h/\mu_0$. The latter quantities are usually larger than unity, particularly in the case of (weak)

Langevin recombination, in which case

$$\mu_0 = L(\mu_e + \mu_h) \tag{3.21}$$

with L a prefactor, with L $\ll$ 1 for weak recombination. Equation 3.20 provides a compact solution to the problem, once the boundary conditions (B_a and B_c) are unambiguously defined. In Parmenter and Ruppel[67] it was conjectured that in order to minimize the voltage over the device, B_a should be minimized and B_c maximized under the constraint of a maximum (injection-limited) value J_{sc} for the electron current at the cathode, and a similar J_{sa} for the hole current at the anode. In the case of two nonsaturated blocking contacts (i.e., ohmic contacts; $J < J_{sa}$ and $J < J_{sc}$), one obtains $B_a = 0$ and $B_c = 1$—representing the situation of a pure hole current at the anode and a pure electron current at the cathode. In that case, and under the approximation $v_e \gg 1$ and $v_h \gg 1$, Parmenter and Ruppel derived their famous result:

$$J = \frac{9}{8} \varepsilon \varepsilon_0 \mu_{eff} \frac{V^2}{d^3} \tag{3.22}$$

where the effective mobility, μ_{eff}, is given by

$$\mu_{eff} = \frac{2}{3} \sqrt{2\pi \frac{\mu_e \mu_h}{\mu_0} (\mu_e + \mu_h)} \tag{3.23}$$

Thus one obtains the same functional dependence on V as unipolar SCL devices (Equation 3.18), but with a current that is enhanced roughly by the square root of the ratio of the smaller carrier mobility to the recombination mobility ($\min[\mu_m, \mu_M]/\mu_0$). For Langevin recombination (Equation 3.21), the result simplifies to

$$\mu_{eff} = \frac{2}{3} \sqrt{2\pi \frac{\mu_e \mu_h}{L}} \tag{3.24}$$

The above results allow for an analysis of the resulting magnetoresistance effects in pure bipolar transport. In the case of Langevin recombination, using Equations 3.22 and 3.24, and adopting the concept of magnetomobilities, leads to

$$\frac{\Delta I}{I} = C \left(\frac{\Delta \mu_e}{\mu_e} + \frac{\Delta \mu_h}{\mu_h} \right) \tag{3.25}$$

with $C = 1/2$. Similarly, assuming a constant μ_0 that is not a function of μ_e and μ_h (instead of Langevin recombination), one finds the same expression with $C = 1$. Thus, in the present case of two ohmic contacts, only the *relative* magnetomobilities matter. As a consequence, even in cases where the total current is entirely dominated by one type of carrier ($|\mu_M| \gg |\mu_m|$ or $|\mu_M| \ll |\mu_m|$), with respect to the magnetoconductance the role of minority and majority carriers is balanced:

$$NMC_M = NMC_m = C \tag{3.26}$$

After having discussed the well-known case of two ohmic contacts, we address the intermediate case with only one ohmic contact (without loss of generality assumed to be the cathode) and the other contact (the anode) being injection limited. In this case there is one parameter, the (saturated) minority carrier current at the anode (J_{sa}), which is not fully specified yet. Note that J_{sa}, and thereby $B_a = 1 - J_{sa}/J$, will be in general a function of the local field at the anode, $E_a = E(d)$. To include this boundary condition quantitatively, one needs to introduce an explicit dependence of the injection current on the local field, $J_{sa}[E_a]$.

Here, we continue by rewriting the integral expression for E_a from Parmenter and Ruppel[67] in terms of incomplete Bessel functions:

$$\beta_z(a,b) = \int_0^z t^{a-1}(1-t)^{b-1}\,dt \tag{3.27}$$

which leads to

$$E_a = \sqrt{\frac{2Jd}{v_e v_h \mu_0 \varepsilon \varepsilon_0} \frac{B_a^{v_e/2}(1-B_a)^{v_h/2}}{\sqrt{\beta_{1-B_a}(v_h,v_e)}}} \tag{3.28}$$

Having compact expressions for E_a (Equation 3.28) and J (Equation 3.20), the final solution, $J[V]$, is obtained by solving J, J_{sa}, and E_a from

$$E_a = E_a^{SCLC}[J_{sa}, J] \tag{3.29}$$

$$J = J^{SCLC}[J_{sa}, V] \tag{3.30}$$

$$J_{sa} = J_{sa}^{model}[E_a] \tag{3.31}$$

for any arbitrary value of V. The superscript SCLC refers to using the functional form in terms of indicated parameters according to the theory of SCL current, while the label "model" refers to a certain model for the functional dependence of the injection current at the anode as a function of the electric field at the anode. The function $J_{sa}^{model}[E_a]$ specifies the so-far failing boundary condition that is needed to solve the set of equations. Numerically, one can do so for any function $J_{sa}^{model}[E_a]$, which should have the property that for low fields the injected minority carrier current is absent or extremely small, while this current becomes significant above a certain threshold field. Since results for different models are qualitatively similar, we concentrate here on three examples:

$$J_{sa}^{model}[E_a] = 0 \quad for \quad E_a < E_0 \quad else \quad J_{sa}^{model}[E_a] = \infty \tag{3.32}$$

$$J_{sa}^{model}[E_a] = 0 \quad for \quad E_a < E_0 \quad else \quad J_{sa}^{model}[E_a] = J_0 \left(\frac{E_a - E_0}{SE_0}\right)^2 \tag{3.33}$$

$$J_{sa}^{\text{model}}[E_a] = J_0 \exp\left[\frac{-(1+S)}{S}\right]\left(\exp\left[\frac{E_a}{SE_0}\right] - 1\right) \qquad (3.34)$$

where E_0 is the threshold field, and $J_0 = \frac{1}{2}\varepsilon\varepsilon_0\mu_e E_0^2/d$ is the single-carrier current right at the threshold field. Furthermore, S is a dimensionless parameter that controls the sharpness of the onset, defined such that $J_{sa}[(1 + S) E_0] = J_0$. Note that Equation 3.32, describing an ideal on/off switch, is just a limiting case of Equation 3.33 for $S \rightarrow 0$ and needs no separate treatment. Equation 3.33 is of particular interest because it provides the opportunity to fully solve the problem analytically in the limit $v_e \gg 1$ and $v_h \gg 1$.[69] The third case, Equation 3.34, was introduced in Bloom et al.[66] This case is not analytically solvable, but is of interest because it results in a smooth onset of the minority carrier injection, which is more in line with experimental results.

The solution corresponding to Equation 3.33 can be written in a compact form[69]:

$$J = J_0\left(\frac{V^2}{V_0^2} + \frac{\Delta\mu_\infty \Delta V^2}{\mu_e V_0^2} + \frac{4\Delta\mu_\infty^2 - 2\Delta\mu_\infty\sqrt{2\Delta\mu_\infty^2 + D_0\Delta\mu_\infty\Delta V^2}}{D_0\mu_e V^2}\right) \qquad (3.35)$$

where $V_0 = 2E_0/3$ is the onset voltage, $\mu_\infty = \mu_{\text{eff}}[V \rightarrow \infty]$ (with μ_{eff} the effective mobility for the device considered), $\Delta\mu_\infty = \mu_\infty - \mu_e$, $\Delta V^2 = V^2 - V_0^2$, and $D_0 = (d^2\mu_{\text{eff}}/dV^2)_{V_0}$ is the second derivative of μ_{eff} at $V = V_0$. Expressions for μ_∞ and D_0 in terms of v_e and v_h are derived in Bloom et al.[69]

Some results, again assuming Langevin recombination, are sketched in Figure 3.19. Panel (a) shows the current vs. voltage for different values of S, and keeping $\mu_e/\mu_h = 10$ and $L = 0.001$ fixed. Below and well above V_0 a V^2 behavior is observed. It is seen, however, that for increasing values of S the pure bipolar regime ($\mu_\infty = \mu_{\text{eff,bip}}$ according to Equation 3.23; black dashed line) is never reached, and the current saturates to a V^2 behavior described by $\mu_\infty < \mu_{\text{eff,bip}}$ (dotted lines), while saturation is postponed to larger voltages. For $S \gg 1$ the increase of the current after the minority carrier injection vanishes completely.

Figure 3.19b shows for the same parameters the development of NMC$_e$ (dashed lines) and NMC$_h$ (solid lines) as a function of V. For intermediate values of S (e.g., $S = 0.2$) NMC$_h$ becomes negative right after minority carrier injection, *fully confirming our qualitative predictions about the sign change*. For increasing voltage, however, NMC$_h$ passes a minimum, after which it rises to a value close to the limiting value known for the pure bipolar regime, NMC$_h$ = 0.5 (Equation 3.26). For a high enough saturation parameter S (green and blue lines), the negative region of NMC$_h$ becomes more pronounced and the current reaches its asymptote before the minimum in NMC$_h$ is reached. As a consequence, the upturn to positive values is suppressed, and NMC$_h$ saturates at a negative value for $V \rightarrow \infty$. In the opposite regime of a very low S (red line), we observe a complete quenching of the negative region of NMC$_h$.

Results for different Langevin factors, L, while fixing $\mu_e/\mu_h = 10$ and $S = 0.2$, are shown in Figure 3.20. These results demonstrate that the weaker the Langevin

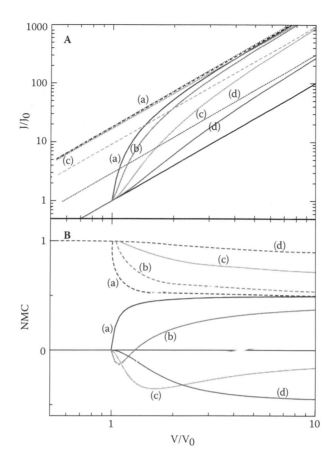

FIGURE 3.19 Results for the current density, $J(V)$, and the normalized magnetoconductivity, NMC(V), for different saturation parameters S, while fixing $\mu_e/\mu_h = 10$ and $L = 10^{-3}$. (A) $J(V)/J_0$ for $S = 0.02$ (red (a)), 0.2 (orange (b)), 0.7 (green (c)), and 2 (blue (d)). The black dotted line represents the extrapolation of the single-carrier current, the black solid line represents the pure bipolar limit, and the colored dotted lines represent the large-bias saturation currents for the corresponding values of S. (b) NMC$_e$ (dashed lines) and NMC$_h$ (solid lines), using the same color code as in (A).

recombination is, the more negative NMC$_h$ becomes, again *in line with our intuitive model*. For a large value of L, no negative NMC$_h$ is observed at all (e.g., $L = 0.1$, red). When reducing L, a negative region develops ($L = 0.01$, orange), which is getting more pronounced upon further reduction ($L = 10^{-3}$, 10^{-6}; green, blue), while the upturn is delayed toward higher voltages.

As an example of a boundary condition that does not allow for a fully analytical treatment, we briefly compare results using the exponential $J_{sa}^{model}[E_a]$ of Equation 3.34. Figure 3.21a and b shows results for $S = 0.1$ and $\mu_e/\mu_h = 2$.[66] Clearly, this model for $J_{sa}^{model}[E_a]$ results in a more gradual onset of minority carrier injection, as expected. A region with negative NMC$_h$ is again obtained around the onset. For comparison, results for the analytically solvable injection model using identical

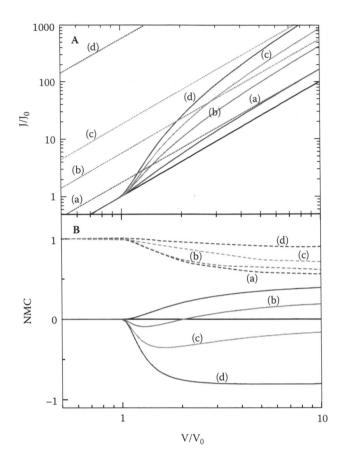

FIGURE 3.20　Results for $J(V)$ and NMC(V) for different Langevin factors, L, while fixing $\mu_e/\mu_h = 10$ and $S = 0.7$. (a) $J(V/J_0)$ for $L = 0.1$ (red (a)), 10^{-2} (orange (b)), 10^{-3} (green (c)), and 10^{-6} (blue (d)). The black solid line represents the extrapolation of the single-carrier current, while the colored dotted lines represent the bipolar current limit for the corresponding values of L. (b) NMC$_e$ (dashed lines) and NMC$_h$ (solid lines), using the same color code as in (A).

parameters are displayed in Figure 3.21c and d. Although detailed dependencies discussed before depend in a subtle way on the injection function, we find that *the specific trend of a negative NMC$_h$ at the onset of hole injection is a general property and is not specific for a special injection function.*

Although the foregoing analysis unambiguously demonstrates the occurrence of a negative NMC$_h$ for an ideal, drift-limited SCL device without traps, one may wonder whether this behavior also exists in realistic devices with a finite density of traps, and when carrier diffusion is no longer neglected. In Bloom et al.[66] explicit modeling for such a more realistic device is presented. In that work, the drift and diffusion equations are solved numerically using the principles laid out by Malliaras and Scott,[70] extending the approach to include trapping in the majority charge carrier (electron) channel. For more details of this modeling, refer to Bloom et al.[66] Figure 3.22a shows

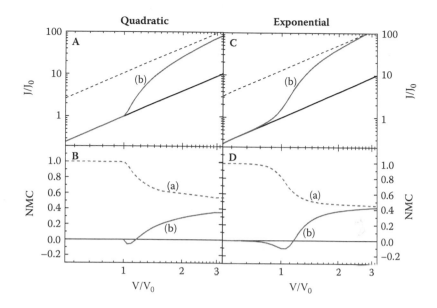

FIGURE 3.21   Results for $J(V)$ and NMC(V) for the quadratic (A, B) and exponential (C, D) injection models, fixing $\mu_e/\mu_h = 1$, $L = 0.025$, and $S = .070$. In (A, C) the red line (b) indicates the modeled $J(V)$ while the black solid line represents the extrapolation of the single-carrier current, and the black dotted line represents the bipolar current limit. In (B, D) NMC$_e$ is represented by red dashed lines (a) while NMC$_h$ is represented by red solid lines (b).

the numerically calculated $J(V)$ with $L = 0.01$. Like in the analytical model, we observe that at low voltages there is a unipolar power law behavior, $J \propto V^n$ (black line in Figure 3.22a, calculated with an ohmic cathode and an injection-limited anode), with the power $n \approx 4.5$ ($n > 2$, due to trapping[71]). We observe a deviation from the power law behavior once minority charge carrier injection starts, like in the ideal device. At high voltage the $J(V)$ behavior saturates to bipolar conduction (blue line in Figure 3.22a, calculated with two ohmic contacts).

With a magnetomobility in the minority channel, the onset of MC occurs with the onset of minority charge injection just like in the analytical model (see red line in Figure 3.22b). We observe a negative NMC$_h$ at the onset of bipolar behavior, and as the voltage increases NMC$_h$ increases, eventually changing sign, which is the same qualitative behavior as in the analytical model and is already significant for much lower values of L. One major difference is that the negative NMC resulting from the magnetomobility in the minority channel is much greater in the numerical calculations than in the analytical model. This is due to the presence of majority traps. As seen in Figure 3.22c, by removing the traps from the majority channel, the negative NMC becomes much smaller. The negative NMC results from a change of the coulomb repulsion in the majority channel by indirectly changing the minority carrier density via the minority carrier mobility. Therefore, it seems reasonable that an increase of the coulomb repulsion by adding traps to the majority channel increases the strength of the negative NMC.

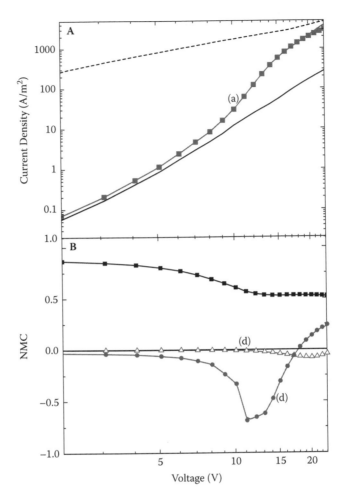

FIGURE 3.22 (A) The red line (a) represents the model calculation of $J(V)$ with traps in the majority (electron) channel. The black solid line represents the extrapolation of the single-carrier current, while the dashed black line represents the pure bipolar current. (B) The corresponding NMC_e (black) and NMC_h (blue solid symbols (d)) including trapping in the majority channel. The NMC_h without trapping (blue open symbols (d)). More details can be found in Bloom et al.[66]

In conclusion, in this section it was shown that a negative NMC due to a magnetomobility of the minority carriers should be a quite common feature for SCL devices above the onset of minority carrier injection. In realistic devices with traps, the effect may be even stronger than predicted for the analytically solvable ideal devices. Whether this negative NMC translates to a sign change in the overall MC depends on the relative strength of the electron and hole magnetomobility, which both contribute according to Equation 3.13. This situation will be described in more detail in the next section, where a case study with sign changes in OMAR will be approached from both an experimental and a theoretical perspective.

3.2.4 Sign Changes in OMAR

3.2.4.1 Review of Recent Experiments

As mentioned in Section 3.2.1.4, it has been observed that one property of OMAR is that the sign of the MC can change depending on device operating conditions such as temperature[33,72] and voltage.[46,73–75] The sign of OMAR can also be changed by the structure of the device, such as the active layer thickness[76] or the injecting electrodes.[75,77] The sign change is puzzling because it is nontrivial to explain why the sign of the MC changes within the currently proposed models for OMAR. Therefore, understanding why the sign of OMAR can change should shed light on the mechanism(s) responsible for OMAR. In this section we will discuss the different types of sign changes that have been observed, and how they have been explained within the context of the proposed microscopic mechanisms of OMAR.

3.2.4.2 Unipolar-to-Bipolar Sign Change

There have been several observations of OMAR changing sign when a change in the operating conditions of the device causes a transition from (mostly) unipolar to bipolar conduction[73–75] (these experimental observations will be discussed in greater detail in Section 3.2.4.5). The (mostly) unipolar regime always corresponds to the current decreasing with magnetic field (negative MC), and the bipolar regime corresponds to the current increasing with magnetic field (positive MC). It has also been observed that the negative and positive MC(B) have different line widths, B_0, and the resulting line shape at the transition from negative to positive MC is a superposition of the negative and positive MC component.[74] Several different explanations for this sign change have been proposed within the context of the microscopic models previously discussed, and they fit into three separate categories.

1. The first category applies to unipolar models. It has been proposed that OMAR is an effect acting on the charge carrier mobility. If the magnetic field alters the charge carrier mobility, it could affect the mobilities of electrons and holes in opposite ways (increasing the mobility of one carrier while decreasing the mobility of the other). This is possible in the bipolaron model for magnetoresistance (Section 3.2.2.3). Also, Rolfe et al.[78] suggested that electrons and holes could scatter differently from triplet excitons, causing the magnetic field to increase the mobility of holes and decrease the mobility of electrons. What is postulated is that when the device is unipolar, the charge carrier transporting the charge has a small negative magnetomobility, giving a small negative MC, while the other charge carrier has a large positive magnetomobility, which dominates the total OMAR response when the device is bipolar, resulting in a large positive MC.[74] However, in the context of these models, it is not understood why electrons and holes would have different signs of the magnetomobility.

2. The second category explains the sign change using two microscopic mechanisms: one microscopic mechanism explains the negative MC in the regime of (mostly) unipolar transport, and a separate mechanism explains the positive MC in the regime of bipolar transport. Hu and Wu[77] propose

that OMAR is a result of a competition between the triplet exciton-charge reaction and the electron-hole recombination (both are described in Section 5.2.2.2). The authors propose that in the mostly unipolar regime the triplet exciton-charge reaction dominates, resulting in a negative MC. In the bipolar regime the electron-hole recombination mechanism dominates, resulting in a positive MC. According to the authors, when the charge transport in the device is mostly unipolar, the triplet exciton-charge reaction dominates, due to the relatively long triplet exciton lifetimes (note that it is necessary for the device to be slightly bipolar in order to have enough triplet excitons to obtain this effect). As the current becomes more bipolar the electron-hole recombination mechanism becomes relatively more important and the MC changes sign. Both the triplet exciton-charge reaction and the electron-hole recombination mechanism rely on the premise that the magnetic field can alter the ratio between singlet and triplet excitons. This is not observed in either charge-induced absorption[79] or fluorescence-phosphorescence measurements,[52] which were already mentioned in Section 3.2.2.2. Also, it would be expected that if triplet excitons cause the negative MC, this effect should have a significant temperature dependence, due to the strong influence of temperature on the triplet exciton lifetime. However, experiments show that the negative MC is only weakly affected by temperature.[72]

Wang et al.[75] studied how OMAR in 2-methoxy-5-(2'-ethylhexyloxy) 1,4-phenylene-vinylene (MEH-PPV) is effected by blending it with PCBM. It is well known from work on organic solar cells that the resulting blend phase separates into MEH-PPV-rich and PCBM-rich phases. This results in the holes and electrons being transported separately in the MEH-PPV and PCBM, respectively. The reason the authors made this blend was that the resulting separation of the electrons and holes greatly reduced electron-hole interactions, and therefore bimolecular recombination was greatly suppressed. In their experiments on the unblended MEH-PPV they observed the typical negative MC in the unipolar regime and the positive MC in the bipolar regime.[75] When a blend of 1:1 MEH-PPV:PCBM was made, in order to decrease the effects of charge recombination, the positive MC in the bipolar regime was quenched, while the negative MC in the unipolar regime remained. From this the authors concluded that the negative MC observed in the unipolar regime is caused by the magnetic field effect on bipolaron formation (Section 3.2.2.3), and that the positive MC observed in the bipolar regime must be due to the magnetic field effect on electron-hole recombination, since this recombination is strongly reduced in the blend. However, adding PCBM to MEH-PPV does not only affect recombination, but also vastly changes the charge carrier mobilities. The electron mobility in the blend is many orders of magnitude higher, due to the lower energetic disorder for electrons in the PCBM. In this case the bipolaron model would also predict that the positive MC should decrease because the MC decreases with decreasing disorder in the bipolaron model (see Figure 3.16).

3. The third category explains the sign change by device physics. It was shown in Section 3.2.3 that decreasing the minority carrier mobility can result in an increase in the current. Using this framework, the sign change can then be explained if the magnetic field only acts to decrease the mobilities of the charge carriers. Therefore, when the device is unipolar, this results in a negative MC, and once the device starts to inject minority carriers and becomes injection-limited bipolar, the sign changes to a positive MC, because the current reacts oppositely to the decrease in minority carrier mobility.[66] As a consequence, it is not necessary to assume that the mobilities of electrons and holes have an opposite response to the magnetic field.

3.2.4.3   High-Voltage Sign Change

Another sign change that has been observed is one that occurs at high voltages, where as a function of increasing voltage the sign changes from positive to negative MC. The line shape of the MC(B) curves is almost the same for both signs of the magnetoconductance, and the magnetitudes of the positive and negative MC are fairly similar.[16] Since the line shape of the positive and negative MC is the same, this sign change is likely to be solely due to device physics, and one does not need two different contributions to OMAR to explain the results.

The model by Bergeson et al.[46] can explain this high-voltage sign change within the context of the electron-hole pair model (Section 3.2.2.2). They argue that the current can react in opposite ways to changes in the rate of charge recombination, depending on the operating conditions of the device. In the bipolar SCLC regime (Equation 3.22), a decrease of the recombination mobility, μ_0, leads to a further interpenetration of charge densities of opposite sign, resulting in a decrease of the space charge and an increase in the current. In this case, $J \propto \mu_0^{-1/2}$; see Equation 3.23. The authors claim that the charge densities of either sign cannot keep increasing as the recombination decreases, because of several factors, like dielectric relaxation. The authors assume that for very weak recombination the current increases with the recombination mobility as $J \propto \mu_0$. Therefore, the electron-hole pair mechanism, in which an increasing magnetic field reduces the recombination rate (i.e., μ_0 decreases), gives a positive MC in the SCLC regime and a negative MC in the regime of very weak recombination. The authors claim that at low voltages the devices show SCLC behavior, yielding a positive MC, whereas they claim that with increasing voltage the charge density becomes too high for SCLC conditions and the device will follow the behavior of very weak recombination, resulting in a negative MC. The high-voltage sign change can also be explained by a unipolar OMAR mechanism affecting the charge mobility in the devices. It was shown in Section 3.2.3 that if there is a magnetomobility in the minority channel, a sign change from negative to positive MC can be induced by forcing the minority charge injecting contact from injection-limited to ohmic behavior by increasing the voltage. This happens at relatively high voltages, and only requires a magnetomobility in the minority channel, so only one line width would be observed. Hence, a unipolar OMAR mechanism may also explain this sign change.

3.2.4.4   Sign Change with Changing Device Thickness

It has also been observed by Desai et al.[76] that the sign of the MC can be changed by changing the thickness of the device. In this experiment they observed in devices with an Alq_3 active layer that samples with a large thickness exhibit a positive MC, while if the thickness decreases below a critical thickness, the device exhibits a negative MC. They attributed the different signs to different reactions involving triplet excitons, which depend on the device thickness. At large thicknesses the authors claim that triplet exciton-polaron quenching dominates, resulting in a positive MC, since they claim the magnetic field acts to reduce the number of triplet excitons (see Section 3.2.2.2). They claim that as the thickness of the Alq_3 layer decreases, it is more likely that a triplet exciton can diffuse to the cathode, which would result in the triplet exciton dissociating into a positive and negative polaron, resulting in a negative MC. The authors point to the fact that when a 20-nm-thick exciton blocking layer of 2,9-dimethyl-4,7-diphenyl-1,10-phenanthroline (BCP) is grown on top of the Alq_3, only a positive MC is observed. Therefore, by not allowing the triplet exciton to go to the cathode to dissociate, they claim that there can no longer be any triplet exciton dissociation, and thus no negative MC.

3.2.4.5   Case Study: Correlation between Sign Change and Onset of Minority Carrier Injection

In order to understand the properties of the sign change we conducted separate experiments on devices with Alq_3[74] and MDMO-PPV[73] active layers.

Our first experiments were performed on an ITO/PEDOT:PSS/Alq_3 (100 nm)/ LiF/Al. In these devices we observed a negative MC at low voltages and a positive MC at high voltages (Figure 3.23). These experiments were done at several temperatures, and in Figure 3.23 we show the MC(B) behavior at 220 K. We observe that the MC(B) line widths are very different for the different signs of MC, with the full width at half maximum of the negative MC (41 mT) much wider than that of the positive MC traces (11 mT). Near the voltage, V_{tr}, where the MC switches sign there is an anomalous behavior, as shown by the local maximum at $B = 0$ mT for the 9 V trace.

Figure 3.24 shows that at low voltages the log(J) vs. log(V) plot is linear, indicating a power law behavior with $J \propto V^6$ (Figure 3.24a to c). This behavior is likely due to SCLC in the trap-filling regime.[71] Investigation of the $J(V)$ characteristics near V_{tr} reveals an interesting trend. At all measured temperatures log(J) vs. log(V) starts to deviate from the power law exactly at V_{tr} (see vertical lines at $V = V_{tr}$ in Figure 3.24). We also noticed that V_{tr} shifts to lower voltages as the temperature increases. The fact that V_{tr} is temperature dependent explains observations that the sign of the MC can change as a function of temperature.[16,46,72]

The increase of the slope of log(J) vs. log(V) is likely due to minority charge carrier (hole) injection. The onset of minority charge carrier injection causes a large increase in the amount of space charge in the device due to charge compensation. To confirm this, we did low-frequency capacitance measurements that showed an increase in inductive space charge at V_{tr}. The reason for the increase of V_{tr} as the temperature decreases is likely due to hopping injection of holes at the anode-Alq_3 interface.[80]

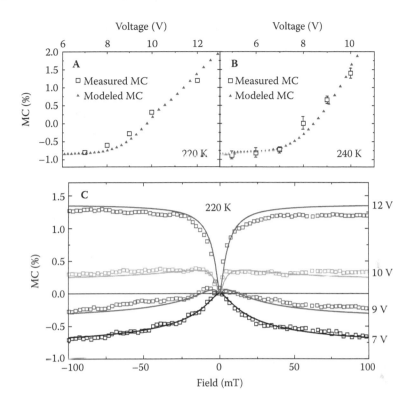

FIGURE 3.23   Modeled and measured MC(V) at (a) 220 K and (b) 240 K. (c) Modeled MC(B) (lines) and measured MC(B) (open squares) at different voltages. The modeled data were calculated according to Equation 3.36 using the values $MC_1 = -0.8\%$ and $MC_2 = 3.7\%$ for 220 K, and $MC_1 = -0.8\%$ and $MC_2 = 4.6\%$ for 240 K. (From Bloom et al., *Phys. Rev. Lett.* 99, 257201 [2007]. With permission.)

To further understand what is happening in the device at V_{tr}, we made ITO/PEDOT:PSS/MDMO-PPV/Ca/Al devices. Unlike the Alq$_3$ devices, these devices show a sign change at room temperature, and show better OLED characteristics, like a lower turn-on voltage. These devices show a sign change at 1.7 V (see Figure 3.25b), and if we look at the $J(V)$ characteristics, we again see that at $V = V_{tr}$ there is a deviation from power law behavior. To get a better understanding of what happens at V_{tr}, we performed measurements of the electroluminescence (EL) current efficiency (see Figure 3.25a, blue line). The EL current efficiency is an indication of how balanced the populations of electrons and holes are with respect to one another. We see that right at V_{tr} there is an onset in the EL efficiency, indicating that above V_{tr} the device begins to become bipolar. The fact that the EL efficiency is approximately 0 below V_{tr} shows that the current is highly unipolar in this region.

We have now established that in the (mostly) unipolar transport regime a negative MC is observed, while in the bipolar transport regime a positive MC is observed. Questions remain, however, since at the transition from positive to negative MC we see in Figure 3.23 that the MC(B) behavior does not correspond to the universal

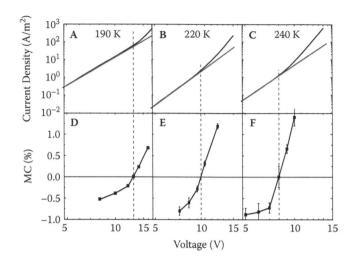

FIGURE 3.24 (a–c) Log(J) vs. log(V) and (d–f) corresponding MC vs. log (V) at 190 K, 220 K, and 240 K. The thin solid red line represents a power law fit to $J(V)$. The vertical dashed lines indicate the transition voltage, V_{tr}, where the MC switches from negative to positive.

OMAR traces discussed in Section 3.2.1.3. We will show that the MC(B) traces are the result of a superposition of two OMAR effects of opposite signs, instead of being the result of an OMAR mechanism that changes sign and shape as a function of voltage.

To demonstrate this superposition we assigned a negative MC contribution to the current in the power law regime (J_1) and a positive MC to the current in excess of the power law (J_2). By fitting the power law regime to $A \cdot V^n$, the power law proportion of the current is $P_1(V) = (A \cdot V^n)/J_{tot}$, where $J_{tot} = J_1 + J_2$. Assuming the current to be a superposition, the magnetoconductance as a function of voltage should be given by

$$MC(V) = P_1(V)MC_1 + (1 - P_1(V))MC_2 \tag{3.36}$$

where MC_1 and MC_2 are the voltage-independent magnitudes of the negative and positive MC, respectively. By only using MC_1 and MC_2 as fitting parameters, we are able to fit the experimental MC(V) data very well to Equation 3.36, as shown in Figure 3.23a and b.

From the analysis of MC(V) we know the magnitudes of MC_1 and MC_2, and the relative proportions of positive and negative MC. Therefore, from the line shapes of the positive and negative MC traces we can calculate the MC(B) curves at several voltages *without any further free parameters*. We see that this calculation exactly matches the experimental MC(B) behavior (see Figure 3.23c). Hence, the sign change is indeed due to a superposition of two OMAR effects, a positive MC effect in the (mostly) unipolar regime and a negative MC effect in the bipolar regime.

3.2.4.6 Discussion

The simplest explanation for the sign change discussed in the previous section is that the minority and majority charge transports have different responses to the magnetic

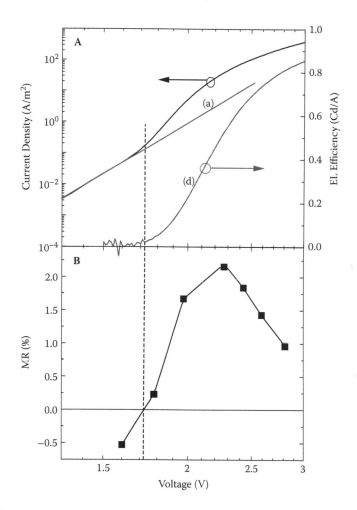

FIGURE 3.25   (A) Log(J) (black) and EL efficiency (blue (d)) vs. log(V). The thin red line (a) in (A) represents a power law fit to the log(J) vs. log(V) relation in the low-voltage regime. (B) MC vs. voltage. The vertical dotted line represents the transition voltage (V_{tr}) from negative to positive MC.

field, which may have opposite signs. Since the sign change occurs at the onset of minority charge carrier injection, one could assume that OMAR in the unipolar regime is dominated by the magnetic field response of the majority carriers, while in the bipolar regime the response is dominated by the minority carriers. It may seem odd that in the bipolar regime the magnetic field response can be dominated by the minority carriers, since due to their low mobility their contribution to the charge transport is negligible. However, we have shown in the case of bipolar SCLC (Equation 3.22) in Section 3.2.3 that it is the relative mobility change of each carrier that determines the change in current (Equation 3.13). This is due to the oppositely charged carriers compensating the coulomb repulsion in the device. Therefore, a magnetic field response of the mobility of the slow minority charges can dominate the response of the MC in the bipolar regime. Also, it has been shown experimentally in unipolar devices that OMAR is significantly stronger in the minority channel than in the majority channel,[75,81] so in the bipolar regime it is likely that the response of the minority channel dominates.

The bipolaron model shows that the line width of OMAR is strongly influenced by (carrier-dependent) material parameters,[50] and as such could indeed be different for electrons and holes. This could explain why the positive and negative MC have different line widths. It is well known that the wavefunctions of electrons and holes of organic semiconductors are spatially different, which may result in electrons and holes experiencing different hyperfine interactions with the nuclear spin of the hydrogen atoms. In electrically detected magnetic resonance experiments the broadening of the resonance is attributed to the random hyperfine fields experienced by the charge carriers. It has been observed in Alq_3 that the electrons have a significantly broader resonance than the holes,[82] indicating that electrons experience a stronger hyperfine field. Since in all the models proposed for OMAR the line width of the $MC(B)$ scales with the hyperfine field, this may explain why we observe a wider $MC(B)$ in the (mostly) unipolar regime than in the bipolar regime. This matches well with our experiments on Alq_3, which indicate that the magnetic field response is dominated by electrons in the unipolar regime and by holes in the bipolar regime. However, it should be noted that in the bipolaron model the line width is determined not only by the hyperfine fields but also by other parameters that may be of dominant importance.[50,61]

The fact that the current reacts oppositely to a magnetomobility in the minority channel may be important in resolving apparent inconsistencies between experiments and the bipolaron model. The bipolaron model predicts both a positive magnetomobility ($d\mu/d|B| > 0$) and negative magnetomobility ($d\mu/d|B| < 0$).[50,61] According to this model the maximum magnitude of the negative magnetomobility is larger than that of the positive magnetomobility (see Figure 3.17). However, the largest MCs that have been observed are positive, which is inconsistent with the bipolaron model, unless the current can react oppositely to the change in the mobility. We have shown the current can react oppositely to the change in the mobility in detail in the device models in Section 3.2.3; therefore, these models may resolve this inconsistency.

More strongly, by using the device models outlined in Section 3.2.3, the sign change behavior in literature can be explained with a universal negative magnetomobility for both the minority and majority charge carriers. Therefore, there is no need for an *ad hoc* assignment of different signs of magnetomobility to different carriers or mechanisms. Two types of sign change behavior have been observed in literature. In one case (described in Section 3.2.4.5), the MC changes from negative to positive with increasing voltage, which occurs at the transition between unipolar and bipolar behavior.[73-75] The resulting line shape is a superposition of two contributions of opposite sign and different field widths, which may be a result of separate magnetic field effects on electrons and holes.[74] In the other case, the sign change occurs at high voltage and goes from positive MC to negative MC with increasing voltage, with a line shape that remains unchanged (Figure 4 in Mermer et al.[16]). This result is consistent with the high-voltage sign change for a magnetomobility in the minority channel, which results from the minority contact becoming less injection limited as the voltage increases.[66]

To summarize, it has been experimentally observed that the MC is negative when the device is mostly unipolar, becoming positive when the device turns (injection-limited) bipolar, and negative again when it is forced toward good bipolar injection. We model the case where both carriers have a negative magnetomobility, and the relative magnetomobilities of the carriers are given by $5\Delta\mu_{maj}/\mu_{maj} = \Delta\mu_{min}/\mu_{min}$. This ratio has been chosen since OMAR has been shown to have a much stronger effect on the minority channel,[75,81] i.e., $|\Delta\mu_{maj}/\mu_{maj}| < |\Delta\mu_{min}/\mu_{min}|$. The results of the model are shown in Figure 3.26. The model predicts exactly the same signs of MC for the different transport regimes as experimentally observed. In both the model and experiment we observe a negative MC when the device is mostly unipolar, then the MC becomes

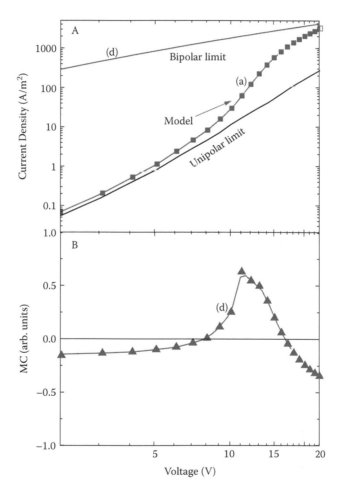

FIGURE 3.26 (A) Numerically determined $J(V)$ of a device with traps in the majority channel and $\mu_{maj}/\mu_{min} = 2$, represented by model (red (a)). The upper and lower limits of the current are given by the bipolar limit (blue (d)) and unipolar limit (black) lines, respectively. (B) Normalized MC vs. voltage in the case of magnetomobility in both the electron and hole channels combined, with $5 \Delta\mu_{maj}/\mu_{maj} = \Delta\mu_{min}/\mu_{min}$.

positive when the device turns (injection-limited) bipolar, and finally, the MC turns negative again when the device is forced toward good bipolar injection.

Of course, observing both sign changes within a single device would provide conclusive experimental evidence that these device models are in fact applicable. However, observing the two sign changes in one device may be difficult, since, as seen in Figure 3.26, the numerical modeling shows that these sign changes occur at currents that are separated by several orders of magnitude, so either the current density is too small to observe the low-voltage sign change or the current density is too large to observe the high-voltage sign change, resulting in device failure. However, it is common to observe a peak in the MC(V),[49,73,75,83] like we observe in the device models (if one considers a negative magnetomobility) when the device becomes less injection limited. Moreover, we also observed that the second sign change is moved to higher voltages or even completely eliminated for a $J_{a_{min}}(E_a)$ dependence that does not allow the device to fully saturate to bipolar SCLC at high voltages (Figure 3.20). This may also explain why both sign changes have not been observed in a single device.

In conclusion, a clear correlation between the sign change in OMAR and the onset of minority carrier injection was observed. Using the treatments outlined in Section 3.2.3, we showed that by assigning a magnetic contrast of the *same* sign to the mobilities of electrons and holes, one can explain both the sign change in the MC and its magnitude. This provides strong evidence that OMAR is an effect acting on the charge carrier mobility.

3.3 MAGNETORESISTANCE IN SPIN VALVES

3.3.1 REVIEW OF EXPERIMENTAL RESULTS

With the rise of (inorganic) magnetic tunnel junctions and growing interest in semiconductor-based spintronics in the 1990s, organic semiconductors and insulators started to be seen as potentially interesting alternatives for spacer layers in spin valves. Apart from chemical tunability and compatibility with organic electronics, the low atomic number (Z), and thereby low spin-orbit coupling,[5] has been a strong motivation to explore organic spin valves (OSVs).[84–87]

After early reports on organic magnetic tunnel junctions (OMTJs), e.g., based on Langmuir-Blodgett (LB) films separating ferromagnetic electrodes,[88] the first promising results in spin injection and detection based on organic semiconductors (OSCs) were reported by Dediu et al.[7] in 2002. The authors used a planar geometry with half-metallic $La_{0.67}Sr_{0.33}MnO_3$ (LSMO) electrodes and sexithienyl (T_6) as OSC, and reported 30% MR at room temperature, while they claimed a spin diffusion length of 200 nm. Because of the incomplete control of the magnetic state, however, no switch between parallel and antiparallel magnetized electrodes could be obtained, but only a transition from a disordered (multidomain) state at zero applied field to the parallel configuration at high fields. Thus, the MR loops were symmetric in the applied magnetic field rather than displaying a characteristic OSV behavior with clear switching fields and hysteretic behavior.

In 2004, Xiong et al.[8] reported the first successful OSV in a perpendicular configuration, using an LSMO bottom electrode and a Co top electrode, sandwiching an Alq_3 spacer layer prepared by shadow-mask evaporation. At low bias and 4 K a negative MR of up to 40% and clear spin valve behavior were observed. Varying the thickness of the Alq_3 spacer layer and measuring the corresponding decay of MR, a spin diffusion length of 45 nm was concluded on. However, devices thinner than 100 nm were "shorted" due to an ill-defined layer of up to 100 nm that may contain pinholes and Co inclusions.

In the past few years, OSV behavior for relatively thick organic spacers (typically >50 nm; thought to be clearly beyond the tunneling regime) has been reproduced by several groups. Attempts using Alq_3 have been most successful,[89–93] but also other small molecules (such as tetraphenyl porphyrin [TPP][94]) and polymers (such as P3HT[95–98]) have been reported. Among ferromagnetic materials, LSMO became a very popular choice as electrode material. It has a high spin polarization, and as an oxide bottom electrode, it is reusable,[7,8] while there may be other issues, e.g., related to the electronic structure of the LSMO/OSC interface, that need to be further explored. For devices with an LSMO electrode and a ferromagnetic 3d transition metal counterelectrode, generally a negative MR is obtained,[8,89,91] which means that the configuration with the parallel magnetic alignment (i.e., at high field) has a higher resistance. OSVs using two 3d-electrodes have also been reported[90,92,99] and tend to yield a positive MR. This inversion has to be assigned somehow to the relative spin polarization of the two magnetic electrodes, but is presently still debated.[89] The role of interdiffusion has been explored in more detail, and a correlation between the degree of interdiffusion and magnitude of MR has been reported.[92] Insertion of a barrier consisting of LiF or Al_2O_3 has been suggested to reduce interdiffusion and thereby limit the formation of an ill-defined layer,[89] while possibly also contributing to reduce the conductivity mismatch[100]; see below. Another option to limit the role of interdiffusion is by exploiting a planar configuration.[7,101]

Although reproduced by several groups, the robustness of the results, as well as their fundamental interpretation, is still being questioned intensively. It has been argued that even for thicker organic spacers most results may be due to direct tunneling through thinner regions,[102,103] while also stray fields may cause hysteretic characteristics that erroneously could be interpreted as a fingerprint of spin valve behavior.[93] In passing we note that in experiments on thin nanowires between electrodes[104] such stray fields may easily become the dominant effect, and that such results should be analyzed with utmost care. Furthermore, it has been indicated recently that care also has to be taken not to confuse OSV behavior with anisotropic magnetoresistance,[105] which can give rise to MR even when only using a single ferromagnetic (single-crystalline) electrode.

Such doubts about the essence of OSVs have fed the desire to directly probe the presence of spin-polarized carriers inside the OSC spacer layer. Unfortunately, magneto-optical techniques that have played a crucial role in the development of inorganic (III-V) semiconductor-based spintronics[106,107] do not work in organic materials, because of insignificant spin-orbit effects and lower symmetries of the molecular structures considered. Very recently, two-photon photoemission[108] and muon spin resonance[90] have been proposed as candidates to probe spins inside organics, and first results were reported indicating that injection of spin-polarized carriers indeed takes place.

In other experiments, the tunneling regime has been intentionally explored by using thinner spacer layers. Clear OSV behavior with positive MR typically up to 10% was reported for OMTJs with two 3d-electrodes, and Alq_3[11] or rubrene[109] as OSC spacer (in combination with a subnanometer Al_2O_3 layer, claimed to promote smooth growth of the OSC). Similar results with negative MR up to 20% were reported for LSMO-based OMTJs with Alq_3 and TPP as spacer layer.[94] Note that the signs of MR (MR < 0 with one LSMO electrode, MR > 0 without) are consistent with the thicker spin injection type of devices discussed above. As to a potential intermediate case, we refer to an interesting transition from single-step OMTJs to two- and more-step tunneling, as discussed in Schoonus et al.[110] Tunneling may also be the dominant effect in large MR effects observed in blends of Alq_3 with Co nano-particles.[112] Finally, we mention more recent results on OMTJs with a spacer layer prepared by the LB technique. Exchange-biased OSVs with 3-hexadecyl pyrrole as the spacer layer were reported to show GMR as large as 20% at room temperature, for single LB layers.[113] The role of pinholes will remain an outstanding question for these types of devices.

While clearly a final assessment of experiments and their interpretation is impossible at the present stage, we make some notes to guide further discussion:

1. Despite recent progress, fabrication of "clean" OSVs remains a technological challenge. Since the current in a device will always select spots with the lowest resistance, even a very small density of imperfections may easily dominate the resistance, and thereby the MR.
2. All reports on successful OSV behavior are based on experiments at low bias, usually less than 1 V. This should be considered an anomalous regime for OLED type of devices that usually need several volts to be in a well-defined transport regime. Thus, all results are obtained in the ohmic (sometimes tunneling-dominated) regime, with a bias typically below the built-in voltage. In this regime the current is often interpreted as a leakage current and regular space-charge-limited behavior is not relevant. It is of particular interest to note that so far no coexistence of OSV behavior and OMAR in a single device and under identical conditions has been observed.
3. In fact, based on present knowledge of inorganic semiconductor-based spin valves, one may argue that OSVs using disordered, low-mobility OSCs should not work at all because of the conductivity mismatch problem.[100] The high resistance of the spacer layer would prevent a significant spin accumulation, a crucial ingredient to develop a sizable MR. Although at the injecting electrode the band alignment of the OSC with the metallic electrode could introduce a barrier, at the collecting electrode such a barrier would be absent, unless a tunnel barrier is inserted explicitly. The successful reports on OSVs even without additional tunnel barrier may point at the role of mechanisms—possibly related to the intrinsic hopping character of the transport, or to the polaronic nature of the charge carriers—not yet understood.

4. A decisive experiment to unambiguously demonstrate successful spin injection and detection would be provided within a Hanle configuration.[114] In such an experiment, MR is measured for a field perpendicular to (or at a certain angle with) the magnetization of the magnetic electrodes. A damped oscillatory MR as a function of field should then be observed, being a signature of the coherent precession of spins while traversing the organic spacer. However, devices based on disordered OSCs investigated so far typically have transit times of microseconds, which would translate to fields of only μTs to observe separable oscillations. The fields thus required are orders of magnitude smaller than the random hyperfine fields due to the hydrogen atoms, so that one may worry about the possibility to measure any MR at such small fields. The potential role of random hyperfine fields in causing a loss of spin coherence, thereby affecting MR, provides a first motivation to investigate spin relaxation in the presence of hyperfine field coupling.

5. A second motivation to undertake a study of spin relaxation in the presence of hyperfine coupling is the fact that the origin of spin diffusion in disordered OSCs is not known. OSCs are characterized by insignificant spin-orbit coupling,[5] and thereby long spin relaxation times of the order of microseconds.[115] Therefore, it may be argued that hyperfine fields play a dominant role in spin relaxation.

A recent theoretical investigation of spin diffusion by hyperfine field coupling[116] will be discussed in the next section. We stress that the results should be taken with utmost care, particularly when comparing with specific sets of experimental data, which, as we have shown, are far from fully understood. Nevertheless, the theoretical study provides a clear and quantitative prediction about the effect of random hyperfine fields on spin transport in OSVs.

3.3.2 THEORY FOR SPIN DIFFUSION IN DISORDERED ORGANIC MATERIALS

As already mentioned in Section 3.2.2.3, charge transport in disordered organic semiconductors takes place by incoherent hopping of charge carriers between localized sites, where the site energy is a random variable, distributed according to a Gaussian DOS. It was also argued in Section 3.2.2.3 that the hyperfine coupling at each localized site, i, can, in good approximation, be treated as a classical, random, and static magnetic field, $\mathbf{B}_{hf,i}$. In the presence of an external magnetic field, $\mathbf{B}_i$, this then leads to an effective magnetic field, $\mathbf{B}_i = \mathbf{B}_{hf,i} + \mathbf{B}$, at each site. The accurate description of OMAR line shapes by considering only the hyperfine field strongly indicates that these hyperfine fields are the main source of spin randomization in organics. Hence, spin diffusion in these materials can be described by a combination of hopping of a carrier in a Gaussian DOS, with standard deviation σ, together with precession of its spin around a local effective magnetic field, which changes after each hopping event. In Figure 3.27 we pictorially indicate this mechanism for spin diffusion. There is an interesting analogy with the D'yakonov-Perel' (DP) mechanism of spin diffusion in crystalline inorganic materials,[4,117] where the charge carrier spin precesses around

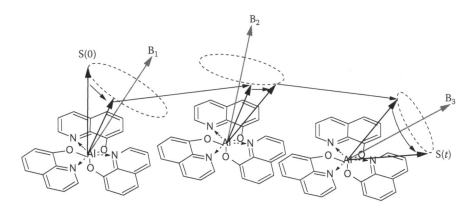

FIGURE 3.27 Mechanism for spin diffusion, in this case for the molecular semiconductor Alq$_3$. Hopping of a carrier takes place between adjacent molecules, with precession of its spin, $S(t)$, around the local effective field $\mathbf{B}_i$ while residing on site i.

an effective magnetic field caused by spin-orbit coupling, pointing in the propagation direction of the charge carrier: the hopping in real space from one site to another is the analogue of the scattering in reciprocal space from one propagation direction to another, and the hyperfine-induced random effective field at a site is the analogue of the spin-orbit-induced effective field in the DP mechanism.

We will consider the situation that a charge carrier (electron or hole) with a fixed spin direction is injected by an electrode into the organic material and moves to the opposite electrode under the influence of an electric field with strength E. We will consider the case relevant for experiments on organic GMR devices,[7–10] where the external magnetic field, $\mathbf{B}$, is applied (anti-)parallel to the initial spin direction. Like in Section 3.2.2.3, we assume that nearest-neighbor hopping takes place by a thermally assisted tunneling process with a rate given by Equation 3.1. It has been shown that positional disorder is much less important than energetic disorder,[53] so for simplicity we will neglect positional disorder and take a fixed distance, a, between nearest-neighbor sites.

A very important parameter is the ratio $r \equiv \omega_{hop}/\omega_{hf}$ between the prefactor, ω_{hop}, in the expression Equation 3.1 for the hopping frequency and the hyperfine frequency, ω_{hf} (~10 ns). If r is large, the influence of the hyperfine field is small and large spin diffusion lengths can be expected, while the opposite holds if r is small. For derivatives of the familiar π-conjugated polymer poly-($para$-phenylene-vinylene) (PPV) we obtain an estimate of 10^9 to 10^{11} s^{-1} for ω_{hop}.[60] Hence, for this class of organic semiconductors r is of the order of 10 to 1000, but the large variation even within this class shows that very different values can be expected for different organic semiconductors. Therefore, we treat r as a parameter.

We start with the consideration of the equation of motion for the spin operator at a site i:

$$\frac{d}{dt}\mathbf{S}(t) = \gamma\mathbf{B}_i \times \mathbf{S}(t) = \gamma\mathbf{B}_{hf,i} \times \mathbf{S}(t) + \gamma\mathbf{B} \times \mathbf{S}(t) \tag{3.37}$$

where the first term is due to the random hyperfine coupling and the second term is due to the Zeeman coupling with the external magnetic field, **B**, pointing in the z-direction, which is also the direction of the initial spin polarization of the injected charge: $\langle \mathbf{S}(0) \rangle = (0, 0, \hbar/2)$. We call $p = 2\langle S_z(t_0) \rangle / \hbar$ and $p' = 2\langle S_z(t_0 + \tau) \rangle / \hbar$ the average spin polarizations of a charge at the moments of arrival, t_0, and at departure, $t_0 + \tau$, from site i, respectively, where $\langle \cdots \rangle$ denotes a quantum mechanical and hyperfine field average. The waiting time τ is exponentially distributed according to a distribution $\omega_i \exp(-\omega_i \tau)$, where $\omega_i = \sum_j \omega_{ij}$ is the total hopping frequency to all possible neighbor sites j. Solving the equation of motion, Equation 3.37, and performing the average over the hyperfine field[58] and over the waiting times leads to a decay of the polarization, $p' = \lambda_i p$, where

$$\lambda_i = \frac{1}{\sqrt{2\pi}} \int_0^\infty \rho \, d\rho \int_{-\infty}^\infty d\zeta \, \exp\left[-\frac{\rho^2 + \zeta^2}{2} \right] \frac{r_i^2 + (\zeta + b)^2}{r_i^2 + \rho^2 + (\zeta + b)^2} \tag{3.38}$$

with $r_i \equiv \omega_i / \omega_{hf}$ and $b \equiv B/B_{hf}$. In the limit of infinite waiting time and zero field ($r_i = b = 0$) we find $\lambda_i = 1/3$. In the limit of small waiting time, $r_i \gg \max(1, b)$, we find asymptotically $\lambda_i \approx 1 - 1/(2r_i^2)$, and in the limit of large field, $b \gg \max(1, r_i)$, we find asymptotically $\lambda_i \approx 1 - 1/(2b^2)$.

It is instructive to consider the situation of a one-dimensional chain of sites with lattice constant in the presence of a large electric field, E, so that only down-field hopping occurs with a rate $\omega_i = \omega_{hop}$. This leads to the same $\lambda_i = \lambda$ for all hops, which is obtained by replacing r_i by $r = \omega_{hop}/\omega_{hf}$ in Equation 3.38. Repetitive application of the above analysis to each site in the chain leads to an exponentially decaying polarization, $p(x) = \exp(-x/l_s)$ (x is the distance from the injection point), with spin diffusion length $l_s = -a/\ln\lambda$. The integral in Equation 3.38 can be evaluated numerically in a straightforward way. The resulting l_s can be rather well approximated by $l_s \approx a[1/\ln(3) + r^2/2 + b^2/2]$. This expression is exact if $r = b = 0$ and has the right asymptotic behavior for $r \to \infty$ and $b \to \infty$. The increase of l_s with increasing b and r can readily be understood qualitatively: the spin Hamiltonian contains a coupling to the hyperfine fields and a Zeeman coupling to the external field. With increasing applied magnetic field, the Zeeman coupling becomes increasingly dominant over the hyperfine coupling and the carrier spin becomes effectively pinned (see, e.g., the appendix of Sheng et al.[35]). The quadratic increase with r results from "motional narrowing," well known in magnetic resonance spectroscopy: the spin relaxation of a carrier moving in a random magnetic field is slower than that of an immobile carrier.

For the three-dimensional situation with a small or moderate electric field the situation is much more complicated, since carriers do not move in a straight line. For this situation we have performed Monte Carlo simulations for hopping on a cubic lattice of sites with lattice constant a, while simultaneously solving the time-dependent Schrödinger equation for the spinor part of the wavefunction. The sizes of the lattices were typically $N \times 50 \times 50$ sites, where N is adapted to the specific situation. In Figure 3.28 we display the results for l_s for a range of values of r and relative

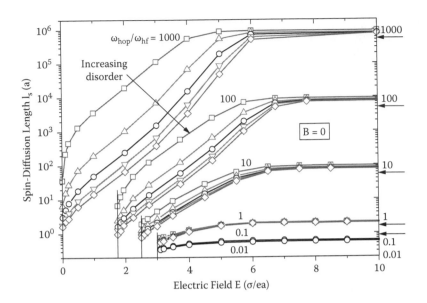

FIGURE 3.28 Spin diffusion length, l_s, vs. applied electric field, E, at zero applied magnetic field, for several ratios $r = \omega_{hop}/\omega_{hf}$ between the hopping and hyperfine frequency and several ratios $\hat{\sigma} = \sigma/kT$ between the disorder strength and the thermal energy: 2 (squares), 3 (up triangles), 4 (circles), 5 (down triangles), and 6 (diamonds). For clarity, an indicated shift along the horizontal axis is applied to the rightmost sets of curves. The arrows at the right axis, labeled with the corresponding ratios ω_{hop}/ω_{hf}, indicate the results of the one-dimensional chain model.

disorder strength $\hat{\sigma} \equiv \sigma/kT$, as a function of the electric field strength, E, for $B = 0$. The arrows at the right axis indicate the results obtained for the one-dimensional chain. These results follow the Monte Carlo results at large E qualitatively and agree with those within a factor of about two. In agreement with the chain model, l_s at large E increases quadratically with r for large r. In Figure 3.29 we display the results for the case of a large relative hopping frequency, $r = 1000$, for a range of values of B and different disorder strengths. In agreement with the chain model ($b \ll r$) the B dependence at large E disappears. In Figure 3.30 we display the results for the case of a small relative hopping frequency $r = 0.01$. In agreement with the chain model, l_s at large E increases quadratically with B for large $b(b \gg r)$. We also analyzed the situation $r = 1$ and found the results to be similar to $r = 0.01$.

A striking feature of all the results displayed in Figures 3.28 to 3.30 is the relatively weak dependence of l_s on the disorder strength for strong disorder ($\hat{\sigma} \geq 4$). For large electric field ($E \gg \sigma/ea$) the dependence on the disorder vanishes almost altogether, which can be readily understood from the chain model, in which disorder plays no role. Some influence of the disorder is found in the three-dimensional Monte Carlo simulations, because of the lateral motion of the carrier, which is not captured by the chain model. In order to provide an explanation for this weak dependence for small electric field ($E \ll \sigma/ea$), we have analyzed the distribution $P(\tau)$ of waiting times τ as experienced by a carrier hopping through the lattice at $E = 0$. The concept of a

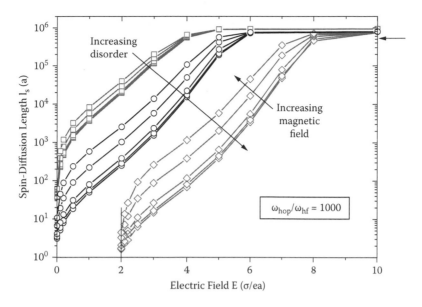

FIGURE 3.29   Spin diffusion length, l_s, vs. applied electric field, E, for a ratio $r = \omega_{hop}/\omega_{hf} =$ 1,000 and different disorder strengths (symbols as in Figure 3.28). The different curves are for different ratios $b = B/B_{hf}$ between the external magnetic field and the strength of the hyperfine field: 0, 1, 2, 5, and 10. A shift along the horizontal axis is applied to the rightmost set of curves. The arrow at the right axis indicates the result of the chain model.

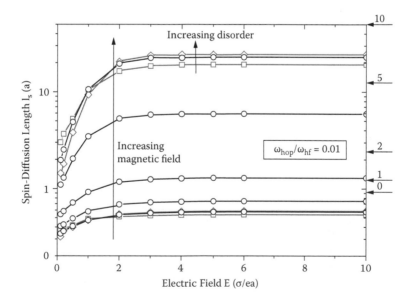

FIGURE 3.30   The same as in Figure 3.29, but for a ratio $r = \omega_{hop}/\omega_{hf} = 0.01$. For $\hat{\sigma} = \sigma/kT = 2$ and 6 (symbols as in Figure 3.28), the result is only given for $b = B/B_{hf} = 0$ and 10. The arrows at the right axis, labeled with the corresponding ratios B/B_{hf}, indicate the results of the chain model.

waiting time distribution has been introduced by Scher and Montroll[118] in the context of their study of dispersive transport, but here we use this concept for the following reason. If τ of a carrier at a site is much smaller than the precession time, its spin will essentially be conserved during its stay on that site: no spin relaxation will occur. In the opposite case, spin relaxation does occur. In Figure 3.31a we display $P(\tau)$ for various values of the disorder strength. The results appear to converge to a universal curve for strong disorder. One can in fact prove (we will not give the proof here) that for large τ the distribution converges to a universal algebraic distribution in the limit of infinite disorder, with $P(\tau) \sim \tau^{-3/2}$; see the dashed line in Figure 3.31a. We note that the waiting time distribution for a Gaussian DOS has been studied before,[119] but for too small disorder strength ($\hat{\sigma} \leq 4$) to observe this universal behavior.

The weak dependence of l_s on disorder appears to be at odds with the fact that the charge carrier mobility depends very strongly on the disorder strength, as shown in Figure 3.31b. The issue is resolved by noting that with increasing disorder the tail of the waiting time distribution will contain an increasing amount of sites with large τ. Sites with increasing τ have an increasing influence on the mobility, since they cause a longer and longer delay in the motion of the carrier. Regarding the spin diffusion, however, the situation is distinctly different. Let us, for example, consider the case $r = 1000$ and $B = 0$. At sites with τ to the left of the solid line in Figure 3.31a essentially no spin relaxation occurs, while at sites with τ to the right of this line almost immediate spin relaxation occurs, regardless of the distance to the line. For large disorder the fraction of the latter sites, obtained by integrating $P(\tau)$ from the dotted line to the right, converges to $\sim 1/r^{1/2}$. This means that on average the spin polarization disappears in $\sim r^{1/2}$ steps. Since at small electric field diffusion of carriers is dominant over drift, one expects $l_s \sim r^{1/4}$. Such a law is indeed observed, as shown in Figure 3.31c. For intermediate electric fields we also studied the waiting time distribution and found a similar convergence to a universal distribution for growing disorder, where the power law behavior gradually changes to an exponential behavior, $P(\tau) \sim \exp(-\omega_{hop}\tau)$, with growing electric field.

When the external magnetic field increases far beyond the hyperfine field ($b \gg 1$), the analysis changes. Let us, for example, again take $r = 1000$, but now $b = 10$. For waiting times shorter than the precession time, γB, i.e., to the left of the dash-dotted line in Figure 3.31a, basically no spin relaxation takes place. To the right of this line *partial* spin relaxation takes place by an amount $\sim 1/b^2$. A similar argument as above now leads to the expectation $l_s \sim b^{3/4}$ in the diffusive regime. This law is tested in Figure 3.31d, and it is seen to be very well obeyed. For a small hopping frequency ($r \ll 1$) and large magnetic field ($b \gg 1$) partial spin relaxation by an amount $\sim 1/b^2$ takes place at *all* sites. Accordingly, we expect $l_s \sim b$ and almost no dependence on the disorder. This is exactly what is found in Figure 3.30, and the linear dependence for large B is verified in Figure 3.31e.

3.3.3 COMPARISON WITH EXPERIMENTAL RESULTS AND DISCUSSION

We now undertake a comparison between the theory and experimental results, as far as they are available at the moment. We have no information about the hopping frequencies of the organic materials used in the spin valves of Dediu et al.,[7] Xiong

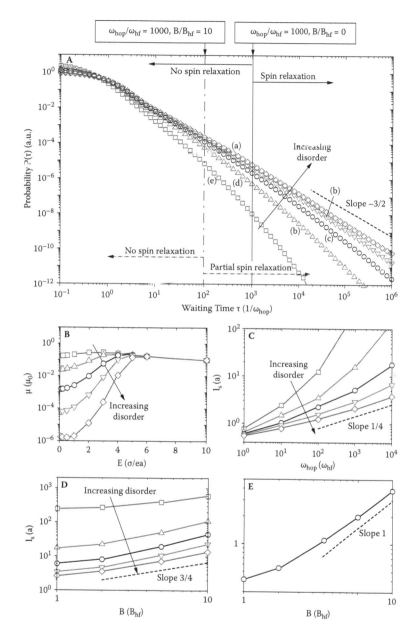

FIGURE 3.31   (a) Waiting time distributions (equally normalized) at $E = 0$ for different ratios $\hat{\sigma} = \sigma/kT$. Solid and dash-dotted lines and arrows: see the main text. (b) Corresponding mobilities μ, in units of $\mu_0 \equiv ea^2\omega_{hop}/\sigma$, as a function of electric field E. (c) Spin diffusion length l_s at $E = 0.1\sigma/ea$ as a function of ω_{hop} at $B = 0$ for $r = \omega_{hop}/\omega_{hf} = 1000$. (d) Spin diffusion length at $E = 0.1\sigma/ea$ as a function of magnetic field B for $r = \omega_{hop}/\omega_{hf} = 1000$. (e) Spin diffusion length at $E = 0.1\sigma/ea$ as a function of magnetic field for $r = \omega_{hop}/\omega_{hf} = 0.01$. The dashed lines in (a) and (c–e) indicate the expected power laws. Symbols as in Figure 3.28. (From Bobbert et al., *Phys. Rev. Lett.* 102, 156604 [2009]. With permission.)

et al.,[8] and Majumdar et al.[9]: T_6, Alq$_3$, and P3HT, respectively. Since the mobilities in these materials are higher than those of the PPV derivatives investigated in Pasveer et al.,[60] we expect that the hopping frequencies are such that $r > 1000$. A calculated value of $\sigma = 0.35$ eV for the energetic disorder of electrons in Alq$_3$[120] leads to $\hat{\sigma} \approx 14$ at room temperature, which is clearly in the strong disorder limit. We can conclude that with a typical value $a \approx 1$ nm the spin diffusion lengths of about 10 to 100 nm found in Dediu et al.,[7] Xiong et al.,[8] and Majumdar et al.[9] and recently confirmed with muon spin resonance studies[90] are compatible with the theoretical results.

Very interestingly, inspection of the experimental GMR traces in Xiong et al.[8] and Majumdar et al.,[9] both using La$_{0.67}$Sr$_{0.33}$MnO$_3$ (LSMO) and Co as electrodes, reveals that in the up- and down-field sweeps the resistance as a function of B changes considerably already *before* the magnetization of the soft layer (with the weaker coercive field) is reversed. This points to a source of magnetoresistance other than the switching of the ferromagnetic layers, and we propose that this is the magnetic field dependence of the spin diffusion length predicted by our theory. To illustrate its effect on the GMR traces, we simulated such traces using a simple phenomenological model.

We describe the organic film by a reservoir for carriers with spin up and down, situated between two ferromagnetic electrodes, 1 and 2, with spin polarizations P_1 and P_2, and restrict ourselves to the low-temperature case, where we can neglect back-hopping from the organic film into electrode 1. A spin flip rate, α, between the two reservoirs is defined, to account for spin relaxation due to the random local hyperfine fields. Next, the occupation of up- and down-spin carriers, N_{up} and N_{do}, respectively, is solved from the rate equations describing hopping from electrode 1 into the organic film, the local spin relaxation, and hopping from the organic film to electrode 2. Doing so for parallel (P) and antiparallel (AP) alignments of the magnetizations of the electrodes provides an expression for the current in the P and AP configurations in terms of model parameters. Then, in a very pragmatic step, spin relaxation in the time domain (described by α) is replaced by a decaying spin density in real space, where $l_s(B)$ corresponds to the experimentally/theoretically determined (magnetic-field-dependent) spin diffusion length. While the result thus obtained depends on the specific way we treat the carriers inside the OSC, we found that the result for the magnetoresistance MR(B) to lowest order in P_1 and P_2 can be written in a generic form (for more details, see the supplementary information to Bobbert et al.[116]):

$$\mathrm{MR}(B) = \mathrm{MR}_{\mathrm{max}} \cdot \frac{1}{2}(R - sm(B))\exp[-d/l_s(B)] + \mathrm{MR}_0 \qquad (3.39)$$

where $m(B) = m_1(B)m_2(B)$ is the product of the magnetizations of the electrodes, $m_i(B)$, measured along the applied field direction and normalized to the saturation magnetization, with $m = +1$ (−1) for P (AP) orientation. The thickness of the organic film is given by d. All details regarding the spin polarization, P_i, and the treatment of carriers in the OSC are contained in the parameters R and s. The case $R = 1$ represents the common case of no spin accumulation for equal polarizations, while $R = 0$ gives a simple Jullière approximation, i.e., direct tunneling between electrodes without carriers accumulating in an intermediate reservoir. For $s = +1$ (−1) these

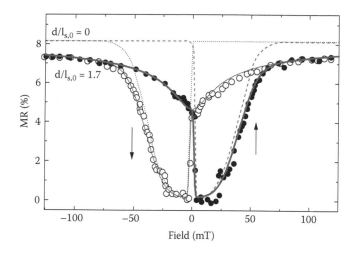

FIGURE 3.32 Full curves: Fit of the model discussed in the main text to the experimental magnetoresistance (MR) traces taken from Figure 3b of Xiong et al.[8] (symbols). Dashed curves: Results neglecting spin relaxation. Thick/thin lines: Up/down field sweep.

polarizations have equal (opposite) signs in parallel alignment of the magnetization of the electrodes. In order to obtain negative MR, as in the case of LSMO/Alq$_3$/Co devices, we need to choose $s = -1$.

We fitted the resulting GMR traces to experimental data of Xiong et al.[8]; see Figure 3.32. In passing we note that fits to other data sets from the same reference have been included in the supplementary information to Bobbert et al.[116] The specific equation used in the fit is obtained from Equation 3.39 when choosing $R = 1$ and $s = -1$:

$$MR(B) = MR_{max} \cdot \frac{1}{2}(1 + m_1(B)m_2(B)) \exp[-d/l_s(B)] \qquad (3.40)$$

We note that the dependence of the resulting curves on R is not strong enough to get an unambiguous fit of this parameter from the present data sets. The magnetic switching, as described by $m_i(B)$, is fitted in accordance with the MR data. In this case we used an error function centered at the coercive field to match the rounded switching of the respective electrodes. For the B- dependence of l_s we assumed

$$l_s(B) = l_{s,0} \left(\sqrt{1 + (B/B_0)^2}\right)^{3/4} \qquad (3.41)$$

with $B_0 = 2.3B_{hf}$. The expression Equation 3.41 accurately fits the magnetic field dependence of the spin diffusion length in the strong disorder limit, as predicted by our theory in the diffusive regime $E \ll \sigma/ea$; see Figure 3.31d. Figure 3.32 shows that this predicted $l_s(B)$ accurately reproduces the shape of MR(B) in the region before switching of the soft layer, with a minimum number of parameters. The fitting procedure yields $l_{s,0} \approx 1.7d$ and $B_{hf} \approx 5.7$ mT, the latter being indeed a typical value for the random hyperfine field. The fact that the curve for $d/l_{s,0} = 0$, which describes the case without spin relaxation, deviates strongly from the experimental data demonstrates that the MR before zero field cannot be explained assuming an injected spin polarization proportional to the measured electrode magnetization.

It should be emphasized that the effect of hyperfine fields on MR(B) range well beyond $B \approx B_{hf}$. More specifically, in the presented case where the precession in between hops is only over a small angle and $B_0 = 2.3B_{hf}$, we find that the effects decay to half their size at a field $B_{1/2} \approx 5.1B_{hf}$ in the limit $l_{s,0} \gg d$. For $l_{s,0} = d$ we obtain $B_{1/2} \approx 7.0B_{hf}$, while $B_{1/2} \to \infty$ for $l_{s,0}/d \ll 1$. As to the specific fitted values of $l_{s,0} \approx 1.7d$ and $B_{hf} \approx 5.7$ mT, we note that $l_{s,0}$ is rather sensitive to the value of R. Although changing R hardly changes the shape of MR(B), the different $l_{s,0}/d$ leads to different B_{hf}. Based on the general behavior just discussed, the value of 5.7 mT should be seen as an upper bound.

As a final remark regarding the fitted results, we emphasize that our theory only predicts a ratio $l_{s,0}/d$. When it comes to the effects of hyperfine fields on MR, an ideal device with a neat spacer of thickness $d \approx 100$ nm and $l_{s,0} \approx 170$ nm will behave similarly to a device where severe interdiffusion causes current to flow predominantly through narrow regions of a few to a few tens of nanometers thick, and a proportionally smaller $l_{s,0}$.

After having shown that rounding of the MR traces before switching of the magnetically soft electrode could indeed be related to hyperfine effects, we continue by comparing temperature and bias dependencies. The theory predicts a rather weak dependence of l_s on the relative disorder strength σ/kT, and hence on temperature, in agreement with experiments.[8,89,121] In the experiments with LSMO as one of the electrodes the GMR effect decreases significantly above $T \approx 100$ K, but this can be fully attributed to a reduction of the spin polarization of the injected current.[89,121] It is important to note that in the experiments the MR rapidly disappears with growing bias voltage[8,89,121] on a voltage scale (~1 V) that corresponds in our theory to electric fields for which $eEa/\sigma \ll 1$. At such fields, l_s has saturated to its value at $E = 0$. Therefore, our view is that the measured bias voltage dependence is not caused by a dependence of l_s on the electric field. This view is supported by experiments that find an asymmetric behavior in the bias dependence of organic spin valves, depending on which of the two unequal electrode materials is positively biased.[8,121] Several possible explanations for the steep decrease of spin valve efficiency with increasing bias have been suggested,[121] but it is clear that this issue requires much further study. If efficient spin injection can be realized with high biases such that $eEa/\sigma > 1$, the theory predicts greatly enhanced spin diffusion lengths of the order of several hundreds of nanometers, up to even millimeters (see Figure 3.28).

Finally, we remark that large spin diffusion lengths can be expected under the condition of fast hopping, i.e., when the hopping frequency is much larger than the hyperfine precession frequency, whereas we have seen in Section 3.2.2.3 that the condition for OMAR is to have slow hopping. It is therefore quite remarkable that a material like Alq_3 is used both in spin valves and in OMAR devices. Possibly, the explanation is that the hopping frequencies in disordered organic semiconductors can vary over many orders of magnitude in the same material. This could mean that while the hopping in Alq_3 is fast enough to guarantee considerable spin diffusion lengths, some critical hops relevant for OMAR are actually slow.

APPENDIX A: POLARON HOPPING

We will first consider the case of polaron hopping in the presence of random hyperfine fields and an external magnetic field $\mathbf{B} = B\hat{z}$. A polaron can be at a localized site α or β; see Figure 3.11. The hyperfine fields experienced by the polaron are $\mathbf{B}_{hf,\alpha}$ and $\mathbf{B}_{hf,\beta}$, leading to effective magnetic fields $\mathbf{B}_\alpha$ and $\mathbf{B}_\beta$ at the sites α and β, respectively. We consider the four-dimensional space spanned by the states $\psi_{\alpha\sigma} \equiv \psi_\alpha \chi_{\alpha\sigma}$ and $\psi_{\beta\sigma} \equiv \psi_\beta \chi_{\beta\sigma}$, where ψ_α and ψ_β are the spatial wavefunctions of a polaron present at α and β, respectively. $\chi_{\alpha\sigma}$ and $\chi_{\beta\sigma}$ are the corresponding spinors, where σ indicates whether the spin is along ($\sigma = 1$) or opposite to ($\sigma = -1$) the local effective magnetic field (thick black and gray arrows in Figure 3.11, respectively). These states are eigenstates of our zeroth-order Hamiltonian, with energies that contain the random site energies $\varepsilon_{\alpha,\beta}$ and the effective Zeeman energies $\Delta_{\alpha,\beta} = -\gamma\hbar B_{\alpha,\beta}/2$ (γ is the gyromagnetic ratio): $E_{\alpha\sigma} = \varepsilon_\alpha + \sigma\Delta_\alpha$ and $E_{\beta\sigma} = \varepsilon_\beta + \sigma\Delta_\beta$. The energies $\Delta_{\alpha,\beta}$ will in general be extremely small ($\sim\mu\text{eV}$) with respect to the typical site energy differences $\varepsilon_\beta - \varepsilon_\alpha (\sim 0.1 \text{ eV})$.

Hopping from site α to β occurs by coupling to a system of phonons, which we don't need to specify here. We will treat the corresponding electron-phonon Hamiltonian as a perturbation. In principle, the phonons should be treated quantum mechanically, but we gain insight by treating the phonon system as a collection of classical modes. The translation to a quantum mechanical treatment is straightforward. The electron-phonon Hamiltonian is supposed to be diagonal in spin space.

We start at $t = 0$ with a polaron at α in the initial state $\psi(t = 0) = \Sigma_\sigma c^0_{\alpha\sigma}\psi_{\alpha\sigma}$, with $\Sigma_\sigma |c^0_{\alpha\sigma}|^2 = 1$. The coupling to a classical phonon mode with frequency ω leads to matrix elements $H'_{\alpha\beta} = V_{\alpha\beta}\cos(\omega t) = H'^*_{\beta\alpha}$ of the perturbing Hamiltonian, where $V_{\alpha\beta}$ is proportional to the amplitude of the mode and the electron-phonon coupling constant. Quite generally, we can write at a later time $\psi(t) = \Sigma_\sigma[c_{\alpha\sigma}(t)\exp(-iE_{\alpha\sigma}t/\hbar)\psi_{\alpha\sigma} + c_{\beta\sigma}(t)\exp(-iE_{\beta\sigma}t/\hbar)\psi_{\beta\sigma}]$. Applying standard time-dependent perturbation theory,[122] we obtain to lowest order in the perturbation:

$$c_{\beta\sigma}(t) \approx -\frac{iV_{\beta\alpha}}{\hbar}\sum_{\sigma'}\frac{\sin[(\omega_0^{\sigma\sigma'} - \omega)t/2]}{\omega_0^{\sigma\sigma'} - \omega}e^{i(\omega_0^{\sigma\sigma'}-\omega)t/2}\pi_{\sigma\sigma'}c^0_{\alpha\sigma'}, \tag{3.42}$$

$$c_{\alpha\sigma}(t) \approx c^0_{\alpha\sigma} + \frac{i|V_{\alpha\beta}|^2}{2\hbar^2}\sum_{\sigma'\sigma''}\left[\frac{e^{i(\omega-\omega_0^{\sigma'\sigma})t/2}\sin[(\omega - \omega_0^{\sigma'\sigma})t/2]}{(\omega - \omega_0^{\sigma''\sigma'})(\omega - \omega_0^{\sigma'\sigma})}\right.$$
$$\left. - \frac{e^{i(\omega_0^{\sigma''\sigma'}-\omega_0^{\sigma'\sigma})t/2}\sin[(\omega_0^{\sigma''\sigma'} - \omega_0^{\sigma'\sigma})t/2]}{(\omega - \omega_0^{\sigma''\sigma'})(\omega_0^{\sigma''\sigma'} - \omega_0^{\sigma'\sigma})}\right]\pi_{\sigma''\sigma'}\pi_{\sigma'\sigma}c^0_{\alpha\sigma'} \tag{3.43}$$

with

$$\omega_0^{\sigma\sigma'} \equiv (E_{\beta\sigma} - E_{\alpha\sigma'})/\hbar, \tag{3.44}$$

and $\pi_{\sigma\sigma'}$ a tensor consisting of the inner products between the spinors at α and β:

$$\pi_{\sigma\sigma'} \equiv \chi_{\alpha\sigma}^{\dagger} \chi_{\beta\sigma'}. \tag{3.45}$$

These spinors can be chosen such that

$$\pi_{11} = -\pi_{-1-1} = \cos\theta/2,$$

$$\pi_{1-1} = \pi_{-11} = \sin\theta/2, \tag{3.46}$$

where θ is the angle between $\mathbf{B}_\alpha$ and $\mathbf{B}_\beta$; see Figure 3.11. We have $\Sigma_{\sigma'} \pi_{\sigma\sigma'}^2 = \Sigma_{\sigma} \pi_{\sigma\sigma'}^2 = 1$.

We now consider the full phonon spectrum with density of states $D(\omega)$, and as usual, we assume incoherent perturbations.[122] We then obtain for the probability at time t that the polaron has made a transition from the initial state to the state with spin σ at site β:

$$p_{\beta\sigma} = \int_{\infty}^{\infty} d\omega D(|\omega|)|c_{\beta\sigma}(t)|^2 . \tag{3.47}$$

For the times relevant for hopping we have $t \gg 1/|\omega_0|$, with $\omega_0 \equiv (\varepsilon_\beta - \varepsilon_\alpha)/\hbar$, and the integrand will be very strongly peaked around $\omega = \omega_0$. Consequently, we can replace $D(|\omega|)$ by $D(|\omega_0|)$. Inserting Equation 3.42 into Equation 3.47 and performing the integral over ω, we find

$$p_{\beta\sigma} = \omega_{\alpha\beta} \left\{ \sum_{\sigma'} \pi_{\sigma\sigma'}^2 \; |c_{\alpha\sigma'}^0|^2 \; t \right.$$

$$\left. + \pi_{\sigma1}\pi_{\sigma-1} 2\,\mathrm{Re}\left[e^{-i(\omega_0^{\sigma1} - \omega_0^{\sigma-1})t/2} \left(c_{\alpha1}^0\right)^* c_{u-1}^0 \right] \frac{\sin[(\omega_0^{\sigma1} - \omega_0^{\sigma-1})t/2]}{(\omega_0^{\sigma1} - \omega_0^{\sigma-1})/2} \right\}, \tag{3.48}$$

with

$$\omega_{\alpha\beta} \equiv \frac{\pi D(|\omega_0|)|V_{\beta\alpha}|^2}{2\hbar^2} = \sum_{\sigma} p_{\beta\sigma}. \tag{3.49}$$

In the evaluation of Equation 3.48, use has been made of the integral

$$\int_{-\infty}^{\infty} dx \frac{\sin(x-a)\sin(x-b)}{(x-a)(x-b)} = \pi \frac{\sin(b-a)}{b-a}. \tag{3.50}$$

Monte Carlo simulations of polaron hopping reproducing the spin-dependent transition rate Equation 3.48 can be performed by using a hopping rate $\omega_{\alpha\beta}$ and "copying" the spinor at the moment of hopping to the new position. That this is a correct procedure can be seen by writing Equation 3.48 in the following form:

$$P_{\beta\sigma} = \omega_{\alpha\beta} \int_0^t dt' \left| \pi_{\sigma 1} c_{\alpha 1}^0 e^{-iE_{\alpha 1}t'/\hbar} + \pi_{\sigma -1} c_{\alpha -1}^0 e^{-iE_{\alpha -1}t'/\hbar} \right|^2,$$

(3.51)

where the integrand is the instantaneous probability that the time-dependent spinor at α has spin σ along the quantization axis at β.

If hopping is very fast compared to precession, $t \ll 2/|\omega_0^{\sigma 1} - \omega_0^{\sigma -1}|$, we can put $\sin[(\omega_0^{\sigma 1} - \omega_0^{\sigma -1})t/2] \approx (\omega_0^{\sigma 1} - \omega_0^{\sigma -1})t/2$ and $e^{i(\omega_0^{\sigma 1} - \omega_0^{\sigma -1})t/2} \approx 1$ in Equation 3.48 and obtain

$$P_{\beta\sigma} = \omega_{\alpha\beta} \left| \pi_{\sigma 1} c_{\alpha 1}^0 + \pi_{\sigma -1} c_{\alpha -1}^0 \right|^2 t, (fast\ hopping).$$

(3.52)

In this case the effect of the hyperfine fields is negligible. On the other hand, if hopping is very slow compared to precession, $t \gg 2/|\omega_0^{\sigma 1} - \omega_0^{\sigma -1}|$, we can neglect the second term in Equation 3.48 with respect to the first, and obtain

$$P_{\beta\sigma} = \omega_{\alpha\beta} \sum_{\sigma'} \pi_{\sigma\sigma'}^2 |c_{\alpha\sigma'}^0|^2 t, (slow\ hopping).$$

(3.53)

Then we can perform Monte Carlo simulations in which we can assume that hopping takes place from eigenspinors at α to eigenspinors at β, with hopping rates proportional to the square of the inner products, $\pi_{\sigma\sigma'}^2$, between these spinors.

Integrating Equation 3.43 over incoherent perturbations and using the properties of $\pi_{\sigma\sigma'}$ leads to the following expression for $c_{\alpha\sigma}(t)$:

$$c_{\alpha\sigma}(t) \approx c_{\alpha\sigma}^0 \left(1 - \frac{1}{2}\omega_{\alpha\beta}t \right),$$

(3.54)

so that within the lowest order of the perturbation the probability at time t to find the polaron with spin σ at site α becomes

$$P_{\alpha\sigma} = |c_{\alpha\sigma}(t)|^2 = \left| c_{\alpha\sigma}^0 \right|^2 (1 - \omega_{\alpha\beta}t).$$

(3.55)

Hence, the increasing probability to find the polaron in the states $\psi_{\beta\sigma}$ is compensated by a decreasing probability to find the polaron in the states $\psi_{\alpha\sigma}$. We have $\Sigma_\sigma(p_{\beta\sigma} + p_{\alpha\sigma}) = 1$. In evaluating Equation 3.54, use has been made of the integral

$$\int_{-\infty}^{\infty} dx \left[\frac{e^{i(x-a)}\sin(x-a)}{(x-a)(x-b)} - \frac{e^{i(b-a)}\sin(b-a)}{(x-b)(b-a)} \right] = i\pi \frac{e^{i(b-a)}\sin(b-a)}{b-a}.$$

(3.56)

We define the density matrix of an ensemble of carriers as

$$\rho_{kl} = \sum_i p^{(i)} \left(c_l^{(i)} \right)^* c_k^{(i)} \exp[i(E_l - E_k)t/\hbar],$$

(3.57)

where i labels the different states $\psi^{(i)}$ of the ensemble and $p^{(i)}$ their probabilities, such that $\Sigma_i p^{(i)} = 1$. The indices k and l label the eigenstates with energies $E_{k,l}$, and $c_{k,l}^{(i)}$ are the expansion coefficients of the $\psi^{(i)}$ with respect to these eigenstates. The form of Equation 3.54 for $c_{\alpha\sigma}$ guarantees that within a density matrix description off-diagonal elements of the density matrix that were initially zero remain zero, because $\rho_{\alpha\sigma,\alpha\sigma'}(t) = (1 - \omega_{\alpha\beta}t)\rho_{\alpha\sigma,\alpha\sigma'}(t = 0)$. This means that the average spin evolution of an ensemble of carriers can be properly described with a Monte Carlo procedure for the hopping of each carrier, together with coherent precession of its spin about the local effective magnetic field.

APPENDIX B: BIPOLARON FORMATION

We now consider the situation that there is polaron at α *and* a polaron at β in Figure 3.11 and assume that bipolaron formation is possible. We will assume that ε_β is very small, but ε_α not, so that bipolaron formation can only occur at β. Furthermore, we will assume that only *singlet* bipolarons can form, because of a large energy splitting between singlet and triplet bipolarons. We now have a total of five eigenstates. We will call the eigenstates with both spins along (black arrows in Figure 3.11, $\tau = 1$) or opposite to (gray arrows, $\tau = -1$) the local magnetic fields the parallel (P) eigenstates, $\psi_{\alpha\tau P}$. Their corresponding energies are $E_{\alpha\tau P} = \varepsilon_\alpha + \varepsilon_\beta + \tau(\Delta_\alpha + \Delta_\beta)$. The eigenstates with one spin along and one spin opposite to the local magnetic fields we will call the antiparallel (AP) eigenstates, $\psi_{\alpha\tau AP}$, and they have corresponding energies $E_{\alpha\tau AP} = \varepsilon_\alpha + \varepsilon_\beta + \tau(\Delta_\alpha - \Delta_\beta)$. The fifth eigenstate is the singlet bipolaron state ψ_β. Its energy is $E_\beta = 2\varepsilon_\beta + U$, where U is the energy penalty for bipolaron formation.

We start at $t = 0$ in the initial state $\psi(t = 0) = \Sigma_{\tau\lambda} c_{\alpha\tau\lambda}^0 \psi_{\alpha\tau\lambda}$, with $\Sigma_{\tau\lambda} |c_{\alpha\tau\lambda}^0|^2 = 1$, and $\lambda = P, AP$. At a later time we have $\psi(t) = \Sigma_{\tau\lambda} c_{\alpha\tau\lambda}(t) \exp(-iE_{\alpha\tau\lambda}t/\hbar)\psi_{\alpha\tau\lambda} + c_\beta(t)\exp(-iE_\beta t/\hbar)\psi_\beta$. Within perturbation theory we find

$$c_\beta(t) \approx -\frac{iV_{\beta\alpha}}{\hbar} \sum_{\tau\lambda} \frac{\sin\left[\left(\omega_0^{\tau\lambda} - \omega\right)t/2\right]}{\omega_0^{\tau\lambda} - \omega} e^{i\left(\omega_0^{\tau\lambda} - \omega\right)t/2} \pi_\lambda c_{\alpha\tau\lambda}^0, \tag{3.58}$$

$$c_{\alpha\tau\lambda}(t) \approx c_{\alpha\tau\lambda}^0 + \frac{i|V_{\alpha\beta}|^2}{2\hbar^2} \sum_{\tau'\lambda'} \left[\frac{e^{i(\omega - \omega_0^{\tau\lambda})t/2}\sin\left[\left(\omega - \omega_0^{\tau\lambda}\right)t/2\right]}{\left(\omega - \omega_0^{\tau\lambda}\right)\left(\omega - \omega_0^{\tau'\lambda'}\right)} \right.$$
$$\left. - \frac{e^{i(\omega_0^{\tau'\lambda'} - \omega_0^{\tau\lambda})t/2}\sin\left[\left(\omega_0^{\tau'\lambda'} - \omega_0^{\tau\lambda}\right)t/2\right]}{\left(\omega - \omega_0^{\tau'\lambda'}\right)\left(\omega_0^{\tau'\lambda'} - \omega_0^{\tau\lambda}\right)} \right] \pi_\lambda \pi_{\lambda'} c_{\alpha\tau\lambda'}^0, \tag{3.59}$$

with

$$\omega_0^{\tau\lambda} \equiv (E_{\beta S} - E_{\alpha\tau\lambda})/\hbar, \tag{3.60}$$

and $\pi_{P,AP}$ the spin part of the inner products between $\psi_{\alpha\tau P,AP}$ and ψ_β:

$$\pi_P = \frac{1}{\sqrt{2}}\sin\theta/2,$$

$$\pi_{AP} = \frac{1}{\sqrt{2}}\cos\theta/2. \tag{3.61}$$

We have $\pi_P^2 + \pi_{AP}^2 = 1/2$. Following arguments similar to those in the previous section, we arrive at the following expression for the probability that at time t the bipolaron state has formed:

$$
\begin{aligned}
P_\beta = \omega_{\alpha\beta}\Bigg\{ &\sum_{\tau\lambda} \pi_\lambda^2 \left|c_{\alpha\tau\lambda}^0\right|^2 t \\
&+ \sum_{\substack{\tau\lambda,\tau'\lambda' \\ \tau\lambda\neq\tau'\lambda'}} \pi_\lambda \pi_{\lambda'} 2\,\mathrm{Re}\left[e^{-i\left(\omega_0^{\tau\lambda}-\omega_0^{\tau'\lambda'}\right)t/2}\left(c_{\alpha\tau\lambda}^0\right)^* c_{\alpha\tau'\lambda'}^0\right]\frac{\sin\left[\left(\omega_0^{\tau\lambda}-\omega_0^{\tau'\lambda'}\right)t/2\right]}{\left(\omega_0^{\tau\lambda}-\omega_0^{\tau'\lambda'}\right)/2} \Bigg\}.
\end{aligned}
\tag{3.62}
$$

For very fast hopping we obtain

$$P_\beta = \omega_{\alpha\beta}\left|\pi_P\left(c_{\alpha1P}^0 + c_{\alpha-1P}^0\right) + \pi_{AP}\left(c_{\alpha1AP}^0 + c_{\alpha-1AP}^0\right)\right|^2 t, \quad (fast\ hopping), \tag{3.63}$$

while for very slow hopping we obtain

$$P_\beta = \omega_{\alpha\beta}\sum_{\tau\lambda}\pi_\lambda^2\left|c_{\alpha\tau\lambda}^0\right|^2 t, \quad (slow\ hopping). \tag{3.64}$$

After integrating Equation 3.59 over incoherent perturbations we obtain:

$$
\begin{aligned}
c_{\alpha\tau\lambda}(t) \approx\; &c_{\alpha\tau\lambda}^0\left(1 - \frac{1}{2}\pi_\lambda^2\omega_{\alpha\beta}t\right) \\
&- \frac{1}{2}\omega_{\alpha\beta}\sum_{\substack{\tau'\lambda' \\ \tau'\lambda'\neq\tau\lambda}}\frac{e^{i\left(\omega_0^{\tau'\lambda'}-\omega_0^{\tau\lambda}\right)t/2}\sin\left[\left(\omega_0^{\tau'\lambda'}-\omega_0^{\tau\lambda}\right)t/2\right]}{\omega_0^{\tau'\lambda'}-\omega_0^{\tau\lambda}}\pi_\lambda\pi_{\lambda'}c_{\alpha\tau'\lambda'}^0,
\end{aligned}
\tag{3.65}
$$

which in the case of very fast hopping becomes

$$c_{\alpha\tau\lambda}(t) \approx c_{\alpha\tau\lambda}^0 - \frac{1}{2}\omega_{\alpha\beta}t\sum_{\tau'\lambda'}\pi_\lambda\pi_{\lambda'}c_{\alpha\tau'\lambda'}^0, \quad (fast\ hopping), \tag{3.66}$$

and in the case of very slow hopping

$$c_{\alpha\tau\lambda}(t) \approx c_{\alpha\tau\lambda}^0 \left(1 - \frac{1}{2} p_\lambda^2 \omega_{\alpha\beta} t \right), \quad (slow\, hopping). \tag{3.67}$$

It is clear from Equations 3.62 to 3.67 that only in the case of very slow hopping does an initially diagonal density matrix remain diagonal. In this case we obtain for the changes of the matrix elements of the density matrix:

$$\rho_{\beta,\beta}(t) = \omega_{\alpha\beta} t \sum_{\tau\lambda} \pi_\lambda^2 \rho_{\alpha\tau\lambda,\alpha\tau\lambda}(t=0),$$

$$\rho_{\alpha\tau\lambda,\alpha\tau\lambda}(t) = (1 - \omega_{\alpha\beta} t) \pi_\lambda^2 \rho_{\alpha\tau\lambda,\alpha\tau\lambda}(t=0). \tag{3.68}$$

With $p_\beta \equiv \rho_{\beta,\beta}$, $P_{\alpha\lambda} \equiv \Sigma_\tau \rho_{\alpha\tau\lambda,\alpha\tau\lambda}$, and $P_\lambda \equiv \pi_\lambda^2$ we obtain the following rate equations:

$$\frac{dp_\beta}{dt} = \omega_{\alpha\beta} \sum_\lambda P_\lambda P_{\alpha\lambda},$$

$$\frac{dp_{\alpha\lambda}}{dt} = -\omega_{\alpha\beta} P_\lambda P_{\alpha\lambda}. \tag{3.69}$$

It is straightforward to include other possible transitions in these rate equations, such as the unbinding of the bipolaron by hopping of one of the charges back to the site α or to another site in the environment. Furthermore, in addition to bipolaron formation, a rate for hopping of the polaron at α to another site in the environment may be included. This can all be included in the Monte Carlo simulations. In the case of very slow hopping these simulations can then be performed as if hopping takes place between eigenstates. In effect, these simulations then correspond to solving the rate equations for the diagonal elements of the density matrix.

REFERENCES

1. S. A. Wolf, D. Awschalom, R. A. Buhrman, J. M. Daughton, S. von Molnár, M. L. Roukes, A. Y. Chtelkanova, and D. Treger, Spintronics: A spin-based electronics vision for the future, *Science*, 294, 1488–95, 2001.
2. M. N. Baibich, J. M. Broto, A. Fert, F. N. V. Dau, F. Petroff, P. Eitenne, G. Creuzet, A. Friederich, and J. Chazelas, Giant magnetoresistance of (001)fe/(001)cr magnetic superlattices, *Phys. Rev. Lett.*, 61, 2472–75, 1988.
3. G. Binasch, P. Grünberg, F. Saurenbach, and W. Zinn, Enhanced magnetoresistance in layered magnetic structures with antiferromagnetic interlayer exchange, *Phys. Rev. B*, 39, 4828–30, 1989.
4. G. Pikus and A. Titkov, *Optical Orientation*, ed. F. Meier and B. P. Zakharchenya, chap. 3, North-Holland, Amsterdam, 1984.
5. J. Rybicki and M. Wohlgenannt, Spin-orbit coupling in singly charged π-conjugated polymers, *Phys. Rev. B*, 79, 153202, 2009.

6. A. R. Rocha, V. M. Garcia-Suarez, S. W. Bailey, C. J. Lambert, J. Ferrer, and S. Sanvito, Towards molecular spintronics, *Nature Mater.*, 4, 335–39, 2005.

7. V. Dediu, M. Murgia, F. C. Matacotta, C. Taliani, and S. Barbanera, Room temperature spin polarized injection in organic semiconductors, *Solid State Commun.*, 122, 181–84, 2002.

8. Z. H. Xiong, D. Wu, Z. V. Vardeny, and J. Shi, Giant magnetoresistance in organic spin-valves, *Nature*, 427, 821–24, 2004.

9. S. Majumdar, R. Laiho, P. Laukkanen, I. J. Vayrynen, H. S. Majumdar, and R. Österbacka, Application of regioregular polythiophene in spintronic devices: Effect of interface, *Appl. Phys. Lett.*, 89, 122114, 2006.

10. J. R. Petta, S. K. Slater, and D. C. Ralph, Spin-dependent transport in molecular tunnel junctions, *Phys. Rev. Lett.*, 93, 136601, 2004.

11. T. S. Santos, J. S. Lee, P. Migdal, I. C. Lekshmi, B. Satpati, and J. S. Moodera, Room-temperature tunnel magnetoresistance and spin-polarized tunneling through an organic semiconductor barrier, *Phys. Rev. Lett.*, 98, 016601, 2007.

12. E. Frankevich, A. Zakhidov, K. Yoshino, Y. Maruyama, and K. Yakushi, Photoconductivity of poly(2,5-diheptyloxy-p-phenylene vinylene) in the air atmosphere: Magnetic-field effect and mechanism of generation and recombination of charge carriers, *Phys. Rev. B*, 53, p. 4498, 1996.

13. J. Kalinowski, M. Cocchi, D. Virgili, P. D. Marco, and V. Fattori, Magnetic field effects on emission and current in alq3-based electroluminescent diodes, *Chem. Phys. Lett.*, 380, 710, 2003.

14. J. Kalinowski, M. Cocchi, D. Virgili, V. Fattori, and P. D. Marco, Magnetic field effects on organic electrophosphorescence, *Phys. Rev. B*, 70, 205303, 2004.

15. T. L. Francis, O. Mermer, G. Veeraraghavan, and M. Wohlgenannt, Large magnetoresistance at room temperature in semiconducting polymer sandwich devices, *New J. Phys.*, 6, 185, 2004.

16. O. Mermer, G. Veeraraghavan, T. Francis, and M. Wohlgenannt, Large magnetoresistance at room-temperature in small-molecular-weight organic semiconductor sandwich devices, *Solid State Commun.*, 134, 631–36, 2005.

17. S. R. Forrest, The path to ubiquitous and low-cost organic electronic appliances on plastic, *Nature*, 428, 911–18, 2004.

18. R. H. Friend, R. W. Gymer, A. B. Holmes, J. H. Burroughes, R. N. Marks, C. Taliani, D. D. C. Bradley, D. A. D. Santos, J. L. Brédas, M. Löglund, and W. R. Salaneck, Electroluminescence in conjugated polymers, *Nature*, 397, 121–28, 1999.

19. C. D. Dimitrakopoulos and P. R. L. Malenfant, Organic thin film transistors for large area electronics, *Adv. Mater.*, 14, 99–117, 2002.

20. D. J. Gundlach, Y. Y. Lin, and T. N. Jackson, Pentacene organic thin film transistors— Molecular ordering and mobility, *IEEE Electron. Dev. Lett.*, 18, 87–89, 1997.

21. M. Shtein, J. Mapel, J. B. Benziger, and S. R. Forrest, Effects of film morphology and gate dielectric surface preparation on the electrical characteristics of organic vapor phase deposited pentacene thin-film transistors, *Appl. Phys. Lett.*, 81, 268–70, 2002.

22. C. J. Brabec, N. S. Sariciftci, and J. C. Hummelen, Plastic solar cells, *Adv. Func. Mater.*, 11, 15–26, 2001.

23. P. Peumans, S. Uchida, and S. R. Forrest, Efficient bulk heterojunction photovoltaic cells using small-molecular-weight organic thin films, *Nature*, 425, 158–62, 2003.

24. M. Granström, K. Petritsch, A. C. Arias, A. Lux, M. R. Anderson, and R. H. Friend, Laminated fabrication of polymeric photovoltaic diodes, *Nature*, 395, 257–60, 1998.

25. T. Ito, H. Shirakawa, and S. Ikeda, Simultaneous polymerization and formation of poly-acetylene film on the surface of concentrated soluble Ziegler-type catalyst solution, *J. Polym. Sci. Polym. Chem. Ed.*, 12, 11–20, 1974.

26. C. K. Chiang, C. R. Fincher, Y. W. Park, A. J. Heeger, H. Shirakawa, and E. J. Louis, Electrical conductivity in doped polyacetylene, *Phys. Rev. Lett.*, 39, 1098–101, 1977.

27. J. H. Burroughes, D. D. C. Bradley, A. R. Brown, R. N. Marks, K. Mackay, R. H. Friend, P. L. Burns, and A. B. Holmes, Light-emitting diodes based on conjugated polymers, *Nature*, 347, 539–41, 1990.

28. M. Pope, H. Kallmann, and P. Magnante, Electroluminescence in organic crystals, *J. Chem. Phys.*, 38, 2042, 1963.

29. C. W. Tang and S. A. Van Slyke, Organic electroluminescent diodes, *Appl. Phys. Lett.*, 14, 913–15, 1987.

30. Y. Sun, N. C. Giebink, H. Kanno, B. Ma, M. Thompson, and S. R. Forrest, Management of singlet and triplet excitons for efficient white organic light-emitting devices, *Nature*, 440, 908–10, 2006.

31. R. Meerheim, K. Walzer, M. Pfeiffer, and K. Leo, Ultrastable and efficient red organic light emitting diodes with doped transport layers, *Appl. Phys. Lett.*, 89, 061111, 2006.

32. D. Tanaka, H. Sasabe, Y.-J. Li, S.-J. Su, T. Takeda, and J. Kido, Ultra high efficiency green organic light-emitting devices, *Jpn. J. Appl. Phys.*, 46, L10, 2007.

33. Ö. Mermer, G. Veeraraghavan, T. Francis, Y. Sheng, D. T. Nguyen, M. Wohlgenannt, A. Kohler, M. Al-Suti, and M. Khan, Large magnetoresistance in nonmagnetic pi-conjugated semiconductor thin film devices, *Phys. Rev. B*, 72, 205202, 2005.

34. M. Pope and C. E. Swenberg, *Electronic Processes in Organic Crystals*, Clarendon, New York, 1999.

35. Y. Sheng, T. D. Nguyen, G. Veeraraghavan, Ö. Mermer, M. Wohlgenannt, S. Qiu, and U. Scherf, Hyperfine interaction and magnetoresistance in organic semiconductors, *Phys. Rev. B*, 74, 045213, 2006.

36. A. L. Efros and B. I. Shklovskii, *Electronic Properties of Doped Semiconductors*, Springer-Verlag, Berlin, 1984.

37. R. Menon, *Organic Photovoltaics*, chap. 3, pp. 91–117, Springer Verlag, Berlin, 2003.

38. G. Bergmann, Weak localization in thin films, *Phys. Rep.*, 107, 1–58, 1984.

39. X. Wei, B. C. Hess, Z. V. Vardeny, and F. Wudl, Studies of photoexcited states in polyacetylene and poly(paraphenylenevinylene) by absorption-detected magnetic resonance—The case of neutral excitations, *Phys. Rev. Lett.*, 68, 666–669, 1992.

40. S. J. Papadakis, E. P. De Poortere, H. C. Manoharan, J. B. Yau, M. Shayegan, and S. A. Lyon, Low-field magnetoresistance in GaAs two-dimensional holes, *Phys. Rev. B*, 65, 245312, 2002.

41. D. M. Zumbühl, J. B. Miller, C. M. Marcus, K. Campman, and A. C. Gossard, Spin-orbit coupling, antilocalization, and parallel magnetic fields in quantum dots, *Phys. Rev. Lett.*, 89, 276803, 2002.

42. V. Prigodin, J. Bergeson, D. Lincoln, and A. Epstein, Anomalous room temperature magnetoresistance in organic semiconductors, *Synth. Met.*, 156, 757–61, 2006.

43. H. Odaka, Y. Okimoto, T. Yamada, H. Okamoto, M. Kawasaki, and Y. Tokura, Control of magnetic-field effect on electroluminescence in alq3-based organic light emitting diodes, *Appl. Phys. Lett.*, 88, 123501, 2006.

44. Y. Iwasaki, T. Osasa, M. Asahi, M. Matsumura, Y. Sakaguchi, and T. Suzuki, Fractions of singlet and triplet excitons generated in organic light-emitting devices based on a polyphenylenevinylene derivative, *Phys. Rev. B*, 74, 195209, 2006.

45. Y. Wu, Z. Xu, B. Hu, and J. Howe, Tuning magnetoresistance and magnetic-field-dependent electroluminescence through mixing a strong-spin-orbital-coupling molecule and a weak-spin-orbital-coupling polymer, *Phys. Rev. B*, 75, 035214, 2007.

46. J. D. Bergeson, V. N. Prigodin, D. M. Lincoln, and A. J. Epstein, Inversion of magnetoresistance in organic semiconductors, *Phys. Rev. Lett.*, 100, 067201, 2008.

47. E. Frankevich, A. Lymarev, I. Sokolik, F. Karasz, S. Blumstengel, R. Baughman, and H. Hoerhold, Polaron pair generation in poly(phenylene vinylene)s, *Phys. Rev. B*, 46, 9320–24, 1992.

48. J. Kalinowski, M. Cocchi, D. Virgili, P. D. Marco, and V. Fattori, *Chem. Phys. Lett.*, 380, 710, 2003.
49. P. Desai, P. Shakya, T. Kreouzis, W. P. Gillin, N. A. Morley, and M. R. J. Gibbs, Magnetoresistance and efficiency measurements of alq3-based OLEDs, *Phys. Rev. B*, 75, 094423, 2007.
50. P. A. Bobbert, T. D. Nguyen, F. van Oost, B. Koopmans, and M. Wohlgenannt, Bipolaron mechanism for organic magnetoresistance, *Phys. Rev. Lett.*, 99, 216801, 2007.
51. A. S. Dhoot and N. C. Greenham, Triplet formation in polyfluorene devices, *Adv. Mater.*, 14, 1834–37, 2002.
52. M. Reufer, M. J. Walter, P. G. Lagoudakis, B. Hummel, J. S. Kolb, H. G. Roskos, U. Scherf, and J. M. Lupton, Spin-conserving carrier recombination in conjugated polymers, *Nature Mater.*, 4, 340–46, 2005.
53. H. Bässler, Charge transport in disordered organic photoconductors—A Monte-Carlo study, *Phys. Stat. Sol. B*, 175, 15–56, 1993.
54. M. Helbig and H.-H. Hörhold, Investigation of poly(arylenevinylene)s: Electrochemical studies on poly(p-phenylenevinylene)s, *Makromol. Chem.*, 194, 1607, 1993.
55. M. N. Bussac and L. Zuppiroli, Bipolaron singlet and triplet states in disordered conducting polymers, *Phys. Rev. B*, 47, 5493–96, 1993.
56. O. Chauvet, A. Sienkiewicz, L. Forro, and L. Zuppiroli, High-pressure electron-spin dynamics in disordered conducting polymers, *Phys. Rev. B*, 52, R13118–21, 1995.
57. H. M. McConnel, *J. Chem. Phys.*, 24, 764, 1956.
58. K. Schulten and P. Wolynes, Semiclassical description of electron spin motion in radicals including the effect of hopping, *J. Chem. Phys.*, 68, 3292, 1978.
59. A. Miller and E. Abrahams, Impurity conduction at low concentrations, *Phys. Rev.*, 120, 745, 1960.
60. W. F. Pasveer, J. Cottaar, C. Tanase, R. Coehoorn, P. A. Bobbert, P. W. M. Blom, D. M. de Leeuw, and M. A. J. Michels, Unified description of charge-carrier mobilities in disordered semiconducting polymers, *Phys. Rev. Lett.*, 94, 206601, 2005.
61. W. Wagemans, F. Bloom, P. Bobbert, M. Wohlgenannt, and B. Koopmans, A two-site bipolaron model for organic magnetoresistance, *J. Appl. Phys.*, 103, 07F303, 2008.
62. U. E. Steiner and T. Ulrich, *Chem. Rev.*, 89, 51–147, 1989.
63. G. Brocks, π-Dimers of oligothiophene cations, *J. Chem. Phys.*, 112, 5353–63, 2000.
64. S. Karabunarliev and E. R. Bittner, Spin-dependent electron-hole capture kinetics in luminescent conjugated polymers, *Phys. Rev. Lett.*, 90, 057402, 2003.
65. B. Movaghar and L. Schweitzer, Bipolaron transport in doped conjugated polymers, *J. Phys. C Solid State Phys.*, 11, 125, 1978.
66. F. L. Bloom, M. Kemerink, W. Wagemans, and B. Koopmans, Sign inversion of magnetoresistance in injection limited organic devices, *Phys. Rev. Lett.* 103, 066601, 2009.
67. R. H. Parmenter and W. Ruppel, Two-carrier space-charge-limited current in a trap-free insulator, *J. Appl. Phys.*, 30, 1548, 1959.
68. B. G. Martin, The effect of contact barriers on space-charge-limited current injection in impurity-band-conduction, *J. Appl. Phys.*, 75, 4539, 1994.
69. F. L. Bloom, W. Wagemans, and B. Koopmans, Device modeling magnetoresistance behavior in space-charge limited devices with one injection limited contact, to be published.
70. G. G. Malliaras and J. C. Scott, The roles of injection and mobility in organic light emitting diodes, *J. Appl. Phys.*, 83, 5399–403, 1998.
71. P. W. M. Blom, M. J. M. De Jong, and J. J. M. Vleggaar, Electron and hole transport in poly(p-phenylene vinylene) devices, *Appl. Phys. Lett.*, 68, 3308–10, 1996.
72. F. L. Bloom, W. Wagemans, and B. Koopmans, Temperature dependent sign change of the organic magnetoresistance effect, *J. Appl. Phys.*, 103, 07F320, 2008.

73. F. L. Bloom, M. Kemerink, W. Wagemans, and B. Koopmans, Correspondence of the sign change in organic magnetoresistance with the onset of bipolar charge transport, *Appl. Phys. Lett.*, 93, 263302, 2008.

74. F. L. Bloom, W. Wagemans, M. Kemerink, and B. Koopmans, Separating positive and negative magnetoresistance in organic semiconductor devices, *Phys. Rev. Lett.*, 99, 257201, 2007.

75. F. J. Wang, H. Bässler, and Z. V. Vardeny, Magnetic field effects in π-conjugated polymer-fullerene blends: Evidence for multiple components, *Phys. Rev. Lett.*, 101, 236805, 2008.

76. P. Desai, P. Shakya, T. Kreouzis, and W. P. Gillin, The role of magnetic fields on the transport and efficiency of aluminum tris(8-hydroxyquinoline) based organic light emitting diodes, *J. Appl. Phys.*, 102, 073710, 2007.

77. B. Hu and Y. Wu, Tuning magnetoresistance between positive and negative values in organic semiconductors, *Nature Mater.*, 6, 985–91, 2007.

78. N. Rolfe, P. Desai, P. Shakya, T. Kreouzis, and W. P. Gillin, Separating the roles of electrons and holes in the organic magnetoresistance of aluminum tris(8-hydroxyquinoline) organic light emitting diodes, *J. Appl. Phys.*, 104, 083703, 2008.

79. T. D. Nguyen, J. Rybicki, and M. Wohlgenannt, Device-spectroscopy of magnetic field effects in a polyfluorene organic light-emitting diode, *Phys. Rev. B.*, 77, 035210, 2008.

80. V. I. Arkhipov, E. V. Emelianova, Y. H. Tak, and H. Bässler, Charge injection into light-emitting diodes: Theory and experiment, *J. Appl. Phys.*, 84, 848–56, 1998.

81. T. D. Nguyen, Y. Sheng, J. Rybicki, and M. Wohlgenannt, Magnetic field-effects in bipolar, almost hole-only and almost electron-only tris-(8-hydroxyquinoline) aluminum devices, *Phys. Rev. B*, 77, 235209, 2008.

82. C. F. O. Graeff, G. B. Silva, F. Nüesch, and L. Zuppiroli, Transport and recombination in organic light-emitting diodes studied by electrically detected magnetic resonance, *Eur. Phys. J. E*, 18, 21, 2005.

83. U. Niedermeier, M. Vieth, R. Pätzold, W. Sarfert, and H. von Seggern, Enhancement of organic magnetoresistance by electrical conditioning, *Appl. Phys. Lett.*, 92, 193309, 2008.

84. S. Sanvito, Spintronics goes plastic, *Nature Mater.*, 6, 803–4, 2007.

85. W. J. M. Naber, S. Faez, and W. van der Wiel, Organic spintronics, *J. Phys. D Appl. Phys.*, 40, R205–28, 2007.

86. F. Wang and Z. V. Vardeny, Organic spin valves: The first organic spintronics devices, *J. Mater. Chem.*, 19, 1685–90, 2009.

87. Z. V. Vardeny, Spintronics organics strike back, *Nature Mater.*, 8, 91–93, 2009.

88. Y. Ando, J. Murai, T. Miyashita, and T. Miyazaki, Spin dependent tunneling in 80NiFe/LB film with ferrocene and tris(bipyridine)ruthenium derivatives Co junctions, *Thin Solid Films*, 331, 158–64, 1998.

89. V. Dediu, L. E. Hueso, I. Bergenti, A. Riminucci, F. Borgatti, P. Graziosi, C. Newby, F. Casoli, M. P. D. Jong, C. Taliani, and Y. Zha, Room-temperature spintronic effects in Alq$_3$-based hybrid devices, *Phys. Rev. B*, 78, 115203, 2008.

90. A. J. Drew, J. Hoppler, L. Schulz, F. L. Pratt, P. Desai, P. Shakya, T. Kreouzis, W. P. Gillin, A. Suter, N. A. Morley, V. K. Malik, A. Dubroka, K. Kim, H. Bouyanfif, F. Bourqui, C. Bernhard, R. Scheuermann, G. J. Nieuwenhuys, T. Prokscha, and E. Morenzoni, Direct measurement of the electronic spin diffusion length in a fully functional organic spin valve by low-energy muon spin rotation, *Nature Mater.*, 23, 11, 2008.

91. L. E. Hueso, I. Bergenti, A. Riminucci, Y. Zhan, and V. Dediu, Multipurpose magnetic organic hybrid devices, *Adv. Mater.*, 19, 2639, 2007.

92. Y. Liu, S. M. Watson, T. Lee, J. M. Gorham, H. E. Katz, J. A. Borchers, H. D. Fairbrother, and D. H. Reich, Correlation between microstructure and magnetotransport in organic semiconductor spin-valve structures, *Phys. Rev. B*, 79, 075312, 2009.

93. G. Salis, S. F. Alvarado, M. Tschudy, T. Brunschwiler, and R. Allenspach, Hysteretic electroluminescence in organic light-emitting diodes for spin injection, *Phys. Rev. B*, 70, 085203, 2004.

94. W. Xu, G. J. Szulczewski, P. LeClair, I. Navarrete, R. Schad, G. Miao, H. Guo, and A. Gupta, Tunneling magnetoresistance observed in $La_{0.67}Sr_{0.33}MnO_3$/organic molecule/co junctions, *Appl. Phys. Lett.*, 90, 072506, 2007.

95. D. Dhandapani, N. A. Morley, A. Rao, A. Das, M. Grell, and M. R. J. Gibbs, Comparison of room temperature polymeric spin-valves with different organic components, *IEEE Trans. Magn.*, 44, 2670–73, 2008.

96. S. Majumdar, H. Huhtinen, H. S. Majumdar, R. Laiho, and R. Osterbacka, Effect of $La_{0.67}Sr_{0.33}MnO_3$ electrodes on organic spin valves, *J. Appl. Phys.*, 104, 033910, 2008.

97. S. Majumdar, H. S. Majumdar, R. Laiho, and R. Oesterbacka, Organic spin valves: Effect of magnetic impurities on the spin transport properties of polymer spacers, *New J. Phys.*, 11, 013022, 2009.

98. N. A. Morley, A. Rao, D. Dhandapani, M. R. J. Gibbs, M. Grell, and T. Richardson, Room temperature organic spintronics, *J. Appl. Phys.*, 103, 07F306, 2008.

99. F. Wang, Z. Xiong, D. Wu, J. Shi, and Z. Vardeny, Organic spintronics: The case of fe/ alq(3)/co spin-valve devices, *Synth. Met.*, 155, 172–75, 2005.

100. G. Schmidt, D. Ferrand, L. Molenkamp, A. Filip, and B. van Wees, Fundamental obstacle for electrical spin injection from a ferromagnetic metal into a diffusive semiconductor, *Phys. Rev. B*, 62, R4790–93, 2000.

101. T. Ikegami, I. Kawayama, M. Tonouchi, S. Nakao, Y. Yamashita, and H. Tada, Planar-type spin valves based on low-molecular-weight organic materials with la0.67sr0.33mno3 electrodes, *Appl. Phys. Lett.*, 92, 153304, 2008.

102. J. S. Jiang, J. E. Pearson, and S. D. Bader, Absence of spin transport in the organic semiconductor alq(3), *Phys. Rev. B*, 77, 035303, 2008.

103. H. Vinzelberg, J. Schumann, D. Elefant, R. B. Gangineni, J. Thomas, and B. Buechner, Low temperature tunneling magnetoresistance on (la,sr)mno3/co junctions with organic spacer layers, *J. Appl. Phys.*, 103, 093720, 2008.

104. S. Pramanik, C.-G. Stefanita, S. Patibandla, K. Garre, N. Harth, M. Cahay, and S. Bandyopadhay, Observation of extremely long spin relaxation times in an organic nanowire spin valve, *Nature Nanotechnol.*, 2, 216–19, 2007.

105. G. Schmidt, reported at SPINOS 2009, unpublished, 2009.

106. J. M. Kikkawa and D. D. Awschalom, Lateral drag of spin coherence in gallium arsenide, *Nature*, 397, 139–41, 1999.

107. S. Crooker, M. Furis, X. Lou, C. Adelmann, D. Smith, C. Palmstrom, and P. Crowell, Imaging spin transport in lateral ferromagnet/semiconductor structures, *Science*, 309, 2191–95, 2005.

108. M. Cinchetti, K. Heimer, J.-P. Wuestenberg, O. Andreyev, M. Bauer, S. Lach, C. Ziegler, Y. Gao, and M. Aeschlimann, Determination of spin injection and transport in a ferromagnet/organic semiconductor heterojunction by two-photon photoemission, *Nature Mater.*, 8, 115–19, 2009.

109. J. H. Shim, K. V. Raman, Y. J. Park, T. S. Santos, G. X. Miao, B. Satpati, and J. S. Moodera, Large spin diffusion length in an amorphous organic semiconductor, *Phys. Rev. Lett.*, 100, 2008.

110. J. J. H. M. Schoonus, Schoonus: 2008, PhD thesis, Magnetoresistance effects in hybrid semiconductor devices. Eindhoven University of Technology, 2008.

111. J. J. H. M. Schoonus, P. G. E. Lumens, W. Wagemans, J. T. Kohlhepp, P. A. Bobbert, H. J. M. Swagten, and B. Koopmans, Magnetoresistance in hybrid organic spin valves at the asset of multiple-step tunneling, *Phys. Rev. Lett.* 103, 146601, 2009.

112. S. Tanabe, S. Miwa, M. Mizuguchi, T. Shinjo, Y. Suzuki, and M. Shiraishia, Spin-dependent transport in nanocomposites of alq(3) molecules and cobalt nanoparticles, *Appl. Phys. Lett.*, 91, 2007.

113. T. Wang, H. Wei, Z. Zeng, X. Han, Z. Hong, and G. Shi, Magnetic/nonmagnetic/magnetic tunnel junction based on hybrid organic Langmuir-Blodgett films, *Appl. Phys. Lett.*, 88, 242505, 2006.

114. N. Tombros, C. Jozsa, M. Popinciuc, H. T. Jonkman, and B. J. van Wees, Electronic spin transport and spin precession in single graphene layers at room temperature, *Nature*, 448, 571, 2007.

115. D. R. McCamey, H. A. Seipel, S. Y. Paik, M. J. Walter, N. J. Borys, J. M. Lupton, and C. Boehme, Spin rabi flopping in the photocurrent of a polymer light-emitting diode, *Nature Mater.*, 7, 723–28, 2008.

116. P. A. Bobbert, W. Wagemans, F. W. A. van Oost, B. Koopmans, and M. Wohlgenannt, Theory for spin diffusion in disordered organic semiconductors, *Phys. Rev. Lett.*, 102, 156604, 2009.

117. M. I. D'yakonov and V. I. Perel', Spin orientation of electrons associated with interband absorption of light in semiconductors, *Sov. Phys. JETP*, 33, 1053, 1971.

118. H. Scher and E. W. Montroll, Anomalous transit-time dispersion in amorphous solids, *Phys. Rev. B*, 12, 2455–77, 1975.

119. B. Hartenstein, H. Bässler, A. Jakobs, and K. W. Kehr, Comparison between multiple trapping and multiple hopping transport in a random medium, *Phys. Rev. B*, 54, 8574–79, 2000.

120. Y. Nagata and C. Lennartz, Atomistic simulation on charge mobility of amorphous tris(8-hydroxyquinoline) aluminum (alq$_3$): Origin of Poole–Frenkel–type behavior, *J. Chem. Phys.*, 129, 034709, 2008.

121. F. J. Wang, C. G. Yang, Z. V. Vardeny, and X. G. Li, Spin response in organic spin valves based on La$_{2/3}$Sr$_{1/3}$MO$_3$ electrodes, *Phys. Rev. B*, 75, 245324, 2007.

122. D. J. Griffiths, *Introduction to Quantum Mechanics*, 2nd ed., Pearson Prentice Hall, Englewood Cliffs, NJ, 2005.

4 Spintronic Applications of Organic Materials

Jung-Woo Yoo, V. N. Prigodin, and A. J. Epstein

CONTENTS

ABSTRACT

We introduce broad aspects of spintronic applications of organic materials. First, we address applications of organic semiconductors as either a spin-conserved tunneling barrier or spin-transporting spacer in ferromagnet–organic semiconductor–ferromagnet heterojunctions, which have been extensively investigated recently. A new class of noble set of magnets, *organic-based magnets*, is presented for their potential integration into magnetic multilayer devices toward all organic spin valves as well as for other opportunities and new phenomena. Finally, we discuss magnetotransport of organic semiconductors based on the spin conversion model.

4.1   INTRODUCTION AND OUTLINES

Recent years witnessed growing attention on the emerging field of *organic spintronics*, where the organic materials function as a medium for exchanging spin information. Spintronics is an area of research that explores spin degree of freedom in electronic applications, which have been stimulated by the prospect of a new paradigm of electronics (Wolf et al., 2001). The distinguishing and manipulating "spin" as well as the charge properties of electrons add a new dimension to the conventional electronics. The perspective of spintronics has limitless concepts of novel solid-state electronic devices (Žutić et al., 2004).

The principal spintronic device, a spin valve, consists of soft and hard magnetic layers decoupled by nonmagnetic spacer materials to allow the parallel (on) and antiparallel (off) alignment of magnetic layers. The fundamental operation of these devices involves polarizing, injecting transporting, and analyzing spin information in each layer and/or interface.

The initial reports on the spintronic application of organic semiconductors (OSCs) as a spin-transporting layer have garnered considerable attention (Dediu et al., 2002; Xiong et al., 2004). The central motivation of this research lies in the low spin-orbit coupling and low hyperfine interaction in OSCs. The availability of organic-based spintronic devices will open completely new fabrication processes and new applications.

In addition, the spin injection and detection can also be achieved within the organic-based units with the advent of organic-based magnets. The magnetic and electronic properties of these magnets can be tuned to meet the application via organic chemistry (Miller and Epstein, 1994, 1995). Furthermore, a variety of novel phenomena, such as photoinduced magnetism, suggest a plethora of potential applications of these materials (Epstein, 2003).

This chapter is aimed to provide a brief introduction for both spintronics and molecule/organic-based magnets as well as the recent development of organic spintronics and possible integration of organic-based magnets into conventional magnetic multilayer devices. The first part of the chapter is focused on the spintronic applications of organic semiconductor films with a brief introduction of spintronics. Though the significant potential lies in these applications, there have been controversies in study of spin injection and transport into/in organic semiconductors raised by a few recent publications. It is of importance to provide a consistent experimental study as well as to develop a coherent view of spin-polarized carrier injection and transport in applying OSC as a spacer in magnetic multilayer devices to promote further development of this field. Then, we introduce available molecule/organic-based magnets and their fascinating new phenomena, as well as exotic electronic state in a room temperature magnet, V[TCNE]$_x$. Finally, we provide theoretical models for magnetoresistance of organic semiconductors.

4.2 SPIN INJECTION AND TRANSPORT INTO/IN ORGANIC SEMICONDUCTOR

4.2.1 SPINTRONICS BACKGROUND

The concept that electric current in a magnetic metal could be composed of spin-polarized carriers dates back seventy years ago (Mott, 1936a, 1936b). Based on the assumption that spin-up and spin-down itinerant electrons in the exchange-split d-band of transition metal ferromagnets are not intermixing by the scattering, Mott (1936a, 1936b) postulated that total conductivity in a ferromagnet is the sum of the two independent parts of majority and minority spins, and the electrical current in a ferromagnetic metal has a net spin polarization P:

$$P = (J_\uparrow - J_\downarrow)/(J_\uparrow + J_\downarrow) \neq 0 \tag{4.1}$$

4.2.1.1 Tunneling Magnetoresistance

Experimentally, the study of the transport of the spin-polarized carriers began with tunneling through a metal-insulator-superconductor (M/I/S) interface (Meservey et al., 1970). At a temperature T below the superconducting transition temperature T_c, relatively large in-plane fields, $H < H_c$, cause a Zeeman splitting of the quasiparticle density of states, $N_{\uparrow,\downarrow}(E) = N(E \pm \mu_B H)$, where μ_B is the Bohr magneton. When the counterelectrode is a metal such as silver, the tunneling conductance is symmetric about $V = 0$, with four peaks at the Zeeman splitting values, $eV = \pm|\Delta \pm \mu_B H|$ (Δ is the superconducting order parameter), due to the singularity of the density of states near the superconducting gap edges, $E_F \pm \Delta$ (Meservey et al., 1970). When the counterelectrode is a ferromagnetic (FM) metal such as Fe, Ni, or Co, the peaks have asymmetric height (Tedrow and Meservey, 1971), since the d-band densities of states at the Fermi level of ferromagnets, $N_\uparrow(E_F)$, $N_\downarrow(E_F)$, differ for spin-up and spin-down subbands due to Zeeman exchange splitting. Tedrow and Meservey (1971) could deduce the net polarization, P, of the tunnel current crossing the FM/I/S interface, directly proportional to the asymmetry of the density of states at the Fermi level of the ferromagnetic layer,

$$P = [N_\uparrow(E_F) - N_\downarrow(E_F)]/[N_\uparrow(E_F) + N_\downarrow(E_F)]. \tag{4.2}$$

This experiment had a broad impact in the fact that the polarized current could tunnel across a barrier at a boundary of FM and maintain its polarization. Soon after, Julliere's model for spin-dependent tunneling recognized that a second ferromagnetic film, FM2, could replace the superconducting film, S, as a detector of spin-polarized carriers (Julliere, 1975). This was the foundation for a magnetoresistive device, *magnetic tunnel junction*, one of the prototype spintronic devices. Here, the tunneling magnetoresistance (TMR) refers to the change of device resistance between parallel and antiparallel configurations of magnetic layers in FM1/I/FM2 magnetic tunnel junction (MTJ) and is derived by Julliere as

$$\text{TMR} = \frac{\Delta R}{R_P} = \frac{R_{AP} - R_P}{R_P} = \frac{2P_1 P_2}{1 - P_1 P_2} \tag{4.3}$$

where R_P and R_{AP} are the device resistance for parallel and antiparallel configurations of magnetization between two magnetic layers, respectively, and P_1 and P_2 are the spin polarizations for each ferromagnetic layer. The change of device resistance by the relative orientation of magnetic multilayers gives birth to a new concept: *nonvolatile memory device*, a device that maintains its memory even without power.

Figure 4.1 shows a schematic view of the change of device resistance between parallel (Figure 4.1a) and antiparallel (Figure 4.1b) alignments of a magnetic trilayer of FM1/I(N)/FM2. Figure 4.1c and d illustrates corresponding ferromagnetic band structures of Stoner's model in both FM layers decoupled by an insulating barrier. With the assumption of spin-conserved tunneling, the electron will tunnel between the same spin subbands as depicted by arrows. The tunnel current will be proportional to the product of the same orientation of spin subbands in both sides of ferromagnetic layers. A higher tunneling rate occurs when the majority spin subbands of both magnetic layers are parallel. Based on Mott's two-fluid model (Mott, 1936a,b), the total device resistance can also be described in a simple resistor network model (Figure 4.1e and f). Here, the up-spin and down-spin electrons are proposed to travel on independent pathways, as depicted in Figure 4.1e and f. Then, the total device resistance is the parallel combination of two channels and is higher for antiparallel alignment of two magnetic layers. Julliere's model formulated the change of magnetoresistance by assuming a constant tunneling rate for all matrix elements and ignoring spin flip scattering at the interface (Julliere, 1975). A realistic model that accounts for and explains a variety of experimental results, such as temperature and bias dependence, especially various dependences of material combinations, would be too complicated to be simplified. Generally, the TMR devices are extremely sensitive to the interfacial quality. Subtle fluctuation of the wavefunction density and symmetry at the interface could strongly suppress/change the MR of devices. The

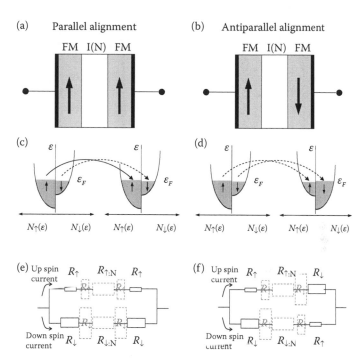

FIGURE 4.1 (a, b) Schematic view of parallel and antiparallel alignment of magnetic tri-layer, consisting of two magnetic layers separated by a nonmagnetic spacer (insulator/metal). (c, d) Band structure of the magnetic layers for parallel and antiparallel alignment. The solid (dashed) arrow depicts the tunneling pathway, given the assumption of spin-conserved tunneling. The tunneling current is proportional to the product of density of state (DOS) for each spin orientation of subbands. (Reproduced with permission from Naber et al., *J. Phys. D. Appl. Phys.* 40, R205 [2007]. Copyright © 2007 IOP Publishing.) (e, f) Two-resistor network model for magnetization of parallel and antiparallel alignment originally proposed by Mott (1936a) and extended two-channel model (illustrated in gray color) with the consideration of a nonmagnetic spacer. (Reproduced with permission from G. Schmidt and W. Molenkamp, *Semicond. Sci. Technol.* 17, 310 [2002]. Copyright © 2002 Institute of Physics. Reproduced with permission from Naber et al., *J. Phys. D. Appl. Phys.* 40, R205 [2007]. Copyright © 2007 IOP Publishing.)

research on MTJ for the application of memory devices has been resurrected after the discovery of room temperature TMR (Moodera et al., 1995; Miyazaki and Tezuka, 1995). Remarkable advances have been accomplished since then. A recent theoretical study suggested the epitaxial magnetic trilayer could produce much higher TMR values with selective wavefunction decay in the barrier (Butler et al., 2001; Mathon and Umerski, 2001). Recent improvements have been reached to over a few hundreds percent of TMR effects using a spin-selective tunneling barrier (Parkin et al., 2004; Yuasa et al., 2004, 2006; Djayaprawira et al., 2005). In these crystalline magnetic trilayers, tunneling coefficients of matrix elements highly depend on the orientations of individual layers due to wavefunction symmetry matching. Therefore, by engineering crystal orientations of individual layers, the decay rates of majority and the minority spin in the barrier can be controlled to significantly improve TMR values.

4.2.1.2   Giant Magnetoresistance

In the late 1980s, two groups (Baibich et al., 1988; Binasch et al., 1989) independently reported huge magnetoresistance in multiple spin valve superlattices. These reports of giant magnetoresistance (GMR) brought a strong impetus for the application of magnetic multilayers and were recognized by the Nobel Prize in Physics in 2007. Historically, GMR first referred to huge magnetoresistance in superlattices of Fe/Cr magnetic multilayers. But the term GMR often is used to distinguish it from TMR, in that the spin-polarized carriers are injected into and transported through the nonmagnetic spacers instead of tunneling in MTJ. The spin diffusion into metal from ferromagnetic contact introduces an imbalance of chemical potential for up- and down-spin in metal spacers. This modification of chemical potential can be detected by a second ferromagnetic contact, as first demonstrated by Johnson and Silsbee (1985) in the bulk Al wire with thin-film ferromagnetic permalloy (Py) as an injector and detectors. Consideration of a nonmagnetic spacer leads to more complex problems, as the up- and down-spins undergo scattering, and so does intermixing. The simple two-independent channel model becomes more irrelevant. The quality of the interface between the nonmagnetic spacer and the ferromagnetic layer also plays an important role, as the efficient spin injection into a spacer becomes more critical. Here, the spin diffusion length (λ_s) in a nonmagnetic spacer becomes a crucial parameter in the magnetoresistance in the GMR devices. The simple extension of the Julliere model with additional input λ_s often applied for the magnetoresistance of the GMR device as follows:

$$\text{GMR} = \frac{\Delta R}{R_P} = \frac{2P_1 P_2 \exp(-d/\lambda_s)}{1 - P_1 P_2 \exp(-d/\lambda_s)} \tag{4.4}$$

Some important insight on the additional features in GMR devices can still be obtained with the consideration of the resistor network model, as described in Figure 4.1e and f. Here, the resistance of a nonmagnetic spacer is supposed to be independent of up- and down-spins, $R_{\uparrow,N} \sim R_{\downarrow,N}$. Then, the simple derivation (Bandyopadhyay and Cahay, 2008) of the spin injection efficiency in this resistor network model shows a maximum value when the spin polarization of the ferromagnet is 100%, *half metal*. Here, the spin injection efficiency strongly depends on the relative conductivity of ferromagnetic metal, σ_F, and nonmagnetic spacer, σ_N, and contact resistance, R_I, as studied by Johnson and Silsbee (1987, 1988). If the nonmagnetic spacer is a semiconducting material, which normally has a ratio of conductivity of $\sigma_{SC}/\sigma_F \sim 10^4$, the spin injection efficiency is nearly diminishing. This is the so-called conductivity mismatch problem (Schmidt et al., 2000), and it has been one of the main stimulants for the intensive effort in developing a room temperature magnetic semiconductor. The other means of circumvention, suggested by Rashba (2000), is an insertion of a tunnel barrier between the ferromagnet and the semiconductor layers, as illustrated in Figure 4.1e and f. In a sample description, the tunneling barrier functions as a spin selective scattering boundary of the interface and eventually introduces a change of total device resistance corresponding to parallel and antiparallel alignments. The conductivity mismatch problem would also be a very critical problem in applying an organic semiconductor for the spin-transporting spacer in a conventional spin valve.

Typical GMR device geometries are generally classified by whether the current flows parallel to the layered structure (current in the plane [CIP]) or perpendicular to the layered structure (current perpendicular to the plane [CPP]), as depicted in Figure 4.2a and b. The CIP geometries were first applied due to its more readable fabrication (Baibich et al., 1988; Binasch et al., 1989). The underlying physical mechanism for the observed giant magnetoresistance is the same for both CIP and CPP geometry. In a CIP device, the electron mean free path in the spacer material is the critical length scale, as the spin-dependent scattering at the interface, depending on the relative orientation of carrier spin and magnetization of the ferromagnetic layer, is essential for the observed giant magnetoresistance. However, in a CPP device, the spin diffusion length in the spacer is the critical length scale, in order to analyze the spin-dependent carrier at the second magnetic layer.

There are a number of phenomena that produce MR in small magnetic layered devices, such as anisotropic magnetoresistance (AMR), local Hall effects, Lorentz magnetoresistance (LMR), etc. (see review of Naber et al., 2007; Žutíc et al., 2004). These effects can easily be mistaken to be spin valve effects and hardly be excluded in local conventional GMR geometry devices. Figure 4.2c shows schematic nonlocal geometry, which is similar to a quasi-1D picture of Johnson and Silsbee (1985;

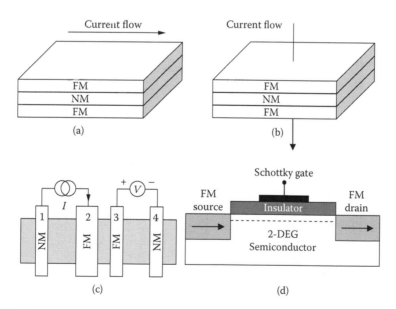

FIGURE 4.2 (a, b) Schematic view of current-in-the-plane (CIP) and current-perpendicular-to-the-plane (CPP) GMR device structure. (c) Schematic view of nonlocal geometry for measuring spin accumulation, which is isolated from current pathway. (Reproduced with permission from Naber et al., *J. Phys. D. Appl. Phys.* 40, R205 [2007]. Copyright © 2007 IOP Publishing.) (d) Schematic view of Datta and Das's (1990) spin field effect transistor (spin-FET). (Reproduced with permission from Datta and Das, *Appl. Phys. Lett.* 56, 665 [1990]. Copyright © 2000 Amerian Institute of Physics.)

Johnson, 2002). Other geometries, such as van der Pauw, were also employed by Jedema et al. (2001). These nonlocal measurements can easily exclude spurious effects by collecting pure spin accumulation and have been widely applied to observe the Hanle effect (Jedema et al., 2002; Lou et al., 2007; Tombros et al., 2007). In the geometry of Figure 4.2c, when the spin-polarized carrier is injected from ferromagnetic electrode 2, the spin current diffuses in both directions and can be analyzed by ferromagnetic electrode 3 and metallic contact 4. Here, the measured voltage is not influenced by the charge current, and the signal is purely due to spin accumulation.

The spintronic devices have already been launched in various commercial electronic applications, especially in memory units. Further application of spintronic devices will be promoted if the active spin-based devices, such as spin transistors, have been achieved. The concept of spin field effect transistor (spin-FET) was introduced by Datta and Das (1990). Figure 4.2d shows a schematic view of spin-FET. The device structure has nearly the same configuration as the conventional field effect transistors (FETs), except the ferromagnetic source/drain contact. The other main difference is the function of gate voltage, which creates an effective field to control precession of the spin current. The two ferromagnetic contacts (source/drain) function as an injector and analyzer of electron spin. When the spin-polarized electrons are injected from the source, they undergo precession, which is controlled by the effective magnetic field from the gate voltage. The device is ON when the spin of electron enters the drain with the same orientation of magnetization as the drain, and OFF when vice versa. It should be mentioned that a fully functional spin FET recently achieved using high mobility InAs semiconducting channel (Koo et al., 2009).

4.2.2 ORGANIC ELECTRONICS

The application of organic materials in electronic devices seems intuitively odd, as the typical organic materials are close to electrical insulators with a large band gap and typically form ill-defined films. Though the initial approach of various electronic applications of organic materials dates back nearly a half century, early devices suffered from low mobility, low liability of devices, sample degradation, poor control of film morphologies and contacts, etc. Much of the initial research was devoted to improve the charge carrier mobility of the organic films. Despite a number of obstacles in this application, the availability of organic-based devices opened up a completely new fabrication process and new application.

The discovery of high conductivity (in comparison to typical organic semiconductors of large band gap) in oxidized, iodine-doped polyacetylene generated a strong impetus for the electronic application of organic materials (Shirakawa et al., 1977; Chiang et al., 1977). A number of novel process technique have been developed, such as soft lithography (Xia and Whitesides, 1998), ink jet printing (Sirringhaus et al., 2000), solution-based methods (Sirringhaus et al., 1998), and vapor phase deposition (Burrows et al., 1995; Shtein et al., 2001). The organic electronic devices include organic light-emitting diodes (OLEDs) (Tang and van Slyke, 1987; Burroughes et al., 1990), photovoltaic cells (Tang, 1986; Yu et al., 1995), and field effect transistors

(FETs) (Sirringhaus et al., 1999). Over the last decades, there have been significant improvements in the performance of those organic electronic devices. Especially, the display based on OLEDs has been commercialized in various appliances.

The organic materials for the electronic applications can be categorized as polymer, oligomer (limited number of repeating monomer units), small molecule, and dendrimer (repeated branch of small molecules). These organic materials are typically π-conjugated systems (alternating single and double bonds between neighboring carbons). This dimerization produces a large Peierl's gap (typically ~1.5 to 3 eV) between the highest occupied molecular orbital (HOMO) and lowest unoccupied molecular orbital (LUMO) levels. Due to their large band gaps, these materials are generally closer to the insulators (often called semi-insulators instead of semiconductors).

The HOMO and LUMO levels are generally delocalized within the molecules through π bonding. The mobility of electrons within the molecule or polymer chain is relatively high. The charge transport in these films strongly relies on the hopping between the adjacent molecules or chains. The hopping between molecules/polymer chains depends on the structural order. The crystallinity of these films increases the overall mobility and often can be improved by various processing, such as annealing, solution processing, and adding a self-assembled seed monolayer (DiBenedetto et al., 2009). The ultra-high-purity single crystal also was developed via physical vapor deposition (PVD) (Laudise et al., 1998), reaching a maximum mobility of ~30 cm²/Vs for pentacene (Jurchescu et al., 2004) and ~20 cm²/Vs for rubrene (Podzorov et al., 2004).

In organic semiconductors, the p-type (or n-type) refers to the holes (or electrons) that are more easily injected from the contacts, and the mobility of the electron/hole in the bulk of OSCs is generally significantly different. The carrier injection in the metal-OSC interface depends on various factors, such as impurities, film morphology, dipole barrier, etc. Due to a large energy gap between HOMO and LUMO levels, there is significant energy barrier height at the metal-OSC interface. The Schottky barrier at the interface is often directly scaled with the metal work function. But, the presence of a dipolar barrier at the interface significantly modifies this level alignment. Another important aspect in organic electronic devices is the space charge limit. Due to a high dielectric constant of the organic semiconductors, there is a continuum of distributed space charge in the bulk of the organic semiconductor layer when the bias is applied from the cathode/anode. In the steady state, the space charge limited current typically shows the characteristic behavior of $J \propto V^2/d^3$.

Recently, there have been increasing research activities in the spintronic application of organic materials. Despite the broad availability of materials and fabrication processing, which have been applied to other organic electronic devices, the current research in organic spintronics has mostly been limited to the application of sublimed amorphous small-molecule films for a non-magnetic spacer layer, indicating research in this field still remains in the infancy stage. In this section we introduce recent work done at The Ohio State University regarding the issue of spin polarized carrier injection and transport into and in OSCs and present both the tunneling magnetoresistance and giant magnetoresistance using amorphous organic small-molecule film spacers in conventional CPP spin valve devices (Yoo et al., 2009, 2010).

4.2.3 ORGANIC SPINTRONICS

The spintronic applications of organic semiconductors (OSCs) were initially motivated by long spin lifetime in OSCs. Dediu et al. (2002) first employed a small-molecule film (α-sexithiophene (α-6T)) as a spin-transporting channel in the planer structure of $La_{2/3}Sr_{1/3}MnO_3$ (LSMO) electrodes separated by $d \sim 100$ nm for the study of the spin injection and transport in organic semiconductors. The observation of MR in that device was ambiguous since the device structure couldn't define parallel (on) and antiparallel (off) alignment of magnetic layers. In addition, the LSMO electrode itself displays linear magnetoresistance (Xiong et al., 2004). Subsequently, a report of the spin valve effect by Xiong et al. (2004) using thick tris(8-hydroxyquinolinato) aluminum (Alq_3) spacers ($d > 100$ nm) brought considerable attention to the spintronic application of OSCs. They employed Alq_3 spacers in large-scale vertical devices with LSMO and Co as soft and hard magnetic layers and reported surprisingly large negative magnetoresistance at low temperature (Xiong et al., 2004). The observed negative spin valve effects were initially attributed to the negative polarization of Co at the Fermi level, as suggested in a previous study of LSMO/Al_2O_3/Co trilayer devices (Teresa et al., 1999). Recently, Santos et al. (2007) demonstrated a positively polarized tunnel current by employing a spin-polarized tunneling study of the Tedrow and Meservey method using Al/Al_2O_3/Alq_3/Co heterojunction. It should be mentioned that this experiment is limited at low temperature and low bias ($T < T_c$; $V_b \sim \backslash \Delta \sim T_c$) and address the polarization tunnel current only at low bias voltage.

Several succeeding studies display negative MR (Wang et al., 2005, 2007; Xu et al., 2007; Dediu et al., 2008) as well as positive MR (Majumdar et al., 2006; Santos et al., 2007; Shim et al., 2008; Ikegami et al., 2008; Liu et al., 2009) using OSC as a spacer. A number of reported studies show widely scattered device resistance, suggesting a lack of consistent study on the device characteristic. Recently, two new experimental probes were implemented in order to study spin injection into OSC layers (Drew et al., 2009; Cinchetti et al., 2009). These reports independently confirmed efficient spin injection into OSC layers; one is via muon spin rotation study and the other is via photoemission study. However, the estimated effective spin diffusion length in amorphous OSC ($\lambda_s \sim 10$ nm) is much shorter (Drew et al., 2009; Cinchetti et al., 2009) than that estimated by electrical detection in a previous device. Despite intensive effort devoted to this application, there has been a lack of fundamental understanding on the physical mechanism of how the spin-polarized carrier mediated through the organic semiconductor (OSC) layer, due to lack of reliable and comprehensive device work. This brought serious challenges (Xu et al., 2007; Jiang et al., 2008; Vinzelberg et al., 2008) to this burgeoning field on the issues of spin transport in organic semiconductor layers, attributing most of the previous reports on spin valve effects to tunneling magnetoresistance through the locally thin areas, as studied for the thin layers of Alq_3 and rubrene ($C_{42}H_{28}$)-based spin valves (Santos et al., 2007; Shim et al., 2008). Therefore, it is very important to deliver a thorough, consistent study on the device characteristics to unravel many contradictions in a number of experimental studies.

Due to light constituting atoms, the spin-orbit coupling in organic small-molecule films is low. The absence of nuclear spin in ^{12}C and the orbital wavefunction of the

π electron carriers lead to weak hyperfine interaction. As a result, the spin lifetime in OSCs will be extremely long compared to that in conventional metals. However, efficient spin-polarized carrier injection and transport in OSC films and electrical detection of MR for OSC-based spin valves are rather complicated. First, there is a large barrier at the metal-OSC interface for carrier injection due to a large band offset between the Fermi level of a typical ferromagnetic metal and the HOMO and LUMO levels of the OSC layer. Typical OSC materials have a band gap of more than 2 eV (for rubrene ~ 2.3 eV; Alq$_3$ ~ 2.8 eV). The dipole interaction and possible chemical interactions at the metal-OSC interfaces introduce a number of defect states, which could interfere with efficient spin-polarized carrier injection. Though the efficient spin injection into HOMO or LUMO levels of organic semiconductor is achieved, the carrier transport in these levels is not band-type conduction unless the film is crystalline. The transport of electrons and holes in the bulk of usual OSC films involves hopping from level to level between neighboring sites through the optimal pathway (variable range hopping [Mott, 1969]), leading to low mobility and interaction with phonons, which could affect spin-polarized carrier transport at high temperature. Finally, there is a conductivity mismatch problem in applying high resistive materials for the spin-transporting spacer (Schmidt et al., 2000). However, the flexibility of organic chemistry has the potential of significant improvement in efficient spin injection and transport in organic semiconductors.

In this section, we present the extensive study of magnetoresistance and device characteristics in LSMO/OSC/Fe heterojunction devices using rubrene (C$_{42}$H$_{28}$) as an OSC spacer (Yoo et al., 2009, 2010). The schematic device structure and rubrene molecule structure are shown in Figure 4.3. Thin rubrene layer for efficient spin-conserved tunneling was initially introduced by Shim et al. (2008) in their TMR devices. A thin layer (1.2 nm) of LaAlO$_3$ (LAO) in this study was employed to improve the interfacial quality between LSMO and rubrene. The thickness of the rubrene layer is varied from the TMR limit ($d \leq 5$ nm) to the giant magnetoresistance limit ($d \geq 20$ nm). The control device, which has only a thin layer of LAO (1.2 nm) as a spacer, displays weak negative MR due to possible exchange coupling and the presence of pinhole channels (Yoo et al., 2010). When the thin layer of rubrene is applied over the LAO to form a hybrid barrier, the device displays

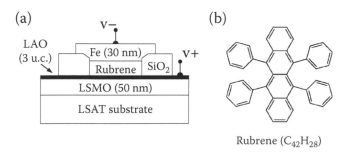

Rubrene (C$_{42}$H$_{28}$)

FIGURE 4.3 (a) Schematic view of device structure. The junction area (200×200 μm) is enclosed by SiO$_2$ (300 nm) layer. (b) Molecular structure of rubrene (C$_{42}$H$_{28}$).

positive TMR over all temperature and bias, similar to the previous work done by Shim et al. (2008). As the thickness of rubrene is increased, the device current strongly relies on carrier injection, resulting in strong temperature-dependent device resistance. The carrier injection in our devices is well described with quantum mechanical transition probability of electrons from the defect states at the interface to the HOMO or LUMO levels of the OSC layer via phonon-assisted field emission. The role of bias field and phonons on the efficient spin-polarized carrier injection and transport into/in OSCs for organic-based spin valve will be discussed. The observed giant magnetoresistance (GMR) in our 20 nm rubrene devices demonstrates the electrical detection of the spin-polarized carrier injection and transport through the HOMO or LUMO levels of rubrene layers.

4.2.4 TUNNELING MAGNETORESISTANCE IN FM/OSC/FM HETEROJUNCTION

Tunneling magnetoresistance with a thin layer of OSC has been extensively studied by the Moodera group (Santos et al., 2007; Shim et al., 2008). Their study of spin-polarized tunneling (Tedrow and Mersevey method) directly suggested a detrimental role of the dipolar barrier for efficient spin-polarized tunneling through the thin layer of OSC. Similar to the approach of previous TMR studies (Santos et al., 2007; Shim et al., 2008), we employed a thin layer of LAO capping on top of the LSMO layer. The capping with a thin oxide layer is primarily for improving metal-OSC interfacial quality for efficient spin injection (Santos et al., 2007; Shim et al., 2008). In addition, applying a thin tunnel barrier will improve efficient spin injection for high, resistive materials (Rashba, 2000). The magnetization-induced second harmonic generation (MSHG) results show that capping with a thin LAO layer displays superior protection of surface polarization of LSMO than does STO (Yamada et al., 2004). The capping with a thin oxide layer could also help avoid unwanted surface contamination of LSMO. It is also reported that capping with a thin oxide layer (Al_2O_3) improves the growth mode of organic small-molecule films (Santos et al., 2007; Shim et al., 2008). Further details of the device fabrication in this study were presented earlier (Yoo et al., 2009, 2010).

Figure 4.4 shows typical MR curves for a device of LSMO (50 nm)/LAO (1.2 nm)/rubrene (5 nm)/Fe (30 nm) at 10 K and 10 mV. The dashed curve is the data recorded with a reverse field scan, as indicated by arrows, and the black curve is vice versa. Arrows indicate the direction of the field scan during the measurement. The device resistance is higher when the two magnetic layers are in antiparallel configuration. The MR steps correspond well to the coercivities of individual Fe (30 nm) and LSMO (50 nm) electrodes. The magnetization curves for individual Fe and LSMO magnetic layers at 10 K recorded by SQUID magnetometry are displayed at the bottom of Figure 4.4.

Figure 4.5a displays *I-V* and d*I*/d*V* curves of a 5 nm rubrene device (LSMO [50 nm]/LAO [1.2 nm]/rubrene [5 nm]/Fe [30nm]) at 10 K. The d*I*/d*V* was measured by standard lock-in technique. As discussed in previous literature (Santos et al., 2007), the absence of zero bias dip in the d*I*/d*V* curve indicates that a thin rubrene layer forms a well-defined tunneling barrier. Figure 4.5b shows the bias dependence of MR determined by the comparison of *I-V* curves for parallel (at $H = 500$ Oe) and

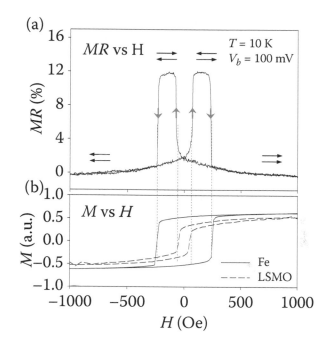

FIGURE 4.4 The prototypical tunneling magnetoresistance in MTJ of LSMO (50 nm)/LAO (1.2 nm)/rubrene (5 nm)/Fe (30 nm) at 10 K and 10 mV. Dashed curve is the data recorded with reverse field scan, as indicated by arrows; black curve is vice versa. Arrows indicate direction of field scan during the measurement. Magnetization curves for individual Fe (30 nm) and LSMO (50 nm) electrodes measured by SQUID magnetometry show excellent correspondence to the steps of MR of the device. (After Yoo et al., 2009, 2010.)

antiparallel (at $H = -150$ Oe) configurations of magnetic layers. The asymmetric MR as a function of V_b is due to different electrodes (Fe, LSMO), similar to previous observations (Xiong et al., 2004; Wang et al., 2007; Xu et al., 2007). The individual MR curves of the 5 nm rubrene device at 10 K are presented in Figure 4.5c for positive biases and Figure 4.5d for negative biases. The steps of MR for all biases show excellent correspondence with each other, as well as the coercivities of Fe (30 nm) and LSMO (50 nm) electrodes.

The T dependence of MR for the 5 nm rubrene device at 10 mV is displayed in Figure 4.6a. The steps of MR curves at each T correspond well to the T-dependent H_c for Fe and LSMO electrodes, as shown in Figure 4.6b. The substantial decrease of MR as T is increased (see Figure 4.6c) is attributed to the T-dependent surface polarization of the LSMO electrode (Figure 4.6d) as probed via spin-resolved photoemission by Park et al. (1998). Similar T dependencies of MR were commonly deserved for various LSMO-based trilayers (Lu et al., 1996; Garcia et al., 2004), as well as organic-based spin valves using LSMO electrodes (Xiong et al., 2004; Wang et al., 2007; Xu et al., 2007; Dediu et al., 2008; Majumdar et al., 2008). The MR of our 5 nm rubrene devices is positive over all T and V_b in contrast to previous reports, which showed negative MR in organic-based spin valves. The difference between

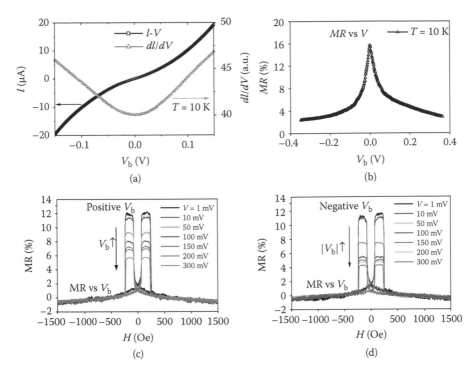

FIGURE 4.5 (a) *I-V* and d*I*/d*V* curves for 5 nm rubrene device (LSMO [50 nm]/LAO [1.2 nm]/rubrene [5 nm]/Fe [30 nm]). (b) MR vs. V_b of 5 nm rubrene device at 10 K determined by the difference in *I-V* curves between parallel ($H = 500$ Oe) and antiparallel ($H = -150$ Oe) configurations for the 5 nm rubrene device. MR vs. V_b at 10 K shows asymmetric decrease for the bias polarity due to asymmetric magnetic electrodes, similar to previous reports (Xiong et al., 2004; Wang et al., 2007; Xu et al., 2007). (c) Positive bias dependence of MR curves for 5 nm rubrene device (LSMO [50 nm]/LAO [1.2 nm]/rubrene [5 nm]/Fe [30 nm]) at 10 K. (d) Negative bias dependence of MR curves for 5 nm rubrene device at 10 K. (From Yoo et al., *Phys. Rev. B.* 80, 205207 [2009]. Copyright © 2006. American Physical Society. With permission.)

our devices and many previous reported organic spin valves using LSMO layers is that we used Fe instead of Co for the second ferromagnetic layer. The variation of MR sign could also be attributed to the highly epitaxial LSMO films as experimentally observed in previous reports for LSMO-based trilayers without OSC spacer (Teresa et al., 1999; Marun et al., 2007). It is also possible that the filamental conduction channel might introduce an irregular change of the MR sign.

4.2.5 GIANT MAGNETORESISTANCE IN FM/OSC/FM HETEROJUNCTION

4.2.5.1 Carrier Injection into OSC Spacer: Thermionic Field Emission

With increasing thickness of rubrene layers, devices display stronger nonlinear *I-V* curves, as shown in Figure 4.7a and b for 20 and 50 nm rubrene devices. The low-bias (~ several mV) resistivity for our typical 20 nm rubrene devices is – higher than GΩ·cm at 10 K, which is substantially higher than the previously reported device

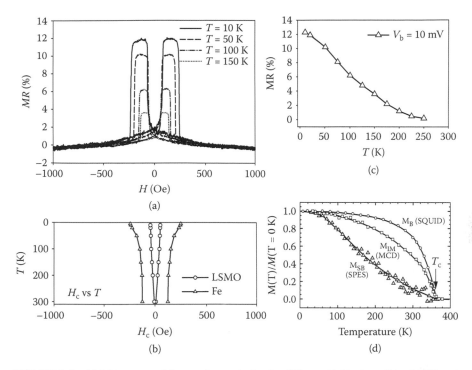

FIGURE 4.6 (a) MR curves of 5 nm rubrene device for different T. (b) H_c vs. T for Fe (30 nm) and LSMO (50 nm) magnetic layers recorded by SQUID magnetometer. (c) T-dependent TMR values for 5 nm rubrene device for $V_b = 10$ mV. Figures 4.6a,b & c are reproduced with permission from Yoo et. al., *Phys. Rev. B.* 80, 205207 [2009]. Copyright© American Physical Society. (d) Comparison of T-dependent magnetization of LSMO films recorded by SQUID, magnetic circular dichroism (MCD) with 50 Å probing depth, and spin-resolved photoemission spectroscopy with 5 Å probing depth. (Reproduced with permission from Park et al., *Phys. Rev. Lett.* 81, 1953 [1998]. Copyright © 1998 American Physical Society. http://link. aps.org/doi/10.1103/PhysRevLett.81.1953.)

works. Figure 4.7c displays T-dependent device currents for the 20 nm rubrene device at bias 1 and 0.1 V. Similar T dependence of device conductance was previously reported for a thin layer of rubrene-based spin valve (Shim et al., 2008). But, the T-dependent change of device resistance in a thin layer (~5 nm) of rubrene-based device, reported by Shim et. al., 2008, is much weaker, suggesting the tunneling current prevails. The resistance of the LSMO electrode, which has strong T dependence (Dediu et al., 2002; Majumdar et al., 2008), can be ignored since it is less than 0.1% of the total junction resistance for the thick rubrene layer devices ($d \geq 20$ nm).

The strong T dependence of organic-based device resistances could be attributed to the T dependence of hopping transport, if the channel distance of the device is long ($d > 100$ nm). For 20 nm of the rubrene channel, the electric field (E) for the applied bias of 1 V could be about ~10^5 (V/cm), and the typical electric field dependent drift mobility (e.g., $\mu \sim \mu_0 \exp\sqrt{E/E_0}$ [Poole-Frenkel form]) in the bulk may be very high and the effect of phonon-assisted hopping could be small, leading to

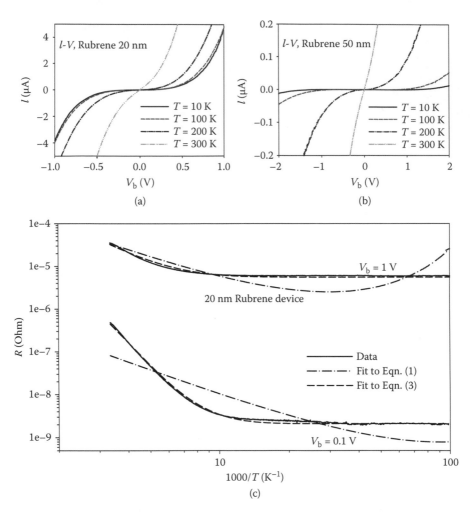

FIGURE 4.7 (a, b) T dependence of I-V curves for 20 nm and 50 nm rubrene devices at 10 K. (After Yoo et al., 2009.) (c) I vs. 1,000/T for the 20 nm rubrene device at $V_b = 0.1$, and 1 V. (After Yoo et al., 2009, 2010.) The device currents decrease exponentially as T decreases and nearly saturates below 100 K. We attribute such T-dependent device currents to metal-OSC interfacial resistances. The dash-dot lines are fit to empirical Equation 4.5 with parameters of $\alpha = 1.3$ and $\phi = 0.767$ eV (for $V_b = 0.1$ V) and 0.763 eV (for $V_b = 1$ V). The dashed lines are fit to Equation 4.6 of Kiveris and Pipinys with parameters of $a = 1.4$, $\alpha\omega = 0.034$ eV, $m^* = 1.2\, m_e$, and $\varepsilon_T = 0.678$ (for $V_b = 0.1$ V) and 0.660 (for $V_b = 1$ V).

weak T-dependent bulk resistance of the rubrene layer. For the case of intermediate thickness ($d \sim 20$ nm), we propose that strong T-dependent device currents mainly originate from activation at the metal-OSC interface barriers rather than bulk resistance (Yoo et al., 2009, 2010).

For a metal-semiconductor junction, the carrier injection across the barrier is generally described with the empirical formula for thermionic field emission as follows

(Brillson, 1993; Vilan et al., 2000):

$$I = I_s \exp\left(\frac{qV}{\alpha k_B T}\right)\left(1 - \exp\left(-\frac{qV}{k_B T}\right)\right) \tag{4.5}$$

where k_B is Boltzmann's constant, and α is the parameter that addresses deviation from ideal thermionic emission ($\alpha = 1$). I_s is the saturation current defined as

$$I_s = \sigma A^* T^2 \exp\left(\frac{-q\phi}{k_B T}\right)$$

where ϕ is the effective barrier height, A^* is the Richardson constant ($4\pi m k_B^2 e/h^3$), and σ is the junction area. The dash-dot (red) lines in Figure 4.7 show fits to Equation 4.5 for T-dependent device currents of 20 nm rubrene device at 1 V and 0.1 V. The fitting parameters are $\alpha = 1.3$ and $\phi = 0.767$ eV (for $V_b = 0.1$ V) and 0.763 eV (for $V_b = 1$ V). Equation 4.5 often well describes the I-V characteristic of hybrid junction devices (Vilan et al., 2000), but does not well explain the whole T range of device currents.

For a metal-OSC interface, a number of factors can introduce substantial change to the injection barrier (Zhan et al., 2007, 2008; Liu et al., 2009). The presence of a dipole barrier at the interface modifies energy-level alignment (Zhan et al., 2007, 2008). Possible chemical reactions and diffusion of metal atoms into the soft OSC layer (Vinzelberg et al., 2008; Liu et al., 2009) create defect states at the interface. Then, the injection of carriers at the metal-OSC interface will depend on release of trapped electrons and holes into the HOMO and LUMO levels of the OSC layer. Figure 4.8 illustrates a schematic view of thermionic field emission at the metal-OSC interface by

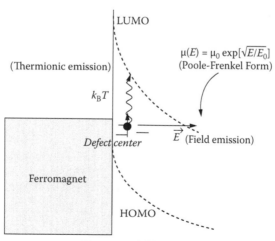

Thermionic field emission

FIGURE 4.8 Schematic view on thermionic field emission at the metal-OSC interface. The electron/holes at the interface are injected into HOMO/LUMO levels of OSC layer by both field emission and thermionic emission. At low T, field emission dominated carrier injection at the metal-OSC by applying high bias. At high T, even small applied bias could induce substantial carrier injection at the metal-OSC interface via thermionic emission. (After Yoo et al., 2009a.)

applying high bias. Injection into HOMO/LUMO can be mediated by phonons, especially at high temperatures. The thermionic emission decreases as the temperature is lowered. At very low temperature, field emission dominates the device currents. The thermionic field emission at the interfaces, together with phonon-assisted hopping in HOMO/LUMO levels, will introduce strong T-dependent device current reflecting thermal activation.

Here, we present an application of the theoretical model developed by Kiveris and Pipinys (Kiveris et al., 1976; Pipinys and Kiveris, 2005) to our organic-based spin valve devices (Yoo et al., 2009a). Their model described carrier injection at the metal-semiconductor interface by equally putting multiphonon activation and field emission of electrons from defect states at the interfaces to the conduction band of semiconductors (Kiveris et al., 1976). The phonon-assisted tunneling rate of electron under the electric field at the metal-semiconductor interface is as follows (Kiveris et al., 1976; Pipinys and Kiveris, 2005):

$$
W_T = \frac{eE}{(8m^*\varepsilon_T)^{1/2}}[(1+\gamma^2)^{1/2} - \gamma]^{1/2}[1+\gamma^2]^{-1/4}
$$

$$
\times \exp\left\{-\frac{4}{3}\frac{(2m^*)^{1/2}}{eE\hbar}\varepsilon_T^{3/2}[(1+\gamma^2)^{1/2} - \gamma]^2\left[(1+\gamma^2)^{1/2} + \frac{1}{2}\gamma\right]\right\},
$$

(4.6)

where

$$
\gamma = \frac{(2m^*)^{1/2}\Gamma^2}{8e\hbar E\varepsilon_T^{1/2}}
$$

Here ε_T is the energetic depth of defect states from the conduction band of semiconductor, $E = V_b/d$ is the applied electric field, $\Gamma^2 = 8a(\hbar\omega)^2(2n+1)$ is the width of the defect states broadened by the interaction with optical phonons, $n = 1/[\exp(\hbar\omega/k_BT) - 1]$, and a is the electron-phonon coupling constant ($a = \Gamma_0^2/8(\hbar\omega)^2$). The dashed (blue) lines in Figure 4.7 show fits to the 20 nm rubrene device current with Equation 4.6. Qualitatively good fit can be achieved over all T and V_b with fixed parameters of $a = 1.4, \hbar\omega = 0.034$ eV, $m^* = 1.2$ m_e $n = 1/[\exp(\hbar\omega/k_BT) - 1]$, and $\varepsilon_T = 0.678$ eV (for $V_b = 0.1$ V) and 0.660 eV (for $V_b = 1$ V). The results show that 20 nm rubrene device currents are controlled by the carrier injection as we anticipated. When the interface resistance dominates, the conductivity mismatch problem would not prohibit electrical detection of MR (Johnson and Silsbee, 1987, 1988; Rashba, 2000). At a high enough electric field in the bulk of rubrene, the spin-polarized drift currents are weakly susceptible to the phonon-assisted hopping, and MR of 20 nm rubrene devices relies more on the spin-dependent carrier injection at the interface. The phonon-assisted carrier injection is negligible below 100 K and starts to increase dramatically above 100 K, and could affect MR of the device at high T. This could also apply for the case of inorganic semiconductor channel spin valves, if the Schotty barrier at the interface dominates the device resistance.

Once the carriers are injected into an OSC layer, the mobility of electrons and holes in the bulk of the OSC layer also strongly relies on the applied electric field (e.g.,

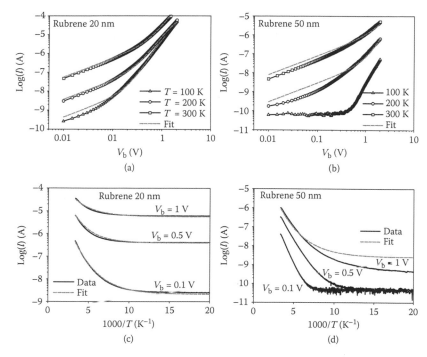

FIGURE 4.9 (a, b) Log$I$ vs. log$V$ plots for 20 and 50 nm rubrene devices for different T. (c, d) I vs. 1,000/T for 20 and 50 nm rubrene devices at V_b = 0.1, 0.5, and 1 V. (After Yoo et al., 2009a.) Dashed lines in (a–d) are fits to Equation 4.6 with fixed a = 1.4, $\hbar\omega$ = 0.034 eV, m^* = 1.2 m_e, and slight adjustment of ε_T = 0.67 ± 0.01 eV for each plot. ε_T = 0.660 (for 100 K in (a)), 0.666 (for 200 K in (a)), 0.674 (for 300 K in (a)), 0.673 (for 200 K in (b)), 0.662 (for 300 K in (b)), 0.678 (for 0.1 V in (c)), 0.674 (for 0.5 V in (c)), 0.660 (for 1 V in (c)), and 0.676 (for 1 V in (d)). Reproduced with permission from Yoo et al., *Phy. Rev. B.* 80, 205207 [2009]. Copyright© 2009 American Physical Society.

$\mu \sim \mu_0 \exp\sqrt{E/E_0}$ [Poole-Frenkel form]). Figure 4.9 shows fitting of Equation 4.6 to *I-V* curves and I vs. 1,000/T curves for rubrene 20 and 50 nm devices. The slight deviation of the fitting at low bias in Figure 4.9a originates from the bulk resistance of OSC, which becomes stronger as T is lowered. As the thickness of the rubrene layer is increased further, the effects of carrier transport in the bulk become more significant in the device current. A larger deviation for our fitting to 50 nm rubrene device currents can be observed at low bias and low T, as shown in Figure 4.9b and d. Fitting for the 50 nm rubrene device was performed with the same fixed parameters of a = 1.4, $\hbar\omega$ = 0.034 eV, m^* = 1.2 m_e, and slight adjustment of ε_T = 0.67 ± 0.01 eV for each plot. Unlike interfacial barrier resistance, which nearly saturates below 100 K for the relatively high bias, the phonon-assisted hopping conductance in the OSC layer continues to drop to a low T following an exponential T dependence. This introduces a strong deviation of our fitting in Figure 4.9d for T < 100 K. In sum, for the thicker rubrene layer device ($d \geq$ 50 nm), higher bias needs to be applied to support efficient field emission and field-driven drift currents (space charge limited regime,

$I \propto V^2/d^3$). Hopping transport over long distances and dephasing of space charge spin in the bulk will affect the polarization of carriers. Finally, a conductivity mismatch problem could make electrical detection of MR more difficult.

4.2.5.2 GMR in FM/OSC/FM Heterojunction

Figure 4.10a displays MR curves of different V_b at 10 K for the 20 nm rubrene device. Resistance of 20 nm rubrene devices at low bias (~ several mV) and low T (10 K) is extremely high due to negligible carrier injection. This suggests that one has to be careful in assessing determined fundamental properties, such as spin relaxation time and diffusion length derived from low-bias (~ several mV) measurements in previous device works with thick OSC layers (Xiong et al., 2004; Pramanik et al., 2007). For a 20 nm rubrene device, clear positive MR signals were detected at 10 K for applied bias of 200 mV above. An increase of MR was observed up to 600 mV, then MR decreased as V_b increased further (Figure 4.10b). For the 30 nm rubrene device, MR at 10 K started to be measured when the V_b was increased up to 500 mV, then became negligible when V_b was as high as 1 V (Figure 4.10b). No MR was observed for 40 and 50 nm rubrene devices over all T and V_b (up to several V). According to a theoretical study of Ruden and Smith (2004), applying very high bias could introduce observation of MR for long OSC channel spin valve devices, since it will promote field emission and increase mobility of carriers and also spin diffusion length.

MR curves of the 20 nm rubrene device for different T at $V_b = 1$ V are presented in Figure 4.10c. The steps of MR curves at different T well correspond to the H_c of Fe and LSMO as presented in Figure 4.6b. The T dependence values of MR for 20 and 30 nm rubrene devices are presented in Figure 4.10d and compared with the T dependence values of TMR for 5 nm rubene device. No clear MR was observed over 150 K for the 20 nm rubrene device. A rapid drop in MR as T increased over 100 K for the 20 nm rubrene device can be associated with the thermionic emission at the interface of the device. For the 30 nm rubrene device, MR disappears above 75 K. The absence of MR for thicker rubrene devices ($d > 30$ nm) could be attributed to hopping transport and thermionic emission, especially at high T, and also a conductivity mismatch problem (Johnson and Silsbee, 1987, 1988; Schmidt et al., 2000; Fert and Jaffrès, 2001).

4.2.5.3 Effects of Thermionic Emission on GMR
in FM/OSC/FM Heterojunction

The spin diffusion lengths (λ_s) in OSCs often were estimated (Xiong et al., 2004; Wang et al., 2007; Ikegami et al., 2008; Han et al., 2006) using Equation 4.3 of extended Julliere's model (1975) with the addition input λ_s. Though the model ignores the effects of interfaces, spin diffusion length could be roughly estimated with the consistent study of channel thickness dependence. However, the nature of strong field-dependent drift mobilities in OSC materials suggests the spin diffusion length (λ_s) is no longer an intrinsic parameter. The spin diffusion lengths of OSC layers are also presumed to have T dependence, because the carrier transport always involves interaction with phonons.

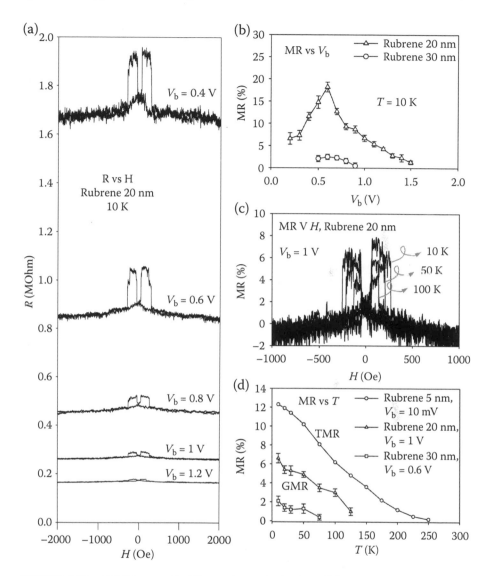

FIGURE 4.10 (a) MR curves for 20 nm rubrene device for a different V_b at 10 K. (b) MR vs. V_b plot for 20 and 30 nm rubrene devices at 10 K. (c) MR curves of 20 nm rubrene device for different T at $V_b = 1$ V. (d) Comparison of T dependence of MR between TMR (5 nm rubrene device) and GMR (20 and 30 nm rubrene devices). Reproduced with permission from Yoo et al., *Phys. Rev. B.* 80, 205207 [2009]. Copyright©, 2006 American Physical Society.

Here, we emphasize the effect of carrier injection at the metal-OSC interface. We ignore the effects of spin depolarization due to transport through the 20 nm rubrene channel at high bias ($V_b = 1$ V). If we assume that the effects of T-dependent spin polarization of magnetic layers on MR (especially surface spin polarization of LSMO) could be subtracted by simply dividing with T-dependent

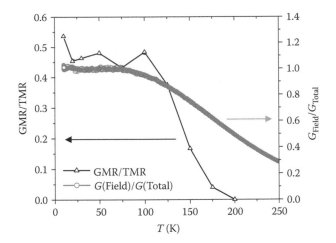

FIGURE 4.11 T dependence of the ratio of GMR (20 nm rubrene device at $V_b = 1$) to TMR (5 nm rubrene device at $V_b = 10$ mV) compared to the ratio of field emission to the total device current at $V_b = 1$ V for the 20 nm rubrene device. We suggest the rapid drop of GMR over 100 K in the 20 nm rubrene device is partly due to spin depolarization at the metal-OSC interface by the thermionic emission. (After Yoo et al., 2009, 2010.)

TMR values, then the T-dependent ratio of GMR to TMR could suggest T dependence of spin depolarization at the metal-OSC interface, which is associated with thermionic emission. The thermionic emission in our 20 nm rubrene device is negligibly below 100 K and starts to increase drastically as T increases over 100 K, whereas the field emission should be T independent. Figure 4.11 displays the T dependence of the ratio of GMR (20 nm rubrene device at $V_b = 1$ V) to TMR (5 nm rubrene device at $V_b = 10$ mV) and ratio of the field emission conductance to the total device conductance at $V_b = 1$ V for the 20 nm rubrene device. The field emission conductance is taken as a constant value of device current at 10 K. The rapid drop of the ratio of GMR/TMR over 100 K could be partly associated with spin depolarization at the interface due to thermionic emission.

4.2.6 SUMMARY AND CONCLUSIONS

In this section, we presented an extensive study on both TMR and GMR in FM/OSC/FM heterojunction using rubrene as a prototype OSC spacer. The thin layer ($d = 5$ nm) of rubrene devices displays efficient spin-conserved tunneling through the OSC layer, as studied previously (Santos et al., 2007; Shim et al., 2008). The TMR behavior for a thin OSC-layered device shows a maximum at near zero bias and exponentially decreases as V_b increases, similar to conventional TMR devices.

As the thickness of the rubrene layer is increased, devices are strongly limited by carrier injection and bulk resistivity. Therefore, low-bias junction resistance for thick OSC-based devices at low T is significantly high. The T dependence of device current is associated with activation leading to strong T-dependent device currents. The activation at the metal-OSC interface dominates the device current up to a certain thickness of OSC layer (for amorphous rubrene, ~ 20 nm). The carrier injection at the metal-OSC interface is associated with the thermionic field emission and can be well described with the phonon-assisted tunneling theory developed by Kiveris and Pipinys. When the interface resistance dominates the device current, the conductivity mismatch problem might not be serious for the electrical detection of MR. However, the thermionic emission at the interface could reduce spin-polarized carrier injection above 100 K. In contrast to TMR behavior, the GMR in an OSC-based spin valve can be measured only at high V_b when T is low, since the device current is limited by carrier injection. The observed GMR shows strong T dependence originating from spin injection and transport in the rubrene layer.

Depending on mobilities of the OSC layers, bulk transport starts to control the device current as d increases over a certain thickness. When the bulk resistance dominates the device resistance, a higher bias should be applied for viable device current (space charge limited regime). The effect of the hopping transport becomes more important in efficient spin transport, and the conductivity mismatch problem would be more serious in electrical detection of MR as d is increased further.

In conclusion, the detailed V_b and T-dependent study of the GMR device clearly demonstrated valid GMR effects via spin-polarized carrier injection and subsequent spin transport into/in the OSC spacer. We explained the carrier injection and transport into/in OSC and their impact on the GMR in an OSC-based spin valve in terms of phonon-assisted tunneling and hopping. Further development of crystalinities of OSC spacer films and mobilities in the bulk of OSC spacer and the interfacial engineering will significantly advance the spintronic application of OSC materials. Observation of the spin precession in OSC layer will signify unambiguous proof of the electrical detection of the spin transport in the OSC layers, which may require substantial improvement of the material's mobility and interfacial quality.

4.3 ORGANIC-BASED MAGNET: NEW PHENOMENA AND SPINTRONIC APPLICATIONS

From the discovery in ancient time, magnets and their applications have been expanded to be ubiquitous in our present life. Various magnetic materials have been primarily used independently in a number of devices. Introduction of a new concept of magnetic devices, such as spin-electronic applications and solid-state quantum computing qubits, derived enormous attention in developing new magnetic materials for desirable magnetic and electronic properties. These new magnetic materials have been formed from alloys of various metals and even organic materials. Magnetism as a field of study is clearly enjoying a renaissance, due to the development of a variety of new materials and the emergence of spintronics.

4.3.1 ORGANIC/MOLECULE-BASED MAGNET

One class, which is both easily processable and magnetic on a microscopic scale, is a novel set of materials known as molecule-based magnets. While traditional magnets owe their magnetization to unpaired electrons in d or f orbitals, molecule-based magnets contain electrons in molecular orbitals, consisting of superpositions of p and even s atomic orbitals, which play a crucial role in magnetic ordering. Organic-based magnets are a specific subdivision of molecule-based magnets that have unpaired spin localized on an organic species, such as 4-nitrophenyl nitronyl nitroxide (Kinoshita, 1994; Tamura et al., 1991) and a complex or fullerene C_{60} with tetrakis(dimethylamino)ethylene $[(CH_3)_2C]_2$ (Allemand et al., 1991), which exhibit magnetic ordering at 0.6 and 16.1 K, respectively.

Due to flexible chemical processing, the magnetic properties of these materials can be tuned to meet the needs of the application, creating "magnets by design" (Miller and Epstein, 1994, 1995). This could open almost endless possibilities for design of materials with desired magnetic properties. Various spin values can be obtained by changing the transition metal ion used or the number of metal ions. By changing the organic ligand, it is possible to obtain a variety of both structural and magnetic lattices, with different dimensionalities and types of spins. This choice of ligand can also affect the relative strength of relevant magnetic interactions, for instance, direct exchange, indirect exchange, and dipole-dipole interactions.

4.3.2 NEW PHENOMENA AND OPPORTUNITIES IN MOLECULE/ORGANIC-BASED MAGNETS

The flexible, low-temperature, and low-cost synthesis of these materials is perhaps their greatest advantage over traditional metal-based magnets. In addition, their low density and biocompatibility make them attractive candidates to replace conventional magnets in certain applications. To be considered for use in certain applications, these materials must meet the demands placed on the new small-scale devices. Indeed, chemical processing also allows for fabrication of magnetic materials in various forms for future use in multilayer structures. The films of $M[TCNE]_x$ magnets were developed by low-temperature chemical vapor deposition (CVD) (Pokhodnya et al., 2000, 2003; de Caro et al., 2000). A variety of organometallic magnetic films were also developed by Langmuir–Blodgett methods (Seip et al., 1997; Petruska et al., 2002; Culp et al., 2003; Talham et al., 2008). In addition, single molecular magnets (Lis, 1980; Sessoli et al., 1993; Gatteschi et al., 1994) have been synthesized that exhibit magnetic behavior at the molecular level. While the processability of organic-based magnets is attractive for application purposes, this chemical tuning also allows these magnets to be used for purely scientific endeavors. The ability to control both the magnetic and structural properties of these materials through organic, organometallic, and coordination metal chemistry helps create an experimental platform to test various theoretical models (Epstein, 2003). These materials can exhibit many interesting phenomena, such as low dimensionality (Miller et al., 1992; Hibbs et al., 2001; Ma et al., 2001; Etzkorn et al., 2002; Pokhodnya et al., 2006; Yoo et al., 2008), magnetic chirality (Coronado et al., 2001, 2002; Inoue et al., 2003; Imai et al., 2004; Gao et al., 2004), and spin frustration (Fujita and Awaga, 1996; Gîrţu et al., 1998, 2000).

A particularly interesting phenomenon in this class of magnets is *magnetic bistability*. Notable examples are spin-crossover complexes, which exhibit high-spin and low-spin thermal transitions (Gütlich and Goodwin, 2004; Kahn and Martinez, 1998); high-spin complexes, which exhibit macroscopic quantum tunneling of magnetization

(Friedman et al., 1996; Thomas et al., 1996); and mixed ferro-ferrimagnetic Prussian blue analogs, which exhibit multiple compensation temperatures (Sato et al., 1996a; Ohkoshi et al., 1999). In addition, their magnetic bistabilities often allow light control (Decurtins et al., 1984; Gütlich and Goodwin, 2004; Kahn and Martinez, 1998; Sato et al., 1996b; Ohkoshi et al., 1997, 2006; Pejaković et al., 2000, 2002; Bleuzen et al., 2000; Tokoro et al., 2003; Park et al., 2004; Hozumi et al., 2005; Yoo et al., 2006, 2007).

During the last decades, the control of magnetic properties in these molecule-based magnets using light has received broad attention due to the needs of modern technology, which searches for "smart," adaptable materials, responsive to external stimuli. In particular, as the information technology's ever-increasing need for faster information processing speeds and higher storage densities requires new ways to write, erase, and read information, the possibility to manipulate information bits at a microscopic (molecular) level by light beams is very attractive.

4.3.3 TCNE-Based Magnet

Tetracyanoethylene [TCNE] has been one of the most important building blocks for organic-based magnets. Among TCNE-based magnets are $[FeCp_2^*]^+[TCNE]^-$ (Cp_2^* = pentamethylcyclopentadyenyl), the first magnet with spins residing in a p orbital (Miller et al., 1985, 1986, 1987, 1988a, 1988b; Chitteppeddi et al., 1987); $[MnCp_2^*]^+[TCNE]^-$ (Yee et al., 1991); and $[MnTPP]^+[TCNE]^-$ (TPP = *meso*-tetra phenylporphyrin) (Miller et al., 1992; Zhou et al., 1993a; Brinckerhoff et al., 1996).

TCNE is a very strong electron acceptor. It can accept an electron in its antibonding π^* orbital, forming charge transfer salts in which it is present as the $[TCNE]^-$ ion. This unpaired π^* electron produces a net spin of 1/2 on $[TCNE]^-$. A single-crystal polarized neutron diffraction study of $[Bu_4N]^+[TCNE]^-$ (Bu_4N = tetra-*n*-butylammonium) paramagnetic charge transfer salt determined the spin density distribution in $[TCNE]^-$ (Zheludev et al., 1994). The spin of 1/2 is delocalized over the entire $[TCNE]^-$ molecular ion, with 33% of the total spin on each of the sp^2 hybridized carbon atoms (central C atoms in Figure 4.12), 13% of the spin on each of the four

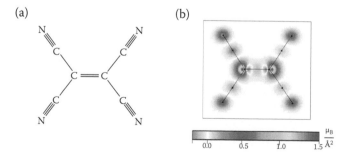

FIGURE 4.12 (a) Schematic structure of tetracyanoethylene [TCNE]. The molecule, as well as its monoanion $[TCNE]^-$, is planar (D_{2h} symmetry). (From Dixon and Miller, *J. Am. Chem. Soc.* 109, 3656 [1987].) (b) The spin distribution of π^* orbital in $[TCNE]^-$ anion examined by neutron diffraction. (Reproduced with permission from Zheludev et al., *J. Am. Chem. Soc.* 116, 7243 [1994]. Copyright © 1994 American Chemical Society.)

nitrogen atoms, and a negative corresponding to about 5% of the total spin on each of the four sp hybridized carbon atoms (see Figure 4.12) (Zheludev et al., 1994; Miller and Epstein, 1995).

The family of molecule-based magnets M[TCNE]$_x$·y(solvent) (M = transition metal) was introduced by the synthesis of V[TCNE]$_x$·y(CH$_2$Cl$_2$), the first room temperature molecule-based magnet, with a magnetic ordering temperature of ~ 400 K (Manriquez et al., 1991). The high magnetic ordering temperature was proposed to result from the direct bonding of the TCNE N atoms to V and a three-dimensional network structure in the solid (Manriquez et al., 1991). This compound is amorphous, chemically unstable, and extremely air sensitive (Manriquez et al., 1991).

Later on, other M[TCNE]$_x$·y(CH$_2$Cl$_2$) (M = Fe, Mn, Co, Ni) magnets have been synthesized (Zhang et al., 1998a). These materials also display relatively high magnetic ordering temperatures, ranging from ~44 K for M = Co, Ni to ~121 K for M = Fe. The infrared absorption spectra in the region of CN stretching vibrations are consistent with metal-coordinated [TCNE]$^-$ for all the compounds. In contrast to the disordered V[TCNE]$_x$·y(CH$_2$Cl$_2$), M = Mn, Fe compounds give relatively sharp x-ray powder diffraction patterns, indicating that they are crystalline. Furthermore, similarity of the diffractograms for these two compounds indicates that they are isostructural (Zhang et al., 1998b).

The structure of M-TCNE class magnet ranges from 1D chain to 3D network structures (Figure 4.13). For example, [MnTPP][TCNE] (TPP = tetraphenyl-porphyrin) exhibits a 1D chain structure of alternating donors (MnTPP) and acceptors (TCNE) with large interchain spacings (Miller et al., 1992; Hibbs et al., 2001). The dipolar interchain interaction in [MnTPP][TCNE] introduces a quasi-1D fractal cluster phase at low temperature (below 4 K) (Etzkorn et al., 2002). Recently synthesized [Fe[TCNE](NCMe)$_2$][FeCl$_4$] shows a 2D layered structure (Pokhodnya et al., 2006). The structure consists of undulating layers composed of Fe^{II} ions with a μ_4-[TCNE]$^-$ bridging within the layer and two axial MeCN coordinations (Pokhodnya et al., 2006). The absence of bridging ligands between the layers suggests novel sets of a 2D Ising layered system (Pokhodnya et al., 2006; Yoo et al., 2008). The solution-prepared Fe[TCNE]$_x$·y(CH$_2$Cl$_2$) powder is the only one of the M[TCNE]$_x$·y(CH$_2$Cl$_2$) magnets whose crystal structure was determined via high-resolution x-ray powder diffraction (XRPD) (Her et al., 2007). As seen in Figure 4.13, the structure of Fe[TCNE]$_x$ consists of six N atoms octahedrally coordinated to Fe^{II} (Her et al., 2007). Each Fe^{II} ion bonds to four μ_4-[TCNE]$^-$ anions in layers, similar to the case of [Fe[TCNE](NCMe)$_2$][FeCl$_4$]. These [Fe^{II}[TCNE]$^-$] layers are interconnected by μ_4-[C$_4$(CN)$_8$]$^{2-}$ ligands with an Fe-N bond length of 2.16(1)° (Her et al., 2007).

4.3.4 Organic-Based Magnetic Semiconductor V[TCNE]$_x$

The powder sample of V[TCNE]$_x$·yS (x ~ 2; S = CH$_2$Cl$_2$, CH$_3$CN, THF) can be prepared in various chemical routes leading to a broad range of T_c (T_c ~ 120 K for CH$_3$CN, T_c ~ 200 K for THF, and T_c ~ 400 K for CH$_2$Cl$_2$) (Brinckerhoff et al., 1995). By coordinating with V, the spinless solvent can directly affect the strength of the effective magnetic coupling between molecular units, leading to a critical

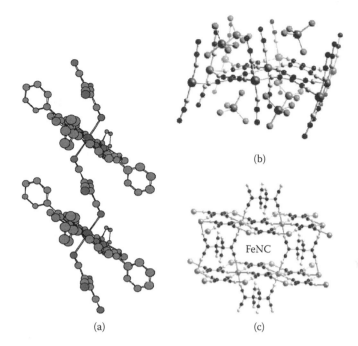

(b)

FeNC

(a)   (c)

FIGURE 4.13 (a) 1D structure of [MnTPP][TCNE] showing [MnTPP] interconnected via trans-μ-[TCNE] coordination. (Reproduced with permission from Hibbs et al., *Inorg. Chem.* 40, 1915 [2001]. Copyright © 2001 American Chemical Society.) (b) 2D structure of [Fe[TCNE](NCMe)$_2$][FeCl$_4$] showing μ_4-[TCNE]$^-$ coordination in plain and axial coordination of MeCN. (Reprinted with permission from Pokhodnya et al., *J. Am. Chem. Soc.* 128, 15592 [2006]. Copyright © 2001 American Chemical Society.) (c) 3D structure of Fe[TCNE]$_x$·y(CH$_2$Cl$_2$) showing μ_4-[TCNE]$^-$ coordination in plain and interlayer coordination of μ_4-[C$_4$(CN)$_8$]$^{2-}$ ligands. (Reprinted with permission from Her et al., *Angew. Chem. Int. Ed.* 46, 1521 [2007]. Copyright © 2007 John Wiley & Sons, Inc.)

temperature ranging from 100 K to 400 K. The presence of solvent can also drastically increase the degree of disorder in the system, leading to a spin-glass-like random magnetic order. Even use of noncoordinating solvents, such as CH$_2$Cl$_2$, PhMe, PhCF$_3$, hexane, and C$_6$H$_{14}$, introduces a change of *T*-dependent magnetic behaviors (Thorum et al., 2006).

Partial substitution of V with Co or Fe in solution-prepared V[TCNE]$_x$·y(CH$_2$Cl$_2$) introduced the flexible ability of controlling magnetic properties (Pokhodnya et al., 2003, 2004). This bimetallic M$_z$V$_{1-z}$[TCNE]$_x$·y(CH$_2$Cl$_2$) (M = Co, Fe; $0 < z < 1$; $x \sim 2$) displays a broad range of magnetic transition temperatures, 150 K < T_c < 400 K, and coercivity, 5 Oe < H_c < 400 Oe, depending on concentration z (Pokhodnya et al., 2003, 2004). Since the V[TCNE]$_x$ is an amorphous structure, the random site occupancy of either Co or Fe within a noncrystalline lattice increases the overall randomness, leading to an increase in H_c (Pokhodnya et al., 2003, 2004). And this random substitution of V with Co or Fe also minimizes phase separation and displays a more gradual transition (Pokhodnya et al., 2003, 2004).

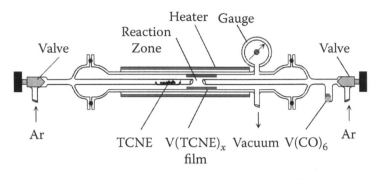

$$x\text{TCNE} + \text{V(CO)}_6 \rightarrow \text{V(TCNE)}_x + 6\,\text{CO}\uparrow$$

FIGURE 4.14 Schematic illustration of the CVD apparatus to prepare thin films of the V[TCNE]$_x$ magnet. (Reproduced with permission from Pokhodnya et al., *Adv. Mater.* 12, 410 [2000]. Copyright © 2000 John Wiley & Sons.)

Thin films of V[TCNE]$_x$ (solvent-free) have been synthesized via the chemical vapor deposition (CVD) method (Pokhodnya et al., 2000), which has a greatly diminished oxygen/moisture sensitivity and increased crystallinity compared to powder samples (Pokhodnya et al., 2000, 2001). The magnetic properties of CVD-prepared V[TCNE]$_x$ film are similar to those of the solution-prepared powder sample, featuring $T_c \sim 400$ K and a small coercive field, $H_c = 4.5$ G, at 300 K (Pokhodnya et al., 2000, 2001). Though the CVD-deposited V[TCNE]$_x$ film still does not have long-range structural order, the local coordination environment of V has been determined by extended x-ray absorption fine structure (EXAFS) analysis (Haskel et al., 2004). The studies show that V ions have a valence state near +2 and are coordinated by 6.04 ± 0.25 nitrogen atoms at room temperature with an average distance of 2.084(5) Å (Haskel et al., 2004).

Based on the surface analysis (XPS, EXAFS, and XANES), IR spectra, elemental composition, and magnetic data, Miller (2009) suggested the simplest formulation of V[TCNE]^-_x[TCNE]$^{2-}_{1-z/2}$ ($1 < z < 2$) for V[TCNE]$_x$ magnet. The materials are supposed to be a glassy structure with a local crystallite order of the combination of two limiting cases ($z = 1$ and $z = 2$). For both $z = 1$ and $z = 2$, the idealized structure is embodied with an undulating 2D layered structure with μ_4-[TCNE]$^-$ coordination with V^{II}, similar to the case of [Fe[TCNE](NCMe)$_2$][FeCl$_4$] and Fe[TCNE]$_x$·y(CH$_2$Cl$_2$). But, the interlayer connection is mediated by μ_4-[TCNE]$^{2-}$ for $z = 1$ and *trans*-μ-[TCNE]$^-$ for $z = 2$, as shown in Figure 4.15.

For the limiting case $z = 2$, V[TCNE]$_2$ is often presumed for describing magnetic and transport study (Prigodin et al., 2002). The magnetic interaction of V(TCNE)$_2$ is provided by the direct exchange coupling between the nearest neighbor of metal ion and the organic ligand via hybrid d-π states delocalized in both V and [TNCE] sites. Then, the antiferromagnetic exchange between two [TCNE]$^-$ ($S = 1/2$) and V^{II} ($S = 3/2$) leaves one uncompensated spin, i.e., ferrimagnet. In this limiting case,

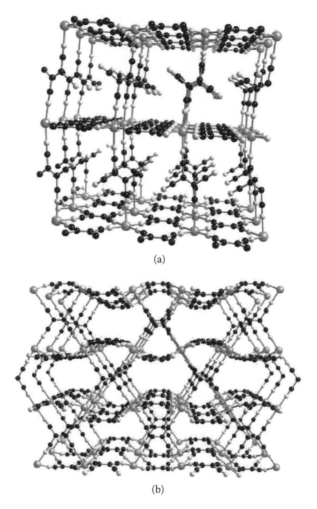

(a)

(b)

FIGURE 4.15 (a) An idealized proposed 3D structure of V[TCNE]$_x$ ($x = 2$) showing layers of μ_4-[TCNE]$^-$ coordination interconnected via *trans-μ*-[TCNE]$^-$. (After Miller, 2009.) (b) An idealized proposed 3D structure of V[TCNE]$^-$[TCNE]$^{2-}_{1/2}$ showing layers of μ_4-[TCNE]$^-$ coordination interconnected via μ_4-[TCNE]$^{2-}$. (After Miller, 2009.) (Reprinted with permission from Miller, J. S., *Polyhedron* 28, 1596 [2009]. Copyright © 2009 Elsevier.)

when all [TCNE] carrying spin 1/2, the coulomb gap induces splitting of π^* levels, leading to fully spin-polarized HOMO and LUMO levels, i.e., half semiconductor (Prigodin et al., 2002), indicating potential application for spintronics.

Recently, Miller (2009) suggested the presence of [TCNE]$^{2-}$ in V(TUNE)$_x$ based on the IR spectra. The presence of spinless [TCNE]$^{2-}$ will introduce saturation magnetization higher than 1 μ_B, which was observed in a sample-to-sample variation (Miller, 2009). In sum, the material was ascribed to be nonstoichiometric and inhomogeneous with two randomly distributed, limiting, small spatial domain structures (Miller, 2009). The idealized geometries were often employed for theoretical study

(Tchougréeff and Dronskowski, 2008; Erdin, 2008; Fusco et al., 2009). These studies support antiferromagnetic coupling between spins in V^{II} and spins in [TNCE]⁻ via direct exchange.

A number of surface studies as well as theoretical analyses have been dedicated to unravel exotic electronic states in $V[TCNE]_{x-2}$. A proposed model by Prigodin et al. (2002) suggested highly spin-polarized valence and conduction bands due to coulomb gap-split π^* levels in [TCNE]⁻. Based on a model system of rubidium intercalated $Rb^+[TCNE]^-$, Tengstedt et al. (2004) estimated the on-site coulomb interaction–induced splitting of π^* orbital of [TCNE]⁻ to be ~2 eV. The d levels of V^{II} lie 1.5 eV above the π^* levels in [TCNE]⁻ based on a combination of photoelectron spectroscopy (PES) and resonant photoemission (RPE) results (Tengstedt et al., 2006). With the consideration of estimated large crystal field splitting, $10Dq \sim 2.3$ eV (de Jong et al., 2007), the LUMO of $V[TCNE]_x$ is ascribed to be $\pi^* + U_c$ levels of [TCNE]⁻ (Tengstedt et al., 2006). Figure 4.16 shows a schematic of a level diagram for $V[TCNE]_x$ suggesting highly spin-polarized HOMO and LUMO levels. The spin-polarized nature of electronic states was also experimentally confirmed via magnetic circular dichroism (MCD) in both HOMO of $V[TCNE]_x$ derived from d levels of V^{II} (Kortright et al., 2008; Tengstedt et al., 2006; de Jong et al., 2007) and π^* levels of [TCNE]⁻ (Kortright et al., 2008). It was also noticed that the HOMOs of $V[TCNE]_x$ are strongly hybridized (Tengstedt et al., 2006; de Jong et al., 2007; Kortright et al., 2008) via covalent σ bonding (Kortright et al., 2008), but the LUMO levels are rather to be localized molecular levels of [TCNE]⁻ (Kortright et al., 2008; Carlegrim et al., 2008a) due to π backbonding (Kortright et al., 2008). Figure 4.16 displays a schematic level diagram of $V[TCNE]_x$ showing spin polarized majority and minority subbands. Recent numerical studies based on density functional theory for $V[TUNE]_x$ (Tchougréeff and Dronskowski, 2008; Erdin, 2008; Fusco et al., 2009) and $[Fe[TUNE](MeCN)_2][FeCl_4]$ (Caruso et al., 2009; Shaw et al., 2009) shows spin polarized valence and conduction subbands suggesting potential applications of these materials for spintronics.

While direct exchange is broadly considered to be dominant in magnetic coupling in $V[TCNE]_x$, other couplings might also affect magnetic properties of this material,

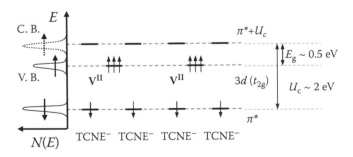

FIGURE 4.16 Schematic-level diagram for $V[TCNE]_x$ (Prigodin et al., 2002). Due to on-site coulomb repulsion, π^* levels of [TCNE]⁻ split into two Hubbard sublevels. For the idealized limiting case, $V[TCNE]_2$, the HOMO and LUMO levels are speculated to be fully spin polarized. (Reproduced with permission from Prigodin et al., *Adv. Mater.* 14, 1230 [2002]. Copyright © 2002 John Wiley & Sons.)

such as double exchange between the spins in the sublattice of V^{II} or $[TCNE]^-$, respectively and superexchange between spins in V^{II} via spinless $[TCNE]^{2-}$. Detailed magnetic study in solution-prepared $V[TCNE]_x \cdot yS$ powder suggests reentrant spin glass behavior (Zhou et al., 1993b; Morin et al., 1993) reflecting the amorphous nature of the material structure. The possible origin of the glass state could be random spatial distribution of magnetic exchange interaction or random distribution of anisotropy axes. While the former is considered a main source of spin glass phase, the latter can also solely induce a glass phase when the anisotropy energy is stronger than the exchange energy (Chudnovsky and Serota, 1982; Chudnovsky, 1986; Kaneyoshi, 1992).

The amorphous nature of $V[TCNE]_x$ introduces both spatial fluctuation of exchange and random magnetic anisotropy. When the anisotropy is comparable to exchange, Chudnovsky predicted a formation of different magnetic phases in the material, depending on the strength of an external magnetic field. Depending on the degree of anisotropy, a correlated spin glass (CGS) phase could appear at low field, while in the intermediate field a small crystallite domain with a locally distributed anisotropy axis, the so-called ferromagnetic wandering axis (FWA) phase, would be formed (Chudnovsky and Serota, 1982; Chudnovsky, 1986, 1989). At high enough field, the system will be near collinear ferromagnetic or paramagnetic phase (Chudnovsky and Serota, 1982; Chudnovsky, 1986, 1989). Here, when the anisotropy is not strong enough, the CGS phase will no longer exist (Chudnovsky and Serota, 1982; Chudnovsky, 1986, 1989).

The behavior of $M(H,T)$ in $V(TCNE)_x \cdot yS$ was successfully explained in the frame of this so-called random magnetic anisotropy (RMA) model (Zhou et al., 1993b; Morin et al., 1993). However, the CVD-deposited solvent-free $V(TCNE)_x$ films show substantial improvement in structural order. Though the nature of amorphous structure suggests the CGS phase as $H \rightarrow 0$, no frequency-dependent ac susceptibility was found in CVD-deposited films of $V(TCNE)_x$ (Pokhodnya et al., 2001).

4.3.5 SPIN-DRIVEN MAGNETORESISTANCE OF $V[TCNE]_x$

In this section, we present magnetotransport studies of $V[TCNE]_x$ films to evaluate its potential applicability for spintronic applications. We report the observation of an increase in resistance with applied magnetic field (positive magnetoresistance) for $V[TCNE]_x$ ($x \sim 2$) at room temperature, up to ~0.7% at $H = 6$ kG. Squeezing of electron wave functions by magnetic field (a mechanism for the positive magnetoresistance in conventional disordered semiconductors) predicts a magnetoresistance three orders of magnitude less than the observed value. We have shown that the observed anomalous value of magnetoresistance in $V[TCNE]_x$ and its variations with magnetic field and temperature are consistent with spin-polarized charge transport. It is critical for spintronic applications to observe that the charge conduction is provided by spin-polarized carriers.

4.3.5.1 Magnetotransport Data and Discussion

The samples fabricated as solvent-free CVD-prepared thin films were characterized magnetically with electronic paramagnetic resonance (EPR). Figure 4.17 shows

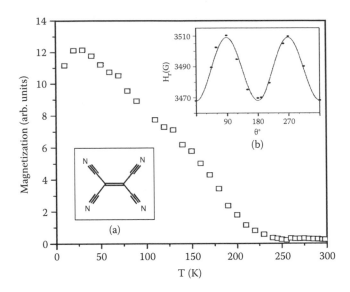

FIGURE 4.17   Magnetization (from EPR) of a V[TCNE]$_x$ film vs T for $H = 3{,}510$ G directed perpendicular to a film. Insets: (a) Structure of TCNE; (b) anisotropy of EPR resonance field H_r at $T = 100$ K. (Reproduced with permission from Prigodin et al., *Adv. Mater.* 14, 1230 [2002]. Copyright © 2002 John Wiley & Sons.)

the magnetization (a result of integrating the EPR signal) of a film vs. temperature. Study of the lower Curie temperature, T_c, sample ($T_c \sim 230$ K) enables us to probe magnetotransport above and below T_c. Below T_c there is anisotropy in the resonance magnetic field (Figure 4.17, inset b) due to the dependence of the demagnetization factor on the orientation of the magnetic field to the film.

Figure 4.18 shows the direct current (dc) conductivity as a function of temperature for a film of $T_c \sim 280$ K. The conductivity temperature dependence varies little with change in T_c. There is a fit to Mott's 3D variable range hopping (VRH) for local-ized charge carriers (Shklovskii and Efros, 1984): $\sigma dc \sim T^{-1/2} \exp[-(T_0/T)^{1/4}]$ with $T_0 = 2.0 \times 10^9$ K. Here $k_B T_0 = B/[N(\varepsilon_F)\xi^3]$, where $B = 21.2$, $N(\varepsilon_F)$ is the density of states at the Fermi level, and ξ is the localization radius. Assuming $\xi \sim 0.5$ nm (size of TCNE), we find $N(\varepsilon_F) \sim 1.0 \times 10^{18}$ (eV)$^{-1}$ × cm^{-3} too low for the charge density obtained from space-filling arguments $n \sim 1/(0.5$ nm$)^3 \sim 10^{22}$ cm^{-3}. Therefore, we replot the T for a simple activation law and estimate the energy gap as $\Delta E \sim 0.5$ eV (Figure 4.18, inset).

Figure 4.19 shows the magnetoresistance of V[TCNE]$_x$, MR = $[\rho(H,T) - \rho(0,T)]/\rho(0,T)$, as a function of temperature for a film with $T_c = 235$ K. The applied field $H = 6$ kG is parallel to the film. The same MR is observed for the magnetic field perpendicular to the film. The maximum of MR(T) is below but close to T_c. The inset in Figure 4.19 illustrates the H dependence of MR for $T = 215$ K and a magnetic field parallel to the film. Similar linear field dependence was obtained for all temperatures below T_c and samples with different T_c. The positive magnetoresistance of conven-tional nonmagnetic semiconductors with $\rho(T) = \rho_0 \exp[(T_0/T)^{1/4}]$ is accounted for by

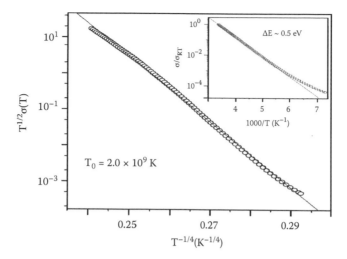

FIGURE 4.18 Conductivity of V[TCNE]$_x$ film (T_c ~ 280 K) plotted as $T^{1/2}\sigma(T)$ vs. $T^{-1/4}$ (3D VRH). Inset: Fit of $\sigma(T)$ to the activated (Ahrrenius) behavior. (Reproduced with permission from Prigodin et al., *Adv. Mater.* 14, 1230 [2002]. Copyright © 2002 John Wiley & Sons.)

a model of squeezing localized wave functions in the presence of the magnetic field. For a weak field the theory yields (Shklovskii and Efros, 1984) $\ln[\rho(H,T)/\rho(0,T)]$ = $A(\xi/\lambda)^4(T_0/T)^{3/4}$, where $A = 5/2016$ and λ is the magnetic length, $\lambda = (c\hbar/eH)^{1/2}$. The regime of weak field is valid so long as $\lambda \gg \xi$. For V[TCNE]$_x$ the localization length ξ ~ 0.5 nm, $\lambda = 33.1$ nm for $H = 6$ kG, and experimentally $T_0 = 2.0 \times 10^9$ K.

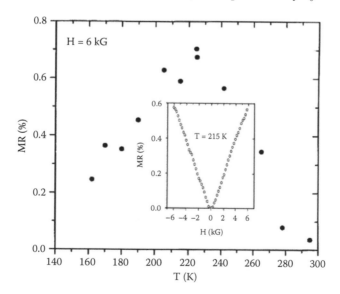

FIGURE 4.19 Temperature dependence of magnetoresistance in V[TCNE]$_x$ film for plane of the film parallel to the applied field $H = 6$ kG for the sample of Figure 4.2. Inset: MR(H) for $T = 215$ K. (Reproduced with permission from Prigodin et al., *Adv. Mater.* 14, 1230 [2002]. Copyright © 2002 John Wiley & Sons.)

For $T = 215$ K and $H = 6$ kG we find $\ln[\rho(H,T)/\rho(0,T)] = 3.6 \times 10^{-6}$; i.e., the MR is very small, $\sim 3.6 \times 10^{-6}$. We tested this conclusion using sulfonated polyaniline (Lee et al., 1999) with hopping transport and no MR was detected at these temperatures. Thus, the observed MR $\sim 6 \times 10^{-3}$ in V[TCNE]$_x$ is larger than that for conventional semiconductors by three orders of magnitude. Also, the experimental field and temperature dependencies of MR are not consistent with the conventional model.

4.3.5.2 Model of a Half-Semiconductor

To explain the anomalous MR of V[TCNE]$_x$ the model of a half semiconductor was suggested (Prigodin et al., 2002; Raju et al., 2003). We consider the electronic structure and the energy diagram of V[TCNE]$_x$. From average stiochiometry and the magnetic saturation moment it is assumed (Miller and Epstein, 1998) that the vanadium is in the V^{II} oxidation state and the [TCNE]$^-$ is the radical anion with its unpaired electron in the π^* orbital. For V^{II} it is anticipated that the spins of three electrons at the 3d level are parallel with a total spin of 3/2. Figure 4.16 represents a one-dimensional "cartoon" for the spatial distribution of energy levels in V[TCNE]$_2$. V^{II} is shown alternating with two [TCNE]$^-$ ions to reflect the stiochiometry. The orientation of electronic spins in the π^* state is opposite of the 3/2 spin of V^{II} due to the large antiferromagnetic exchange coupling, J.

The charge transport involves hopping among [TCNE]s. The π^* orbital of each [TCNE] can accept two electrons with opposite spins, but the energy of this double-occupied [TCNE]$^{2-}$ includes the additional coulomb repulsion, U_c. The relevant model for description of electronic states on TCNEs is the Hubbard model with nearly half filling (Onoda and Imada, 2001). Here the π^* band is split into two nonoverlapping subbands provided $U_c \gg t$, where t is the electronic transfer integral between neighboring TCNEs.

The spins of electrons in the lower filled Hubbard subband should have antiferromagnetic order, but the corresponding exchange constant, $J' = 2t^2/U_c$, is much less than the antiferromagnetic exchange constant, J, between spins of V^{II} and [TCNE]$^-$. Therefore, in the magnetic phase the electronic spins of lower subband are parallel and their orientation is opposite to the spin of V^{II}. According to the Pauli exclusion principle, the upper empty subband may be occupied by electrons with spins polarized in the direction opposite to the lower subband, i.e., along the direction of the V^{II} spins.

The one-site coulomb repulsion U_c or the gap between π^* and $\pi^* + U_c$ subbands in analogy with known organic materials U_c is estimated as 2 eV (Tengstedt et al., 2004). According to Tengstedt et al. (2004, 2006), the 3d level of V is located inside the Hubbard gap. Therefore, for the stiochiometric case the temperature dependence of conductivity is determined by thermal activation of electrons from 3d levels of Vs to $\pi^* + U_c$ levels of TCNEs (see Figure 4.16). The corresponding activation energy, E, besides the energy separations between the two levels, $E_0 = E(\pi^* + U_c) - E(3d) \sim$ 0.5 eV, includes the change of magnetic energy associated with this electronic transition. Taking into account the opposite spin polarizations of π^* subbands and their antiferromagnetic coupling with V^{II} spins, the magnetic contribution to the activation energy, ΔE, is approximated as $\Delta E - \Delta E_0 = -4J\langle S\rangle\langle\sigma\rangle$ (Prigodin et al., 2002; Raju et al., 2003, 2010), where $\langle S\rangle$ is the spin 3/2 polarization of V^{II} and $\langle\sigma\rangle$ is the spin 1/2 polarization of the lower π^* subband. Applying the magnetic field changes the spin polarization, and hence the activation energy, for conduction electrons and the conductivity. In this way, the magnetoresistance can be estimated from the relation $MR = (-4J/k_BT)[<S(H)><\sigma(H)> - <S(0)><\sigma(0)>]$.

In the paramagnetic phase the spin polarizations are induced by an external magnetic field and become noticeable only upon approaching T_c. Within the mean field theory (Yeomans, 1999) $<S> \sim -<\sigma> \sim \chi h$, where χ is the Curie-Weiss susceptibility per spin, $\chi \sim 1/\delta$, with $\delta = (T - T_c)/T_c$ being the "distance" from T_c ($k_BT_c \sim 3.2\ J$). Here $h = \mu_B gH/(k_BT_c) \ll 1$ and $h^{2/3} \ll \delta \ll 1$. As a result, the gap widens with the applied field and $MR \sim (h/\delta)^2$.

In the critical region, $|\delta| \ll h^{2/3}$, the induced polarization obeys the scaling law $<S> \sim -<\sigma> \sim h^{1/3}$ and $MR \sim h^{2/3}$. Below T_c ($|\delta| \gg h^{2/3}$, $\delta < 0$), there is the spontaneous spin polarization $<S> \sim -<\sigma> \sim |\delta|^{1/2}$, with field-dependent corrections $\sim \chi h$, where $\chi \sim 1/|\delta|$. Therefore, in the ferrimagnetic phase the MR is proportional to a magnetic field, $MR \sim h/|\delta|$.

To check this prediction, we decrease the Curie temperature by exposition of a sample on air. As a result, the conductivity of the sample drops by a few orders of magnitude, but its temperature dependence continues to follow up the Ahrrenius law with the same activation energy. This temperature behavior of conductivity confirms the model prediction about the band diagram. Figure 4.20 shows up the magnetic field dependence of magnetoresistance below and above $T_c = 240$ K. Linear and quadratic dependence of magnetoresistance on a magnetic field clearly can be identified. As a function of temperature, the MR has its maximum near T_c. For $H = 6$ kG and $T_c = 230$ K the model predicts the MR maximum $\sim h^{2/3} \sim 2\%$, which is of the order of the observed value.

Concluding with Figure 4.21a, the phase diagram of a half semiconductor in a plane of magnetic field and temperature is presented. Figure 4.21b schematically

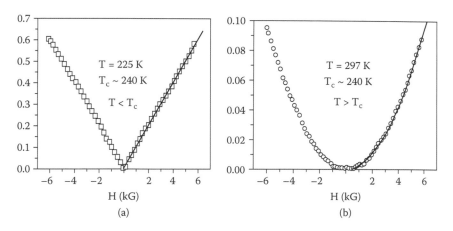

FIGURE 4.20 Variation of magnetoresistance (%) with magnetic field for magnetic semiconductor V[TCNE]$_x$ (a) below T_c (H dependence) and (b) above T_c (H^2 dependence) in agreement with the model prediction for a half semiconductor. (Reproduced with permission from Prigodin et al., *Adv. Mater.* 14, 1230 [2002]. Copyright © 2002 John Wiley & Sons.)

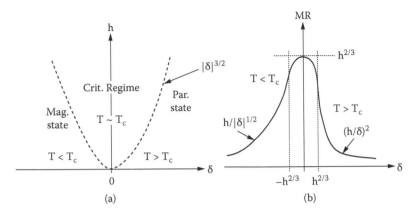

FIGURE 4.21 (a) Mean field phase diagram of ferrimagnet in plane of reduced temperature, $\delta = (T - T_c)/T_c \ll 1$, and reduced magnetic field, $h = g\mu_B H/k_B T_c \ll 1$. (b) Mean field temperature and magnetic field dependence of magnetoresistance for half semiconductor like V[TCNE]$_x$ near of ferrimagnetic transition.

shows the dependence of magnetoresistance as a function of a temperature at a fixed magnetic field at crossing the critical point. All the T and H dependencies agree well with the model predictions. Thus, the experimental results for resistance and MR of organic-based magnet V[TCNE]$_x$ support the existence of spin-polarized subbands—a half-semiconducting state—in which the electron spins in valence and conduction bands have some polarization. This is a very desirable property for utilizing organic-based magnetic semiconductors in spintronic applications.

4.3.6 PHOTOINDUCED MAGNETISM

Since the light-irradiated low-spin and high-spin transition in spin-crossover complexes has been discovered (Decurtins et al., 1884), the photoinduced magnetism has grown as a particular area of research that has attracted considerable attention, due to its facinating phenomena and scientific importance as well as possible application to information technology.

Photoreactivity is a widespread phenomenon in organic and inorganic chemistry. A variety of molecular units may change its structural and electronic properties under the light irraddiation. As a consequence, a phase transformation may occur, with or without hystersis. These photoexcited states are typically long-lived metastable states at low temperature following the Arrhenius law of activation with a lifetime of ~$\exp(E_a/k_B T)$. The molecule/organic-based magnet that has transition ion bonds to molecular/organic ligands may undergo change in magnetic properties through such photoreactivity. This phase transformation takes place locally in the individual complex molecules as well as the short- and long-range cooperative interactions in the lattice.

Figure 4.22 shows a schematic illustration for a mechanism of these phenomena. The diagram displays adiabatic potential curves (U) for a ground state (GS), photoexcited intermediate state (IS), and photoexcited metastable state (MS) with the

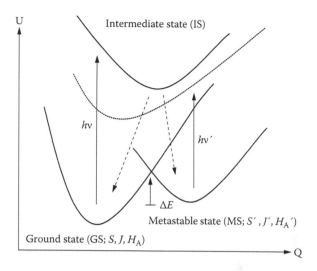

FIGURE 4.22 Schematic illustration of a switching process by the light irradiation.

energy barrier ΔE. The horizontal axis represents the relevant nuclear configuration coordinate Q. The minima of the GS and MS are separated along the configuration coordinate Q. The MS state usually lies higher in energy and usually has weaker metal-ligand (M-L) bonds and, accordingly, longer bond lengths r(M-L) (Q) than the ground state. This is as a consequence of increasing the population of antibonding molecular orbitals and simultaneously decreasing the population of weakly π-back-bonding molecular orbitals. Upon the illumination of light ($h\nu$) the ground state with spin S and orbital momentum J transits to the intermediate state (IS), followed by a radiative or a nonradiative decay into either GS or MS with spin and orbital momentum S' and J'. At temperature $T \ll \Delta E$, the electrons that reach the minimum of the MS potential curve remain trapped in it. The energy barrier between the potential wells governs the lifetime of the metastable state (Arrehenius law). Switching back from the MS to the GS may be possible by the illumination of different frequencies of light $h\nu$. There are various classes of molecule/organic-based complexes, which show light-sensitive structural changes accompanied by drastic changes of their magnetic or optical properties.

The photoinduced change in magnetic properties was initially studied extensively in spin-crossover complexes (Gülitch and Goodwin, 2004) following its discovery (Decurtins et al., 1984). These paramagnetic molecular complexes show high-spin–low-spin transition via temperature, pressure, and light irradiation. A new impetus to this field has been given by discoveries of materials in which photoinduced magnetic (PIM) phenomena coexist with cooperative magnetic behavior, i.e., with magnetic order (Sato et al., 1996a). This coexistence brought a possibility for optical control of the magnetic order, and a number of novel, spectacular, and easily detectable effects, such as optical control of the magnetic ordering temperature, spin freezing temperature, coercivity, etc. These novel materials include Prussian blue analogs (Sato et al., 1996b; Ohkoshi et al., 1997, 2006; Pejaković et al., 2000; Bleuzen et al., 2000; Tokoro et al., 2003; Park et al., 2004; Hozumi et al., 2005), diluted magnetic semiconductors

(Koshihara et al., 1997), doped manganites (Matsuda et al., 1998), organic-based magnet ($Mn(TCNE)_x \cdot y(CH_2Cl_2)$) (Pejaković et al., 2002), and organic-based magnetic semiconductor ($V(TCNE)_x$) (Yoo et al., 2006, 2007). When referring to the photoinduced magnetism, one in fact refers to a variety of distinct physical mechanisms, different for each of these materials.

Recently, dramatic PIM effects were found in $Mn(TCNE)_2 \cdot y(CH_2Cl_2)$ ($y \sim 0.8$) upon excitation with light in the blue region of the spectrum (Pejaković et al., 2002). While many phenomenological aspects of the PIM are similar to the PIM effect in Co-Fe Prussian blue magnets, a detailed analysis suggests a different underlying physical mechanism. This is the first observation of PIM in an organic-based magnet, i.e., magnet with spins that reside on organic species $(TCNE)^-$. Upon excitation with a $\lambda = 488$ nm laser line, the magnetic susceptibility is substantially increased in the entire temperature region below T_c (Pejaković et al., 2002). The photoinduced state has a long lifetime ($>10^6$ s) (Pejaković et al., 2002). Effects of illumination can be erased partially by the excitation of a $\lambda = 514.9$ nm line as well as be erased totally by annealing up to 250 K (Pejaković et al., 2002). Based on extensive optical absorption studies, the photoinduced magnetism is explained as a result of the formation of a metastable state, stabilized by a lattice distortion, in which exchange interaction is locally enhanced (Pejaković et al., 2002).

4.3.7   MULTIPLE PHOTONIC RESPONSE IN ORGANIC-BASED MAGNETIC SEMICONDUCTOR V[TCNE]$_x$, $x\sim2$

In addition to the obvious scientific importance due to the novelty of the phenomenon involved, the phenomenon of photoinduced magnetism has a significant technological potential. Possible applications include information technology, such as magneto-optic devices. Several major breakthroughs should be achieved to meet technological applications. First, the material should exhibit PIM at or near room temperature with a long lifetime. This implies the magnetic ordering temperature should be near or above room temperature, and also the photoexcited state should be sufficiently robust to withstand the vibrational excitations in the material at room temperature. Second, the material should be available in the form of thin films. This, on the one hand, allows their deposition on various substrates, necessary in device fabrication. On the other hand, thin-film configuration allows light to induce effects in the bulk of the material. In addition, designing and developing materials that allow optical control of not only magnetic properties but also electrical properties will provide more functionality for the future application. This will open up new opportunities, along with various concepts of spintronics device application. In this section, we introduce concomitant photoinduced magnetic and electrical phenomena in the films of an organic-based magnetic semiconductor V[TCNE]$_x$, $x \sim 2$ (Yoo et al., 2006, 2007).

Figure 4.23a shows the effect of illumination on the field-cooled magnetization (M_{FC}) of $V(TCNE)_x$, measured at 10 K, in a static magnetic field of 50 Oe. Upon excitation with the $\lambda \sim 457.9$ nm argon laser line (light intensity ~ 20 mW/cm^2), which corresponds to the $\pi \rightarrow \pi^* + U_c$, the magnetization decreases and reaches saturation in about six hours. After turning the laser off, M_{FC} exhibits an additional decrease

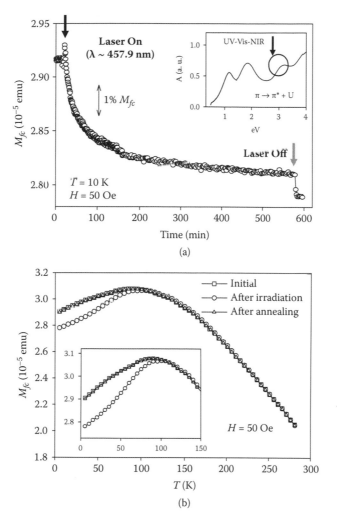

FIGURE 4.23 (a) Effects of argon laser excitation ($\lambda \sim 457.9$ nm, light intensity ~ 20 mW/cm², at 10 K) on the field-cooled magnetization of V(TCNE)$_x$ film. Inset shows UV-Vis-NIR spectra of absorption, indicating $\pi \rightarrow \pi^* + U_c$ excitation around 3 eV. (b) Temperature dependence of field-cooled magnetization (□: before illumination; o: after illumination [$\lambda \sim 457.9$ nm, light intensity ~ 20 mW/cm², at 10 K for 10 h]; △: after annealing sample [to 280 K and waiting for 10 min]). (Reproduced with permission from Yoo et al., *Phys. Rev. Lett.* 97, 247205 [2006]. Copyright © 2006 American Physical Society. http://link.aps.org/doi/10.1103/PhysRevLett.97.247205.)

due to cooling of the sample with an estimated $\Delta T < 5$ K (M_{FC} increases with the increase of T at 10 K [see Figure 4.23b]). The photoexcited state is preserved even in the dark after illumination. For example, the photoinduced magnetization decreases by only ~ 0.5 % after one week at 10 K, which gives a lifetime of $> 10^7$ s by assuming exponential relaxation of PIM. The UV-Vis-NIR absorption of inset in Figure 4.23a shows the $\pi \rightarrow \pi^* + U_c$ excitation band, in the region of 2.5 to 3.5 eV, is the same excitation band that initiates photoinduced magnetism in Mn(TCNE)$_x$ (Pejaković et al., 2002).

The light-induced effects on the conductivity of the CVD-deposited V(TCNE)$_x$ film are displayed in Figure 4.24. The material was illuminated at 120 K by an Ar-ion laser ($\lambda \sim 457.9$ nm, light intensity ~ 20 mW/cm^2). After two hours of illumination, the resistivity (ρ) of the V(TCNE)$_x$ film decreases by a factor of 3 (Figure 4.24a). The shift of resistivity (ρ) after turning off the laser is due to the thermal heating effect

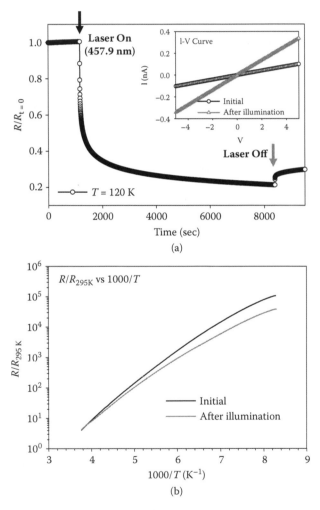

FIGURE 4.24   (a) Normalized resistance of V(TCNE)$_x$ film upon illumination ($\lambda \sim 457.9$ nm, light intensity ~ 20 mW/cm^2, at 120 K). Inset shows *I-V* curves before and after illumination at 120 K. (b) Normalized resistances as a function of temperature for both ground and photoexcited states (black: before illumination; blue: after illumination [$\lambda \sim 457.9$ nm, light intensity ~ 20 mW/cm^2, for 2 h at 120 K]). (Reproduced with permission from Yoo et al., *Phys. Rev. Lett.* 97, 247205 [2006]. Copyright © 2006 American Physical Society. http://link.aps.org/doi/10.1103/PhysRevLett.97.247205.)

during the light irradiation with a $\Delta T < 2$ K. The photoinduced conductivity can be erased by warming the sample above 250 K, similar to the effect of heating on PIM. The inset in Figure 4.24a shows Ohmic *I-V* curves for both ground and photoexcited states at 120 K. After two hours of illumination, the *I-V* slope is increased by a factor of 3.

Detailed conductivity studies for both solution-prepared powders and CVD-deposited films of V(TCNE)$_x$ revealed that the electronic transport for these samples has semiconductor-like behavior with an activation gap of ~0.5 eV at room temperature (Prigodin et al., 2002; Raju et al., 2003). Normalized resistances as a function of temperature for both ground and photoexcited states are shown in Figure 4.24b. After illumination, the temperature dependence of resistivity substantially deviates from the ground state at low temperature. However, $\rho(T)$ values for both ground and photoexcited states are nearly indistinguishable as the temperature increases, and become identical above 250 K. The estimated effective energy gap at 130 K from the linear plot of $\log(R/R_{295K})$ vs. $1{,}000/T$ for the photoexcited state is 0.28 eV, which is 0.04 eV less than the value for the ground state.

The metastability of photoinduced effects for both magnetic and electronic properties of V(TCNE)$_x$ films suggests the presence of light-induced distortion in the overall structure. The mid-IR absorption spectrum in Figure 4.25 shows several features in the 2,050 to 2250 cm^{-1} region (multiple peaks for CN vibrations are associated

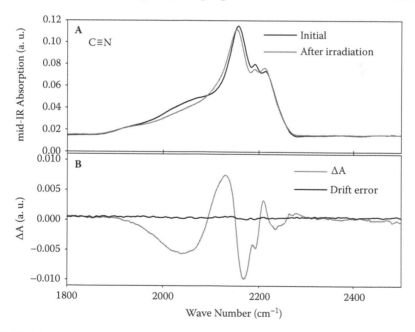

FIGURE 4.25 Mid-IR spectra of CN stretching modes in V(TCNE)$_x$ film recorded at 10 K (black: initial spectrum; gray: recorded after 10 min of light illumination [$\lambda \sim 457.9$ nm, light intensity ~ 30 mW/cm^2 at 10 K]). (b) Photoinduced absorption (ΔA; gray line) compared to the drift error (black line) recorded with 10 min interval. (Reproduced with permission from Yoo et al., *Phys. Rev. Lett.* 97, 247205 [2006]. Copyright © 2006 American Physical Society. http://link.aps.org/doi/10.1103/PhysRevLett.97.247205.)

with different bridging, such as *trans*-μ-N, *cis*-μ-N, μ_4-(TCNE), and possible modulation due to orbital overlap between N and V (Pokhodnya et al., 2000; Miller, 2006), Partial bleaching of CN vibration modes by the illumination ($\lambda \sim 457.9$ nm, light intensity ~ 30 mW/cm^2, at 10 K for 10 min) supports the presence of photoinduced structural distortion. The $\pi \rightarrow \pi^* + U_c$ transition in (TCNE)$^-$ could introduce a structural change, since it can instantaneously weaken the bonding within (TCNE)$^-$. Therefore, light excitation (2.7 eV) can introduce a transition from the ground state to an intermediate state, which can relax back to the original state or a metastable state with a local energy minimum, induced by structural relaxation.

The photoinduced magnetism is proposed to originate from photoinduced change in the orbital overlap of metal-to-ligand bonding, especially spin-carrying orbitals (t_{2g} orbitals of V^{II} and π^* orbitals of (TCNE)$^-$) and photoinduced structural changes. The change in the overlap integral of spin-carrying orbitals induced by changes in (TCNE)$^-$ structures can explicitly modulate kinetic exchange interaction. We speculate the change in exchange energy could be spatially inhomogeneous due to the random structural change resulting in the increase of magnetic anisotropy and random magnetic order. Enhanced magnetic anisotropy in the photoexcited state may induce a decrease in magnetization below reentrant transition, $T_r \sim 90$ K, and lead to a slight shift of T_r toward higher temperature. However, above this increased T_r no significant change is found in the overall magnetization, though the photoexcited metastable state can be fully erased only when warmed to 250 K.

On the other hand, the origin of the photoinduced conductivity is associated predominantly with the charge transfer integral between the orbitals of adjacent (TCNE)$^-$ in addition to the orbital overlap of the V^{II} and (TCNE)$^-$ bond, which are related to carrier mobility and concentration. The photoinduced random structural change in the system could lead to broadening of π^* subbands so that the effective activation energy gap between the valence and conduction bands can be reduced, producing a higher carrier concentration. The increased disorder could also create favorable directions at local sites for electron hopping, resulting in stronger charge transport pathways. In short, the light-induced disorder in the system may result in both a higher carrier concentration, due to reduced activation energy gap, and an increased mobility due to improved charge transfer overlap; however, stronger random magnetic order decreases overall magnetization at low magnetic fields.

In summary, we have presented multiple photonic effects in V[TCNE]$_x$ films through photoinduced magnetic, electric, and spectroscopic studies. Both photoinduced magnetic and electrical properties originate from structural changes triggered by $\pi \rightarrow \pi^* + U_c$ excitation in [TCNE]$^-$. The photoexcited state, which exhibits reduced magnetic susceptibility and increased conductivity, is a long-lived metastable state at low temperature and can be fully recovered to the ground state by warming above 250 K. The organic-based hybrid magnet V[TCNE]$_x$ is a particularly interesting system not only for its room temperature magnetic order but also for its exotic spin-polarized electronic structure and transport properties. In addition, the demonstrated optical control of both magnetic and electrical properties may introduce a new paradigm in developing functional materials and devices associated with spintronics.

4.3.8 PHOTOINDUCED MAGNETISM AND RANDOM MAGNETIC ANISOTROPY IN ORGANIC-BASED MAGNETIC SEMICONDUCTOR V[TCNE]$_x$, $x\sim2$

Further investigations are needed to account for the underlying mechanism for photoinduced magnetism in V[TCNE]$_x$. We employed photoinduced ferromagnetic resonance (PIFMR) to study the mechanism of photoinduced magnetism in V[TCNE]$_x$ films (Yoo et al., 2007). This PIFMR study showed that random magnetic anisotropy (RMA) plays a central role in photoinduced magnetic phenomena in V[TCNE]$_x$. Upon optical excitation ($\lambda \sim 457.9$ nm), the ferrimagnetic resonance spectra display a substantial increase in line width and shift in the resonance field, depending on the orientation of the applied magnetic field. This result shows that the PIM in V[TCNE]$_x$ originates from the enhanced RMA induced by increased structural disorder caused by the light.

The time-dependent evolution of FMR spectra of V[TNCE]$_x$ under illumination is illustrated in Figure 4.26. The angle between the normal to the film and the external field was set to $\theta \sim 54.7°$, where the spectrum collapses to a single resonance as the effects of demagnetization and uniaxial anisotropy of films are essentially eliminated (see Equation 4.7). Upon illumination ($\lambda \sim 457.9$ nm, light intensity ~ 20 mW/cm^2, $T = 30$ K), the FMR absorption spectra became substantially broader and the resonance field was shifted to a lower value. The intensities of the first integration of the FMR spectra fit well to Lorentzian curves. The time dependence of the line width (full width half maximum [FWHM]) and resonance field (obtained from spectra in Figure 4.26) during illumination is displayed in the inset of Figure 4.26. At 30 K, the line width is increased by twofold, and the resonance field is decreased by 10 Oe upon

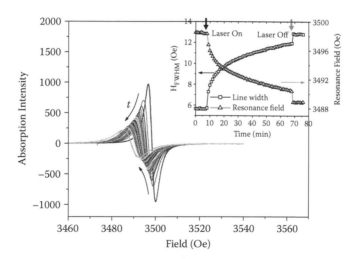

FIGURE 4.26 Evolution of FMR spectrum under light illumination ($\lambda \sim 457.9$ nm, light intensity ~ 20mW/cm^2, at 30 K). Inset shows the line width and resonance field obtained by Lorentzian fit of the FMR spectra. (Reproduced with permission from Yoo et al., *Phys. Rev. Lett.* 99, 157205 [2007]. Copyright © 2007 American Physical Society. http://link.aps.org/doi/10.1103/PhysRevLett.99.157205.)

illumination for 1 h ($\lambda \sim 457.9$ nm, light intensity ~ 20 mW/cm^2). The light-induced FMR spectra were recovered to the initial FMR lines, when the sample was warmed to 250 K and cooled back to initial T. In contrast, the dc magnetization of the film at external field 3,500 Oe is almost saturated and does not show any change due to the illumination. However, the FMR absorption spectra show a substantial change in spin dynamics in the system caused by the illumination. Furthermore, increased line width in FMR spectra suggest slowing down of spin relaxations and a decrease in the effectiveness of exchange narrowing. This could be attributed to either increased fluctuation of exchange coupling or increased RMA.

Figure 4.27 displays typical FMR spectra of CVD-deposited V(TCNE)$_x$ film for various orientations of the external magnetic field with respect to the normal to the plane of film for both ground and photoexcited states at 30 K. For a measurement of the photoexcited state, the sample was illuminated for 2 h with $\lambda \sim 457.9$ nm and light intensity ~ 20 mW/cm^2. When $4\pi M \| H$ (M is the magnetization and H is the applied magnetic field), the magnetization of films lies nearly along the applied field for all orientations. For a small anisotropy field H_A and $4\pi M - H_A \ll H_r$, the angular dependence of the resonance behavior for a planar sample can be expressed by a simple equation (Pechan et al., 1985; Plachy et al., 2004):

$$H \cong H_r + \left(2\pi M + \frac{H_A}{2}\right)(2 - 3\sin^2 \theta) \tag{4.7}$$

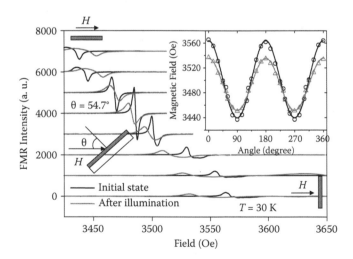

FIGURE 4.27 Effects of illumination on angular dependence of FMR spectra of V(TCNE)$_x$ film at 30 K for the angle ($\theta = 90, 75, 60, 55, 45, 30, 15, 0°$ from top to bottom) from the normal to the plane of the film to the external magnetic field (black line: ground state; gray line: after illumination with $\lambda \sim 457.9$ nm, light intensity ~ 20 mW/cm^2 for 2 h). Inset shows angular dependence of line shift for a particular FMR line of V(TCNE)$_x$ film for both ground (o) and photoexcited ($\triangle$) states. Solid lines are fits to Equation 4.7. (Reproduced with permission from Yoo et al., *Phys. Rev. Lett.* 99, 157205 [2007]. Copyright © 2007 American Physical Society. http://link.aps.org/doi/10.1103/PhysRevLett.99.157205.)

where θ is the angle from the normal to the plane of the film to the external magnetic field, and H_r is the internal resonance field (i.e., the field at which resonance will occur in the absence of any shape effects). The internal resonance field can be described by an effective g-value, $g_{eff} = \hbar \omega \mu_B H_r$, where $\hbar$ is Planck's constant, ω is the spectrometer operating frequency, and μ_B is the Bohr magneton. H_A is a perpendicular uniaxial anisotropy field. As the V(TCNE)$_x$ has an amorphous structure, crystalline anisotropy is negligible and RMA mainly accounts for H_A, which can be assumed to be independent of angle.

One can see that several spectral features in Figure 4.27 can be observed when θ differs from 54.7°. These features correspond to nonuniform regions with distinct $4\pi M - H_A$ values within the sample, and collapse to a single spectrum at $\theta = 54.7°$. Similar angular-dependent FMR spectra in V(TCNE)$_x$ films were reported previously with a larger number of resonances due to inhomogeneity of samples (Plachy et al., 2004). The inset of Figure 4.27 shows the angular dependence of the resonance field for a specific FMR line (the most pronounced resonance spectrum) and fit to Equation 4.7. Note that there is a substantial reduction of angle-dependent line shift after the light irradiation. Such light-induced shift of the resonance field suggests a substantial increase of H_A, because the magnetization at $H \sim 3,500$ Oe is almost identical for both ground and photoexcited states according to the SQUID measurements (Yoo et al., 2006).

The T dependence of H_A can be obtained from the T dependence of the resonance fields for particular orientations, $\theta = 0°, 90°$, and $54.7°$, which reduce to $H(0°, T) = H_\perp \cong H_r + 4\pi M - H_A, H(90°, T) = H_\parallel \cong H_r - 2\pi M + H_A/2$, and $H(54.7°, T) \cong H_r \cong 1/3(H_\perp + 2H_\parallel)$, respectively. $M(T)$ was estimated from the total integrated intensity of FMR spectra with values scaled using dc magnetic measurement results ($M = 17.7$ G at 5 K and 3,500 Oe). Figure 4.28a shows temperature dependence of the resonance field of the most pronounced FMR line for angles $\theta = 0°, 54.7°$, and $90°$. Figure 4.28b exhibits temperature dependence of subtracted $4\pi M(T) - H_A(T) \cong H(0°, T) - H(54.7°, T)$. Figure 4.28c displays obtained $H_A(T)$ for both ground and photoexcited states. The measurements for photoexcited states were made after illumination with $\lambda \sim 457.9$ nm, light intensity ~ 20 mW/cm^2 for 2 h at 30 K. Fitting the Bloch law ($M_s(T) = M_s(0)(1 - BT^{2/3})$) to the total integrated intensity of FMR spectra is employed to subtract $M(T)$. In a phenomenological approach, $H_A = 2K/M$, where K is an anisotropy constant of RMA. The anisotropy constant increases linearly as T is lowered (Figure 4.28c). This behavior is similar to that experimentally observed in a spin glass system with a relation $K(T) = K(0)(1 - (2/3)T/T_f)$, where T_f is spin freezing temperature (Schultz et al., 1980) and qualitatively similar to the theoretical prediction for spin glasses (Becker, 1982). Note that a substantial increase of anisotropy field H_A caused by the illumination at 30 K can be observed over the wide range of temperature. At 10 K, the anisotropy field ($H_A \sim 143$ G) extracted from the most pronounced resonance line increased by 17% after the illumination.

In amorphous magnets, when the anisotropy energy is small compared to the exchange energy, Chudnovsky (1986) predicted a formation of different magnetic phases in the material according to the strength of an applied magnetic field. Depending on the degree of anisotropy, a correlated spin glass (CSG) phase may appear at low

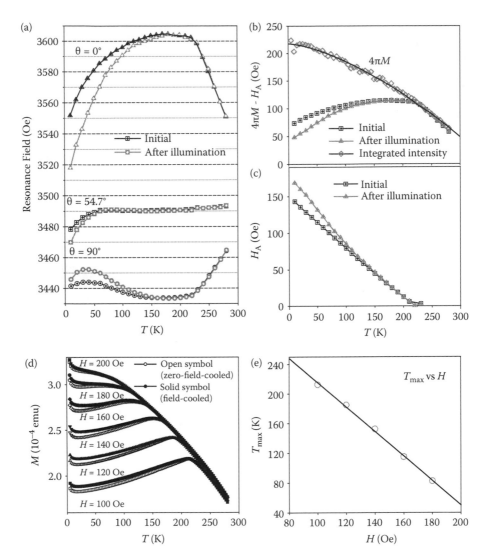

FIGURE 4.28 (a) Light-induced effect on the temperature dependence of the resonance field for the most pronounced FMR line at $\theta = 0°$, 54.7°, and 90° (black line and symbol: ground state; gray line and symbol: photoexcited state; measured after illumination with $\lambda \sim 457.9$ nm, light intensity ~ 20 mW/cm^2 for 2 h at 30 K). (b) Extracted $4\pi M - H_A$ from selected FMR line between $\theta = 0°$ and 90° for both ground and photoexcited states. ◊ symbol shows total integrated intensity of FMR spectra and black line is Bloch function fit to the total integrated intensity of FMR spectra. (c) Temperature dependence of H_A after subtraction with Bloch function fit to the total integrated intensities of FMR absorptions for both ground and photoexcited states. (d) Field-cooled and zero-field-cooled dc magnetization for an applied magnetic field of 100, 120, 140, 160, 180, and 200 Oe. (e) Magnetic field dependence of the zero-field-cooled peak temperatures (T_{max}). (Reproduced with permission from Yoo et al., *Phys. Rev. Lett.* 99, 157205 [2007]. Copyright © 2007 American Physical Society. http://link.aps.org/doi/10.1103/PhysRevLett.99.157205.)

field, while at intermediate field small domains with local correlated anisotropy axis, the so-called ferromagnetic wandering axis (FWA) phase, may be formed. At high enough field, the system will be nearly a collinear ferromagnetic phase.

The main difference between a spin glass and CSG is the origin of spin freezing. In a spin glass, the freezing of spin is a consequence of random magnetic interaction. Mean field theory produces the Almeida-Thouless (AT) instability line, $\delta T_f \propto H^{2/3}$, for Ising spin glasses (de Almeida and Thouless, 1978), and transverse freezing with the Gabay-Toulouse (GT) line, $\delta T_f \propto H^2$, followed by the AT longitudinal freezing line for Heisenberg spin glasses (Gabay and Thouless, 1981). On the other hand, the spin freezing of CSG relies on random magnetic anisotropy.

Figure 4.28d shows the temperature dependence of field-cooled (FC) and zero-field-cooled (ZFC) dc magnetization. At low field, $H = 5$ Oe, M slowly increases as T increases and does not show peaks until 280 K. AC magnetic susceptibility has similar T-dependent behavior with no frequency dependence below 300 K (Pokhodnya et al., 2001). At relatively high fields ($H > 100$ Oe), M decreases monotonically as T increases, suggesting substantial excitation of spin waves with low wave vectors, which is expected for the FWA phase of disordered V(TCNE)$_x$ (Zhou et al., 1993b; Pokhodnya et al., 2001). As T is decreased, M peaks at a temperature, T_{max}, and is suppressed as T is lowered further. The M exhibits weak irreversibility, given by the deviation between field-cooled and zero-field-cooled magnetization below a particular $T_{irr} \sim T_{max}$. As the intensity of the applied magnetic field increases, the T_{max}, where the spins start to freeze, shifts to lower temperatures following a linear relation $T_{max} = a - bH$ (Figure 4.28e), which was earlier reported for another disordered magnetic system (Tejada et al., 1990). This external field dependent shift of peaks may then be a direct consequence of RMA, which increases linearly as T is lowered, and makes CSG distinct from the spin glass, where the position of peaks in ZFC curves follows the AT line and/or GT line.

In short, the persistent PIM in V(TCNE)$_x$ films was probed via PIFMR studies. The responses of line shift and line width of FMR spectra of V(TCNE)$_x$ films to the illumination are well correlated to a light-induced increase of RMA resulting from an increased structural disorder. The results also display the role of magnetic anisotropy in amorphous magnets, providing distinction between spin glass and CSG. The PIFMR study of V(TCNE)$_x$ film demonstrates a new mechanism for PIM in organic or molecule-based magnets, beyond changing spin numbers and exchange couplings, and suggests a different approach for the optical control of magnetic properties in materials.

4.3.9   ANOMALOUS MAGNETIC BISTABILITY AND NUCLEATION OF MAGNETIC BUBBLES IN A LAYERED 2D ORGANIC-BASED MAGNET

Recently the first 2D crystal structure of a metal-TCNE magnet, [Fe(TCNE)(NCMe)$_2$][FeCl$_4$], was reported (Pokhodnya et al., 2006). The structure consists of undulating layers composed of Fe^{II} ions with a μ_4-[TCNE]$^-$ bridgings within the layer and two axial MeCNs coordinations (Pokhodnya et al., 2006). There are no covalent bonds between layers (Pokhodnya et al., 2006). Additional paramagnetic [FeCl$_4$]$^-$ anions occupy sites between the layers of [Fe(TCNE)(NCMe)$_2$]$^+$ but do not contribute to the magnetic ordering (Pokhodnya et al., 2006). The absence of bridging ligands between the layers suggests only that dipolar coupling exists between the layers. In each layer, the magnetic coupling between spin in [TCNE]⁻ ($S = 1/2$) and spins in Fe^{II} ($S = 2$) is suggested to be antiferromagnetic, resulting in ferrimagnetic order (Pokhodnya et al., 2006), similar to other metal-TCNE magnets. Thus, each individual layer is considered an ideal 2D ferromagnetic Ising system. In this section, we present unique properties of magnetic bistability of the 2D layered system [Fe(TCNE)(NCMe)$_2$][FeCl$_4$] (TCNE = tetracyanoethylene) (Pokhodnya et al., 2006; Yoo et al., 2008). The dc magnetization displays anomalous irreversibility between zero-field-cooled (ZFC) and field-cooled (FC) states, which we propose to originate from magnetization reversal of single 2D layers through the nucleation and growth of "bubble domains." We show that the rate of bubble generation, together with the bubble size and characteristic relaxation time, is strongly contingent on the external magnetic field (H) and thermal energy ($k_B T$).

The FC and ZFC magnetizations upon warming at different external fields are shown in Figure 4.29. A strong irreversibility is observed between FC and ZFC states (Pokhodnya et al., 2006). The ZFC magnetization is not only strongly suppressed below the irreversible temperature, T_{irr}, but is also nearly negligible over a wide temperature range below T_c at low field. The ac susceptibility also exhibits the slowing down of spin dynamics as temperature (T) decreases through T_c, which was attributed to spin glass-like behavior (Pokhodnya et al., 2006). The strong irreversibility may be attributed to (1) a metamagnetic phase transition due to the interlayer coupling, (2) a superparamagnetic blocking energy, (3) a strong spin freezing due to spin glass order, or (4) an irreversible magnetic domain formation as the sample is cooled through T_c with an applied field. The observed hysteresis in [Fe(TCNE)(NCMe)$_2$]

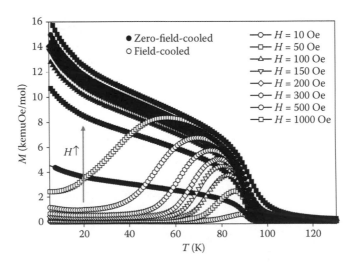

FIGURE 4.29 Field-cooled (solid symbols) and zero-field-cooled (open symbols) magnetization for different applied magnetic fields ($H = 10, 50, 100, 150, 200, 300, 500, 1000$ Oe). (Reproduced with permission from Yoo et al., *Phys. Rev. Lett.* 101, 197206 [2008]. Copyright © 2008 American Physical Society. http://link.aps.org/doi/10.1103/PhysRevLett.101.197206)

[FeCl$_4$] features a large coercive field and remanence in contradiction to the behavior of a metamagnet, whose hysteresis curve shows a step-like feature due to a first-order phase transition (Stryjewski and Giordano, 1977).

The time-dependent relaxation of ZFC, FC, and thermoremanent magnetization (TRM) at 70 K with an external field of 100 Oe is shown in Figure 4.30. It is well known that the canonical spin glass system exhibits so-called aging due to slow growth of the coherence length toward the equilibrium state (Mydosh, 1993). The time-dependent relaxation of the ZFC magnetization displays no obvious aging effects for waiting times $t_w = 10^3$ s and 10^4 s (see Figure 4.30), indicating absence of spin glass order. In addition, another typical signature of spin glass relaxation, memory effect (Mathieu et al., 2001), was not observed in this material.

For superparamagnets, there exists a unique response function, $f(t)$, for magnetic relaxation at a given temperature, as described by $\Delta M(t) = \Delta H \cdot f(t)$, where ΔH is the change of magnetic field at $t = 0$ (Lundgren et al., 1986). Such relaxation can even be extended for a cooperative system like a spin glass with an additional input, t_w, due to memory of historical events of the system (Lundgren et al., 1986). One can readily test this fundamental signature by the principle of superposition, as follows (Lundgren et al., 1986; Mathieu et al., 2001).

$$M_{ZFC}(t_w, t) = M_{FC}(0, t_w + t) - M_{TRM}(t_w, t) \qquad (4.8)$$

where M_{FC}, M_{ZFC}, and M_{TRM} are FC, ZFC, and thermoremanent magnetization, respectively. Figure 4.30 displays a huge difference in the magnetic relaxation between FC and ZFC states. This suggests that the magnetic irreversibility between M_{FC} and M_{ZFC} does not originate from either superparamagnetic blocking or spin glass ordering.

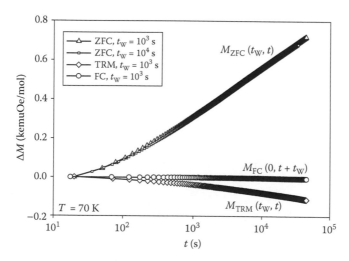

FIGURE 4.30 Time relaxation of M_{ZFC}, M_{FC}, and M_{TRM} at $T = 70$ K, $H = 100$ Oe for waiting time $t_w = 10^3$ and 10^4 s. (Reproduced with permission from Yoo et al., *Phys. Rev. Lett.* 101, 197206 [2008]. Copyright © 2008 American Physical Society. http://link.aps.org/doi/10.1103/PhysRevLett.101.197206)

Here, we propose that due to the large interlayer separation, the individual layers are magnetically weakly interacting with each layer and can be described by the 2D ferrimagnetic Ising plane. Because the polycrystalline powder samples were used for measurement, the easy axis of individual crystallites is randomly oriented. When the sample was cooled in a field below T_c, all layers had a component of their magnetization directed along the field, providing the macroscopic magnetization of the FC phase. However, if the sample was cooled in the absence of a field, each single layer had magnetization randomly either up or down along its own axis, resulting in zero total macroscopic magnetization. The magnetization of a single 2D layer is frozen due to large blocking energy below T_c. Then, the reorientation of magnetization in those layers by a magnetic field will occur through the nucleation of magnetic bubbles (Barbara, 2005).

For a 2D ferrimagnetic Ising system, the mean field spontaneous spin polarization is $S = |\tau|^{1/2}$ below T_c, where reduced temperature $\tau = 1 - T/T_c$. Far below T_c, all spins align up or down perpendicular to the layer. If the magnetic field occurs opposite to the magnetization of a layer, it initiates the nucleation of bubble domains. The energy of a bubble domain with a radius n (in units of repeat cell) is estimated (Kittel, 1949) to be

$$E_n \approx -\mu_B H(2S)(\pi n^2) + (2\pi n)S^2\sqrt{JK} \qquad (4.9)$$

where the first term is the energy gain due to reorientation of the bubble spin along the field. The second term represents the exchange energy loss in the bubble wall of thickness, $w \sim (J/K)^{1/2}$, that depends on the exchange constant (J) and anisotropy constant (K). The maximum bubble energy, E_n, which corresponds to the activation energy, E_b, occurs at the critical radius $n = S(JK)^{1/2} 2\mu_B H$ and $E_b = \pi S^3 JK/2\mu_B H$. Thus, the rate of bubble nucleation determined by the activation energy, E_b, depends on the magnetic field and temperature. A similar description of magnetic reversal with $1/H$ dependence of the energy barrier was extensively studied in metallurgical magnetic systems (Barbara and Uehara, 1976; Barbara, 2005).

The crossover field H_c derived from the experimental data is presented in Figure 4.31a. Here H_c, for ZFC-to-FC crossover, is the field at which the magnetization starts to increase sublinearly and irreversibly as presented in minor hysteresis loops in the inset of Figure 4.31a. Below H_c, $M(H)$ is almost linear with a negligible remanence and coercive field. Above H_c, the hysteresis loops display irreversibility and remain open upon completing the cycle. Alternatively, within the bubble nucleation model, H_c for the ZFC-FC crossover can be defined, at which the bubble nucleation time ($\tau = \tau_0 \exp[E_b/k_B T]$) becomes comparable with the measuring timescale, τ_m. Then, the T dependence of crossover field H_c follows

$$1/H_c = 1/H_0 + \ln(\tau_m/\tau_0)\frac{2\mu_B k_B T}{\pi S^3 JK} \qquad (4.10)$$

where H_0 is introduced following the empirical formula of Barbara and Uehara (1976) to limit H_c as $T \to 0$. H_0 reflects a lower limit of bubble size, $n \geq w$ (Barbara,

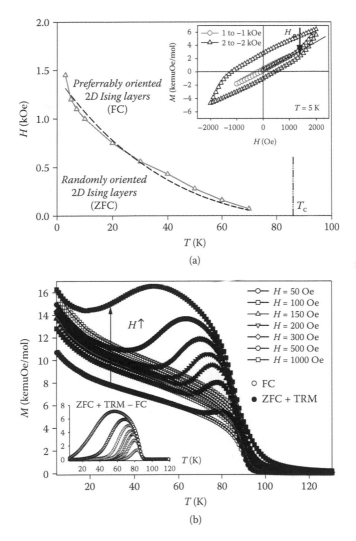

FIGURE 4.31 (a) A T-dependent crossover field obtained from minor hysteresis loop as described in inset. Inset shows minor hysteresis loop for field scan range of ±1,000 and ±2,000 Oe. (b) $M_{FC}(T)$ (open symbols) and $M_{ZFC}(T) + M_{TRM}(T)$ (solid symbols) for different applied fields (H = 50, 100, 150, 200, 300, 500, 1000 Oe). Inset shows $M_{ZFC}(T) + M_{TRM}(T) - M_{FC}(T)$ for each applied field. (Reproduced with permission from Yoo et al., *Phys. Rev. Lett.* 101, 197206 [2008]. Copyright © 2008 American Physical Society. http://link.aps.org/doi/10.1103/PhysRevLett.101.197206)

2005). From the above model, H_0 is estimated as $\mu_B H_0 \sim k_B K$. The black dashed line is the fit of $H_c(T)$ to Equation 4.10 employing $\tau_m/\tau_0 = 10^{11}$. A quantitative fit yields $H_0 = 1431 \pm 6$ Oe and $JK = 54.8 \pm 0.7$ K^2. Taking $J \sim T_c \sim 90$ K, we get $K \sim 0.6$ K, which is close to the value $K \sim 0.2$ K estimated from the obtained H_0. Here, the K values give an order of ~ 10 repeat unit cells for the wall thickness as well as a minimum size of bubble, as $T \to 0$. A similar T dependence for a coercive field due to bubble nucleation was reported in metallurgical magnets (Barbara and Uehara, 1976).

The $M_{ZFC}(T) + M_{TRM}(T)$ and $M_{FC}(T)$ for different applied fields are shown in Figure 4.31b. These data were collected while increasing the T with the same rate of linear T sweep, $\delta T/\delta t = 0.25$ K/min. There exists large differences between $M_{ZFC}(T) + M_{TRM}(T)$ and $M_{FC}(T)$ over a wide T range that depends on the applied field. We attribute this difference to the thermal activation of bubble domains. If the sample is warmed from the ZFC state, the applied field H overcomes H_c at certain T-initiating nucleation of bubbles. Here, the activation of bubbles introduces a strong asymmetric magnetic relaxation between $M_{ZFC}(t)$ and $M_{TRM}(t)$, as shown in Figure 4.31b.

Below T_c, the rate of magnetic response to external magnetic field in the ZFC phase also is determined by the nucleation rate. The relaxation time (τ) is anticipated to behave as $\tau \sim \tau_0 \exp[E_b/k_B T]$. Thus, the relaxation time has the following temperature and field dependencies: $\ln\tau \propto |T|^{3/2}/T$ and $\ln\tau \propto 1/H$. The median relaxation time (τ_c) together with the distribution width of $\tau(\alpha)$ can be determined from the complex linear susceptibility ($\chi_{lin}(\omega) = \chi'(\omega) + i\chi''(\omega)$) through the Cole-Cole analysis (($\chi(\omega) = \chi_s(\omega) + (\chi_0(\omega) + \chi_s(\omega))/(1 + (i\omega\tau_c)^{(1-a)})$, where $\chi_0(\omega)$ and $\chi_s(\omega)$ are the isothermal ($\omega = 0$) and adiabatic ($\omega \to \infty$) susceptibilities, respectively (Cole and Cole, 1941). The Cole-Cole analyses were performed following the procedure of Dekker et al. (1989).

The temperature dependencies of the characteristic relaxation times for different applied fields ($H = 0, 50, 100$ Oe) are displayed in Figure 4.32a. The range of frequencies used for the ac field was 11 to 10^4 Hz with an amplitude of 5 Oe in order to obtain a distribution of relaxation times. All data were collected with increasing temperature from the ZFC state. As the temperature decreases through the transition, the median relaxation time increases, indicating the growth of the correlation length of the spin system. When there is no applied field, the critical slowing down near T_c ($T > T_c$) is characterized as a power law, $\tau \propto \tau_0 |\tau|^{-z\nu}$, where ν is the critical exponent for correlation length ($\xi \sim \tau^\nu$) and z is the dynamic critical exponent ($\tau \sim \xi^z$), respectively. Here, scaling for $z\nu$ varies for a different set of τ_0 and T_c as well as the temperature fitting range. For a fixed value of $\tau_0 = 10^{-9}$ s, the highest confidence values for $z\nu = 2.26 \pm 0.06$ and $T_c = 85.90 \pm 0.05$ K were obtained for the fitting range of 86.2 K $\leq T \leq$ 87.0 K (gray line in Figure 4.32a). The obtained $z\nu$ value is significantly lower than the typical values for spin glass systems (Mydosh, 1993). Taking the exact value of $\nu = 1$ for the 2D Ising model (Patashinskii and Pokrovskii, 1979), the z value is within the range of many theoretical simulations ($2.06 \leq z \leq 2.35$) for the dynamic 2D Ising model (Nightingale and Blöte, 1996). When the field is applied, a substantial decrease of the overall relaxation times is observed.

The evolution of τ_c of spins with increasing applied field at a fixed temperature ($T = 83.0$ and 83.5 K) is displayed in Figure 4.32b. The results exhibit asymptotic behavior of spin relaxation, $\ln\tau \sim 1/H$ as $1/H \to 0$. The dashed lines are fit to the linear region of $\ln\tau$. Here, the slopes correspond to $\pi S^3 JK/(2k_B T\mu_B)$. At $T = 83$ K, the slope is 331.7, which produces $JK = 191$ K^2. The inset in Figure 4.32b displays α as a function of external field. The monotonic increase of α can also be due to domain activation.

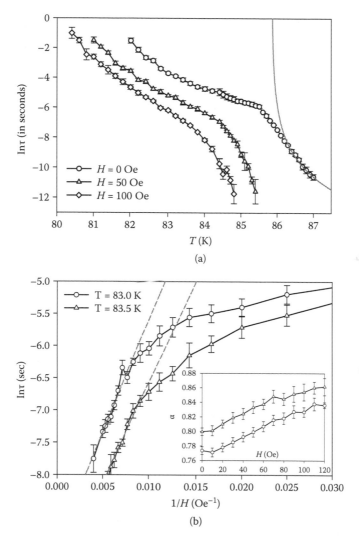

FIGURE 4.32 (a) The characteristic relaxation time $\tau_c(T)$ for dc applied field $H = 0$, 50, and 100 Oe, obtained from complex linear susceptibility. The gray line indicates a power law fit. (b) The characteristic relaxation time, $\tau_c(H)$, for a fixed temperature of $T = 83.0$ K and 83.5 K. Inset displays broadening of width of spin relaxation time distribution as increasing external field. (Reproduced with permission from Yoo et al., *Phys. Rev. Lett.* 101, 197206 [2008]. Copyright © 2008 American Physical Society. http://link.aps.org/doi/10.1103/PhysRevLett.101.197206.)

In summary, we presented an unusual macroscopic magnetic bistability of the FC and ZFC magnetization in 2D layered organic-based magnets [Fe(TCNE)(NCMe)$_2$] [FeCl$_4$]. We propose that this unique magnetic bistability can be explained through the consideration of an ensemble of uncoupled 2D Ising layers and their magnetic reversal initiated by the nucleation of bubbles. Here, the molecule-based systems that can be adapted to different situations via organic methodology provide new materials that display new macro-microscopic phenomena as well as solid-state physics of static/dynamic phase transitions.

4.4 MAGNETOTRANSPORT OF ORGANIC SEMICONDUCTORS

4.4.1 INTRODUCTION

In comparison with inorganic crystalline semiconductors, the organic semiconductors have more a complex and irregular structure (Baldo and Forrest, 2001; Delucia et al., 2002). Their electronic bands feature charged defects such as polaron states whose energy levels are in a gap. Disorder leads to a smooth distribution in polaron energy near valence and conduction band edges. Charge transport in organic semiconductors represents hopping of polarons, and the impact of disorder can be seen in the dispersive character of charge transport as well as in the electric field and temperature dependencies of the charge carrier mobility (Prigodin and Epstein, 2008).

For these reasons the representation of excitons as hydrogen-like atoms is not applicable for organic semiconductors. An electron and a hole appear to be bound by the coulomb attraction potential but stay separated, being localized in the proximate states of the valence and conduction bands. This type of e-h bound pair is also considered as highly excited excitons or exciton precursors, meaning that the bound e-h pair becomes the real exciton of Frenkel type when the electron and hole are on the same molecule. We will call them coulomb pairs. It is important to stress that the lifetime of these coulomb pairs may be very long because the disorder and polaronic effects hold electron and hole spatially separated, while e-h recombination is determined by the direct overlap between wave functions of electron and hole. They can migrate by hopping and interact with other charge carriers. Such e-h bound pairs for organic semiconductors were exploited to explain experiments on fluoresce, in particular, quenching of electroluminescence with an electric field (Morteani et al., 2004; Rothe et al., 2005).

Here we address another inherent feature of coulomb pairs that enables us to control charge transport and light emission in organic semiconductors by a weak magnetic field (Frankevich and Balabanov, 1965; Kalinowski et al., 2003; Francis et al., 2004; Mermer et al., 2005; Prigodin et al., 2006; Xu et al., 2006; Wu and Hu, 2006; Bobbert et al., 2007; Wagemans et al., 2008; Bergeson et al., 2008).

Because the electron and hole of e-h pairs spatially are separated, the exchange energy between their spins remains small. Indeed, the coulomb attraction energy decays with e-h distance as $-eV(r) = -E_0 (r_0/r)$, where E_0 and r_0 are the energy and radius of the Frenkel exciton. But the exchange energy drops with distance much faster, $J(r) = J_0 \exp[-2r/r_0]$. As a result, the remote electron and hole already coulombically may be bound ($|eV(r)| > k_B T$), but in contrast to the conventional exciton, the spin configuration is still flexible ($|J(r)| \ll k_B T$). Besides the exchange interaction, there is the dipolar interaction between e and h spins characterized by the constant $D(r) \sim (g\mu_B)^2/r^3 \sim \alpha^2 E_0(r_0/r)^3$, where $\alpha = 1/137$ is a hyperfine structure constant. The dipolar interaction is much weaker than the hyperfine one at small distances but prevails for large separations where the dipolar interaction becomes negligibly small.

The electron and hole of e-h bound pairs may either dissociate back into free charge carriers or recombine in radiative and nonradiative ways. Because of spin conservation the recombination of e-h bound pairs is anticipated only from their singlet spin state. However, due to degeneracy of the triplet and singlet spin states for the well-separated electron and hole of bound pairs, weak spin perturbations can change their spin state. These perturbations could be the hyperfine interaction and an external magnetic field (Prigodin et al., 2006). As a result of spin interactions, the triplet e-h pairs can convert into singlet ones and recombine. In this case, a weak magnetic field produces a strong effect on the spin dynamics, and hence on the recombination rate of e-h pairs and on the charge current.

We have shown that the recombination current may have a maximum as a function of recombination constant (Bergeson et al., 2008). For high recombination constants the current is space charge limited and decreases with increasing the e-h recombination constant. At a low recombination constant the recombination takes place in the whole volume and the current increases with increasing the recombination constant. The characteristic recombination constant separating those two regimes depends on the thickness of the sample, applied voltage, and charge carrier mobilities. The present model is capable of accounting for a variety of magnetotransport behaviors in organic semiconductors.

In the next section we describe the dependence of injected bipolar current on the e-h recombination constant. Then we study the formation of coulomb pairs and the dependence of their recombination rate on the magnetic field. Discussion of experimental results within the present model is given in the concluding section.

4.4.2 Charge Transport in Organic Semiconductors

Undoped organic semiconductors are good insulators because of their large energy gaps (>2 eV), and they can conduct electricity only due to injected charge carriers. The electric field of injected charge carriers compensates in part for the external electric field and limits the charge transport. This space charge limited transport is studied in the literature (Lampert and Mark, 1970).

In Figure 4.33 an energy diagram of a semiconductor with a unipolar charge transport is shown. We assume that there is the ohmic contact between the semiconductor and the anode, and the hole concentration in the semiconductor nearby the anode is p_0. The holes start to flow through the semiconductor layer if the external potential, V_{ext}, is larger than the built-in voltage, $V_{bi} = \Phi_a - \Phi_c$, where $\Phi_{a,c}$ are the anode and cathode work functions. The voltage, V, which determines the current, is $V = V_{ext} - V_{bi}$.

$I(V)$ characteristics of the device shown in Figure 4.33 are described by the equation

$$V = (8IL^3/9\varepsilon\mu_p)^{1/2}[(1 + L_c/L)^{3/2} - (L_c/L)^{3/2}], \quad L_c = \varepsilon I/(2e^2\mu_p p_0^2) \tag{4.11}$$

According to Equation 4.11, for $V \gg V_c$ the current obeys Ohm's law:

$$I = ep_0\mu_\pi(V/L) \tag{4.12}$$

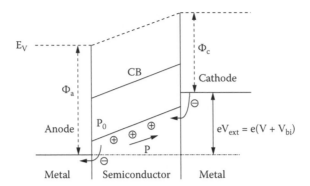

FIGURE 4.33 Energy diagram for metal-semiconductor-metal structure with unipolar (hole) charge transport. p_0 is the hole concentration in the semiconductor nearby the anode. The hole current flows if the external voltage, V_{ext}, is larger than the built-in voltage, $V_{bi} = \Phi_a - \Phi_c$, where $\Phi_{a(c)}$ is the work function of the anode (cathode). The voltage drop across the semiconductor $V = V_{ext} - V_{bi}$ determines the space charge limited current (Equations 4.11 to 4.13).

where $V_c = (e/\varepsilon) p_0 L^2$ is the charging voltage. We note $\varepsilon_0/2e = 2.76 \times 10^5$ (cm $\times$ V)$^{-1}$ so that $V_c = 6$ V for $p_0 = 10^{17}$ cm^{-3}, $L = 10^2$ nm, and $\varepsilon_r = 3$. For $V \ll V_c$ the current (Equation 4.11) follows the universal law (Lampert and Mark, 1970):

$$I = (9/8)\ \varepsilon\mu_p\ (V^2/L^3) \tag{4.13}$$

Experimental $I(V)$ characteristics for the injected current have more complex behavior than that given by Equations 4.11 to 4.13 and are related to the presence of traps and the electric field dependence of charge carrier mobility. In particular, the power law $I \sim V^\gamma$ holds for the exponential distribution of trap depths with characteristic scale T_0 so that $\gamma = T_0/T$ (Blom and Vissenberg, 2000). The dependence of mobility on the electric field often obeys the law $\mu = \mu_0 \exp[(E/E_0)^{1/2}]$. In this case, for large voltages the current follows the same law, $I \sim I_0 \exp[(V/V_0)^{1/2}]$. Another source of strong field dependence is the charge injection through the barrier interface (Khramtchenkov et al., 1996).

At bipolar charge injection (see Figure 4.34) the charges of electrons and holes neutralize each other inside a sample, and at balanced injection, $p_0 = n_0$, the current obeys Ohm's law, i.e.,

$$I = e(\mu_p p_0 + \mu_n n_0)\ (V/L) \tag{4.14}$$

The e-h recombination destroys local electroneutrality and decreases the current. At low recombination and $\mu_p = \mu_n = \mu$ the current decrease, δI, can be estimated by the perturbation and the result reads

$$\delta I/I = -(\mu_r/\mu)\ (V_c/V)\ [1 + (1/3)(V_c/V)], \quad V_c = (e/\varepsilon) p_0 L^2, \quad \mu_r = \varepsilon B/2e \tag{4.15}$$

where B is the recombination coefficient. According to Equation 4.15, the effect of recombination becomes more noticeable at low applied voltages $V \ll V_c$. In the limit of high recombination mobility, $\mu_r \gg \mu$, the electrons and holes become spatially

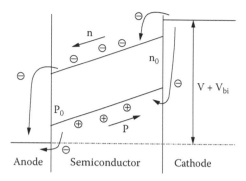

FIGURE 4.34 Metal-semiconductor-metal structure with ohmic contacts for electrons and holes.

separated and the current through the sample is given by Equations 4.11 to 4.13 for the unipolar space charge limited current with replacements $V \to V/2$, $L \to L/2$.

To increase the e-h recombination output in LEDs, the layers for blocking electron transport into the anode and hole transport into the cathode are incorporated (see Figure 4.35). In the case when the current is provided solely due to recombination and in the limit p_0, $n_0 \to \infty$, the exact $I(V)$ characteristics were found by Parmenter and Ruppel (1959). The current I as a function of applied voltage, V, and the film thickness, L, is given again by the space charge limited law:

$$I = (9/8)\, \varepsilon\mu_{\text{eff}}(V^2/L^3) \tag{4.16}$$

where μ_{eff} is the effective mobility determined as

$$\mu_{\text{eff}} = (4/9)\mu_R v_n v_p \Gamma^2(3(v_n + v_p)/2)\Gamma^{-2}(3v_n/2)\Gamma^{-2}(3v_p/2)\Gamma^{-3}(v_n + v_p)\Gamma^3(v_n)\Gamma^3(v_p) \tag{4.17}$$

Here $\Gamma(x)$ is the gamma function and the other parameters in Equation 4.17 are

$$\mu_r = \varepsilon B/(2e), \qquad v_n = \mu_n/\mu_r, \qquad v_p = \mu_p/\mu_r \tag{4.18}$$

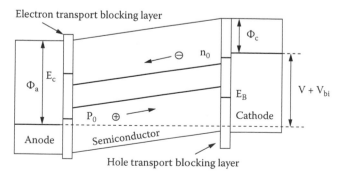

FIGURE 4.35 LED structure with electron and hole transport blocking layers.

Equation 4.17 is valid at any relationship between the mobilities. In the limit of high recombination $v_{n,p} \ll 1$, the effective mobility in Equation 4.17 reads

$$\mu_{\text{eff}} = \mu_n + \mu_p \tag{4.19}$$

In this limit the entire sample is split into two spatially separated parts with electron and hole space charge limited currents. The e-h recombination takes place in the narrow region where electron and hole currents meet. In the opposite limit of high electron and hole mobilities, $v_{n,p} \gg 1$, according to Equation 4.17 the effective mobility in Equation 4.16 becomes

$$\mu_{\text{eff}} = (2/3)[2\pi\mu_n\mu_p(\mu_n + \mu_p)/\mu_R]^{\frac{1}{2}} \tag{4.20}$$

In this case electrons and holes are spread over the whole sample by concentrating near the contacts ($p \sim n \sim 1/(\mu_r x)^{1/2}$ and $p \sim n \sim 1/(\mu_r(L-x))^{1/2}$), where most of charge recombination occurs.

Thus the recombination current (Equation 4.16) monotonically increases with decreasing the recombination constant. This increase is provided by an increase of charge carrier concentrations $p \sim n \sim 1/\sqrt{\mu_r}$. However, charge concentrations inside the semiconductor cannot exceed the contact values p_0 and n_0. Therefore, the dependence (Equation 4.20) becomes invalid at very small μ_r.

The dependence of recombination current in the limit of vanishing recombination coefficient can be found using simple arguments. In the absence of recombination the semiconductor represents the bulk e-h capacitor in which electron and hole concentrations are determined by electrochemical potentials of anode and cathode, $\xi_{a,c}$ (Prigodin et al., 2008) (see Figure 4.36a). The profile of electrical potential across the structure is shown in Figure 4.36b. The electrical field is absent inside of a bulk semiconductor due to screening of the external field by holes at the cathode and by electrons at the anode. The corresponding screening lengths, $r_{n,p}$, are assumed to be less than L. The electrical potential of bulk semiconductor, ϕ_0, is determined by the requirement of electroneutrality. The external potential, $V_{\text{ext}} = V + V_{\text{bi}}$, determines the difference between chemical potentials, $\zeta_{n,p}$, of electrons and holes. So there is the following set of equations for determining the electron and hole concentrations:

$$c(V) \equiv n(\zeta_n) = p(\zeta_p), \qquad \zeta_n = \xi_c - e\varphi_0, \ \zeta_p = \xi_a - e\varphi_0, \qquad \xi_c - \xi_a = e(V + V_{\text{bi}}) \tag{4.21}$$

where $n(\zeta_n) = \int f(E - \zeta_n)N_C(E) \, dE$ and $p(\zeta_p) = \int (1 - f(E - \zeta_p))N_V(E)dE$, with $N_{C,V}(E)$ being the density of states in conduction and valence bands, and $f(E)$ the Fermi function. The recombination current through the device represents a leakage current of a bulk supercapacitor and, in the first approximation along the recombination coefficient B, is

$$I = eBLc^2(V) \tag{4.22}$$

Assuming that density of states of conduction and valence bands can be described by the same function, $N(E)$, Equation 4.20 reads

$$c(V) = \int_0^\infty f(E + E_g/2 - e(V + V_{\text{bi}})/2)N(E)\,\mathrm{d}E$$

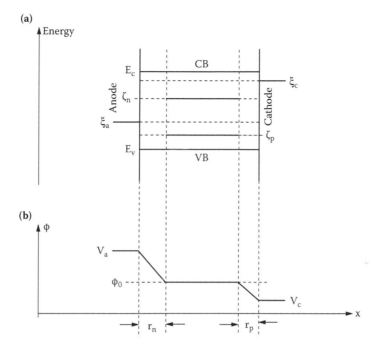

FIGURE 4.36 Semiconductor layer as a supercapacitor. (a) Energy diagram of the LED structure shown in Figure 4.35 for suppressed e-h recombination: the chemical potentials for electrons and holes, $\zeta_{n,p}$, inside the semiconductor are controlled by anode and cathode electrochemical potentials, $\xi_{a,c}$. (b) Profile of electrical potential across the structure: $r_{n,p}$ values are the screening Debye radius due to electrons and holes, ϕ_0 is the constant electrical potential of the semiconductor, and V_a and V_c are the electrical potential of the anode and cathode.

where $E_g = E_c - E_v$ is the gap. For $e(V + V_{bi}) > E_g$ the electrons as well as the holes represent the degenerate gas and we have

$$c(V) = \int_0^{(V+Vbi-Eg)/2} N(E)\,dE \tag{4.23}$$

In the approximation of effective mass, m, the charge concentration is equal to $c(V) = m^{3/2}(3\pi^2\hbar^3)^{-1}(V + V_{bi} - E_g)^{3/2}$. For low applied voltages, $e(V_{bi} + V) < E_g$, when electrons and holes remain nondegenerate, Equation 4.21 becomes

$$c(V) = c_0 \exp[(V + V_{bi})/2k_BT], \quad c_0 = \exp(-E_g/2k_BT) \int_0^\infty N(E)\exp(-E/k_BT)\,dE \tag{4.24}$$

Again in the approximation of effective mass it follows that

$$c_0 = 2(mk_BT/2\pi\hbar^2)^{3/2}\exp(-E_g/2k_BT).$$

Equation 4.22, with Equations 4.23 and 4.24, describes the recombination current at vanishing recombination constant. To get the overall dependence of

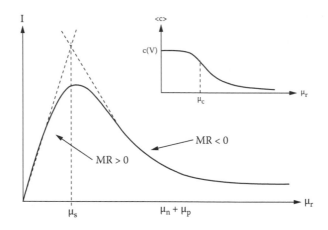

FIGURE 4.37 Schematic dependence of recombination current on the recombination mobility. There are two regimes: recombination ($\mu_r < \mu_c$) and space charge ($\mu_r > \mu_c$) limited currents. Assuming that the recombination coefficient decreases with magnetic field, the magnetoresistance is positive and negative in these regimes. The insert shows the mean over a sample charge carrier density, $<c>$ (electrons or holes), as a function of the recombination mobility. The saturation value, $c(V)$, is determined the external voltage (Equations 4.23 and 4.24). (Reproduced with permission from Prigodin et al., *Synth. Met.* 156, 757 [2006]. Copyright © 2006 Elsevier.)

current on the recombination coefficient at a qualitative level, we can match Equation 4.22 with the current for a small recombination constant given by Equations 4.16 and 4.20. So we find that the current has a maximum at $\mu = \mu_c$ (see Figure 4.37), where

$$\mu_c = (9\pi/32)^{1/3} \, (\mu_n + \mu_p) \, (2 + \mu_n/\mu_p + \mu_n/\mu_n)^{-1/3} \, (V/V_c)^{4/3}, \quad V_c = (e/\varepsilon)_c(V)L^2 \qquad (4.25)$$

As long as the recombination mobility $\mu_r > \mu_c$, the current follows the law

$$I = (3\varepsilon/4)[2\pi\mu_n\mu_p(\mu_n + \mu_p)]^{1/2} \, V^2 L^{-3}\mu_r^{-1/2} \qquad (4.26)$$

At low recombination mobility $\mu_r < \mu_c$, the current is given by Equation 4.22. The insert in Figure 4.37 shows the average charge carrier density in these two regimes.

The parameter that is most sensitive to an external magnetic field is the recombination coefficient B. The dependence of current on the recombination coefficient $B(H)$ eventually leads to the magnetoresistance. We define the magnetoresistance as

$$MR = I(0)/I(H) - 1 = C \, [B(H) - B(0)]/B(0) \qquad (4.27)$$

where the factor C is given:

$$C = d(\ln I(B))/d(\ln B)|_{H=0} \qquad (4.28)$$

As has been seen from Equations 4.16 and 4.22 and Figure 4.37, the sign of C may vary with the voltage.

The recombination of e-h pairs occurs in a few steps, and the magnetic field dependence of recombination constant B is considered in the next section.

4.4.3 COULOMB PAIRS

Electron-hole pairs can be considered to be bound if their mutual attraction energy, $|-eV(r)| = e^2/(4\pi\varepsilon r)$, is larger than $k_B T$. This condition determines the Onsager radius (Albrecht and Bässler, 1995),

$$r_c = e^2/(4\pi\varepsilon k_B T) \tag{4.29}$$

for formation of bound e-h (coulomb) pairs. For $\varepsilon = 3$ and $T = 300$ K we have $r_c = 18.5$ nm. The formation coefficient of coulomb pairs, b, can be defined as the mutual drift flow of electron and hole through the Onsager sphere, i.e.,

$$b = 4\pi r^2 (\mu_n + \mu_p)(dV(r)/dr)|_{r=rc} = (e/\varepsilon) (\mu_n + \mu_p) \tag{4.30}$$

Equation 4.30 represents the Langevin formula (Pivrikas et al., 2005).

Distribution of long-living coulomb pairs over spatial separations is determined by their attraction energy, which also can be presented in the form

$$E(r) = -E_0 r_0/(r + r_0), \quad E_0 = e^2/(4\pi\varepsilon r_0) \tag{4.31}$$

where we introduced such scales as the size of Frenkel exciton, r_0, and its energy, E_0. The probability, $P(r)$, for a pair to have a given e-h separation, r, is proportional to the Boltzmann factor, $\exp[-E(r)/k_B T]$, and the structural factor, $W(r)$, i.e.,

$$P(r) = (1/N) \exp[r_c/(r + r_0)] W(r) \tag{4.32}$$

where N is the normalization factor, so that $_0\int^{rc} P(r)dr = 1$. The function $W(r)$ is the property of semiconductors to arrange long living e-h pairs (Pope and Swenberg, 1999). In polymers, the electron and hole of long-living coulomb pairs are placed on neighboring chains (Morteani et al., 2004; Rothe et al., 2005). In this case $W(r)$ excludes the distances less than the interchain separation. For the homogeneous media it follows that $W(r) = 4\pi r^2$.

In general, the long-living e-h pairs represent the electrons and holes trapped by the potential fluctuations and $W(r)$ is related to the correlation of such traps or localized states. However, one can anticipate that the main dependence of $P(r)$ is due to the Boltzmann factor in Equation 4.32. Therefore, the dissociation rate of e-h bound pairs defined as the diffusive escape rate from the Onsager sphere, $q = (D_n + D_p)$ $(dP(r)/d_r)|r = r_c$, reads

$$q = (\mu_n + \mu_p)(E_0/er_0^2)(E_0/2k_B)\exp[-E_0/k_B T] \tag{4.33}$$

The above equilibrium consideration, Equations 4.32 and 4.33, makes sense if the lifetime of e-h pairs is larger than the drift relaxation time for the coulomb pairs:

$$\tau_{dr} = \int_{r0}^{rc} dr/v(r); \quad v(r) = -(\mu_n + \mu_p)dV(r)/dr \tag{4.34}$$

which also is equal to the time of diffusive spreading within the Onsager sphere:

$$\tau_{dr} = \tau_{dif}, \quad \tau_{dif} = r_c^2/(D_n + D_p) \tag{4.35}$$

Introducing the lifetime for the singlet Frenkel exciton, τ_0, the recombination rate for e-h singlet bound pairs is anticipated to be $k_s = 1/(\tau_0 + \tau_{dif})$ and the corresponding singlet recombination coefficient reads

$$B_s = (1/4)\, b[k_s/(k_s + q)] \tag{4.36}$$

where a factor ¼ is the statistical weight of singlet pairs. Triplet e-h pairs recombine after their conversion into the singlet state, as it is considered in the next section.

4.4.4 Spin Conversion of Coulomb Pairs

In the region $r_a < r < r_c$ the singlet-triplet conversion takes place for $g\mu_B H < a$. At $g\mu_B H > a$ the conversion takes place only between S and T_{+1} levels in the narrow region $\sim r_0(a/g\mu H)$ around r_H.

The spin interaction within an e-h pair can be written as (Prigodin et al., 2006)

$$H = H_0 + H_{S-T} \tag{4.37}$$

$$H_0 = \mu_B g\, (S_n + S_p)H - J(r)\,[1/2 + 2\, S_n S_p] \tag{4.38}$$

$$H_{S-T} = \mu_B \delta\, (S_n - S_p)H + [a_n\, I_n S_n + a_p\, I_p\, S_p] + H_{in} \tag{4.39}$$

Here $S_{n,p}$ represents the spin of hole and electron, and $g_{n,p}$ is the corresponding g-factors. H_0 describes noninteracting singlet and triplet spin states of bound e-h pair, $J(r)$ is the exchange integral, and $g = (1/2)(g_n + g_p)$. The first term of H_0 is the Zeeman splitting of the triplet manifold, and the second term separates the singlet and triplet states as shown in Figure 4.38a.

H_{S-T} in Equation 4.34 mixes the singlet and triplet states. The first contribution to H_{S-T} is due to the difference of g-factors for electron and hole, $\delta = (g_n - g_p)/2$. The next term is the hyperfine interaction where $I_{n,p}$ are the nuclear spins of the host centers for electron and hole. The last term in Equation 4.34 includes an incoherent interaction of electron and hole spins with the environment.

The singlet-triplet mixing produced by the first term in Equation 4.39 (Δg-mechanism) can be studied in a straightforward way. This term describes the transitions between only the T_0 component of the triplet manifold and the singlet state. If the initial spin configuration at the time $t = 0$ is the triplet state, the spin wave function for the next time moment is (Boehme and Lips, 2003)

$$|\Psi(t)\rangle = \{[(E - J)/2E]\exp[iEt/\hbar] + [(E + J)/2E]\exp[-iEt/\hbar]\}|T_0\rangle$$

$$+ [(E^2 - J^2)^{1/2}/2E]\{\exp[iEt/\hbar] - \exp[-iEt/\hbar]\}|S\rangle \tag{4.40}$$

where $E = [J^2 + (\mu_B H \delta)^2]^{1/2}$. Thus, according to Equation 4.40, the probability for the e-h pair to be in the singlet state at time t is

$$P_s(t) = \gamma[1 - \cos(2Et/\hbar)], \quad \gamma = (1/2)\,(\mu_B H \delta)^2/[J^2 + (\mu_B H \delta)^2] \tag{4.41}$$

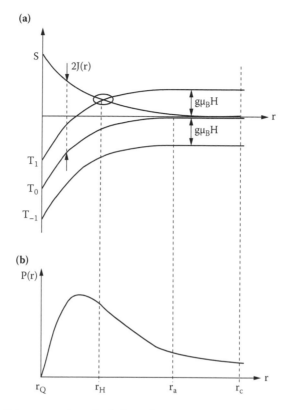

FIGURE 4.38 Spin conversion of e-h bound states. (a) Spin levels (singlet S, triplets T_0 and $T_{\pm 1}$) of e-h pair as a function of e-h separation r and the Zeeman splitting energy $g\mu_B H$. Splitting between S and T_0 is controlled by the exchange interaction between electron and hole spins, $J(r) = J_0 \exp(-2r/r_0)$, where r_0 is the size of Frenkel exciton and J_0 is the energy difference between singlet and triplet levels of Frenkel exciton. (b) $P(r)$ is the distribution of e-h bound pairs over separations. $r_c = r_0(E_0/k_B T)$ is the Onsager radius of formation of bound e-h pair, where E_0 is the energy of Frenkel exciton. The characteristic scales r_a and r_H are determined by equalities $J(r_a) = a$ and $J(r_H) = g\mu_B H$, where a is hyperfine frequency. (Reproduced with permission from Prigodin et al., *Synth. Met.* 156, 757 [2006]. Copyright © 2006 Elsevier.)

Taking into account that the e-h pairs recombine only from the singlet state with the rate k_s, the recombination rate for the initial triplet T_0 state becomes γk_s. Also, bound e-h pairs can dissociate into free electrons and holes with the rate q; therefore, the total e-h recombination constant B_0 for the T_0-channel due to the Δg-mechanism is

$$B_0(H) = (b/4) \, [\gamma k_s/(\gamma k_s + q)] \qquad (4.42)$$

Equation 4.42 enables us to find the relative variation of recombination coefficients as the function of magnetic field, and then the magnetoresistance can be calculated by using Equations 4.27 and 4.28.

The parameter $\delta = (g_n - g_p)/2$ in Equation 4.41 is anticipated to be small because of low atomic numbers for organic materials; therefore, the magnetoresistance described by Equation 4.42 remains very low for moderate magnetic fields. More essential S-T mixing is provided by the hyperfine interaction in Equation 4.39. The comprehensive analysis of this interaction requires consideration of the Liouville stochastic equation (Pedersen and Freed, 1973; Ito et al., 2003), which can be done only numerically. However, qualitatively, the hyperfine mixing can be understood within simple arguments and its effect is anticipated to be similar to the above Δg-mechanism.

One can note that there is the characteristic e-h separation, $r_a = (r_0/2) \ln(J_0/a)$, at which the exchange interaction $J(r) = J_0 \exp(-2r/r_0)$ is comparable to the hyperfine constant a. For separations $r > r_a$ the exchange interaction becomes small, so that the S-T mixing is determined by the relationship between the hyperfine constant, a, and the Zeeman splitting energy, $\mu_B gH$ (see Figure 4.38). In the absence of a magnetic field, all four spin states are degenerate and equally are presented in the spin wave function due to the hyperfine interaction. A weak external magnetic field, $\mu_B gH \ll a$, separates $T_{\pm 1}$ states from S_0 and T_0, and the mixing effect of the hyperfine interaction on these states becomes weaker. Therefore, for low magnetic fields the triplet recombination coefficients can be described as

$$B_{\pm 1} = [1 - c_1(\mu_B gH/a)^2] B_s, \quad B_0 = B_s \tag{4.43}$$

where c_1 is a numerical factor of order of unity. For large magnetic fields the hyperfine mixing $T_{\pm 1}$ with S_0 can be considered as a perturbation and the recombination constants $B_{\pm 1}$ behave as

$$B_{\pm 1} = c_2 [a/g\mu_B H]^2 B_s, \quad B_0 = B_s \tag{4.44}$$

where c_2 is another numerical factor. Equations 4.43 and 4.44 support the suggestion (Francis et al., 2004; Bobbert et al., 2007) that the overall magnetic field dependence of $B_{\pm 1}$ can be approximated by the Lorentzian shape

$$B_{\pm 1}/B_s = a^2/[a^2 + (\mu_B gH)^2] \tag{4.45}$$

According to Equation 4.44 at $\mu_B gH \gg a$, the magnetic field dependence of recombination constants approaches saturation. However, in this range of magnetic fields the spin conversion may occur at a spin-level crossing for pairs with e-h space separation $r < r_a$ (see Figure 4.38). The pair with separation $r_H = (r_0/2) \ln(J_0/\mu_B gH)$ has the degeneracy of a singlet state with one of $T_{\pm 1}$ triplet levels. The level crossing takes place in a very narrow range of separation, $\Delta r \sim r_0 (a/\mu_B gH)$, but the density of the pairs with such separations can be large due to the Boltzmann factor. So one can expect that for these magnetic fields the recombination constant of resonance triplet pairs is

$$B_r = Z B_s, \quad Z = 4\pi r^2{}_a \Delta r \, P(r_a) \tag{4.46}$$

The overall triplet recombination constant $B_t = B_0 + B_{+1} + B_{-1}$ as a function of magnetic field is shown in Figure 4.39.

Flipping of electron and hole spins by phonons included in the last term of Equation 4.39 establishes the equilibrium occupancy (equipartion) for all four e-h spin levels within the spin-lattice relaxation time T_1. The above coherent mechanism of spin conversion prevails if $aT_1 \gg 1$. Because of weak spin-orbit coupling, T_1 is

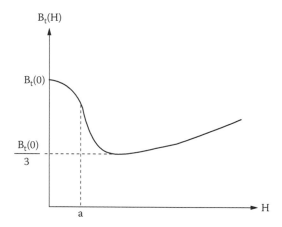

FIGURE 4.39 Schematic dependence of recombination constant on the magnetic field following the spin conversion shown in Figure 4.38. The maximum is associated with a central body of distribution of the e-h pair over size.

anticipated to be as long as microseconds (Joo et al., 1998), while the hyperfine frequency occurs to be $a = (10^7 - 10^8)$ s^{-1}. Therefore, the coherent mechanism of spin conversion for e-h bound pairs is anticipated to be dominant. Also, it is supposed that the singlet recombination rate is fast enough, so that $k_s T_1 \gg 1$.

Besides longitudinal spin relaxation there is the transverse spin relaxation (dephasing) described by T_2. This process provides the equilibrium between S and T_0 and can be included into the corresponding coherent conversion rate determined by the hyperfine interaction constant, a.

In the above consideration of spin conversion we explicitly assume that the hyperfine frequency, a, is larger than the hopping frequency, ω_h, with which electrons and holes change the nuclear surrounding. In the opposite case, the hyperfine interaction causes the small spin changes before charge carrier hops for another molecule. The effect of this spin scattering can be included into the dephasing time, $T_2 = \omega_h/a^2$ (Schulten and Wolynes, 1977).

The hopping frequency, ω_h, varies with temperature and disorder. The disorder suppresses the hopping frequency and keeps the electron and hole separated, and this way makes the coherent mechanism of magnetoresistance more effective.

4.4.5 DISCUSSION

It is now widely accepted that the low magnetic field magnetoresistance of organic semiconductors is related to competition of random precession of charge carrier spins due to hyperfine interaction and aligning of the spins by an external magnetic field (Prigodin et al., 2006; Xu et al., 2006; Bobbert et al., 2007; Wagemans et al., 2008; Bergeson et al., 2008). We suggested that this competition influences the e-h recombination of spatially separated but coulombically bound pairs and have shown that the recombination intensity may control the bipolar current in organic semiconductors (Prigodin et al., 2006).

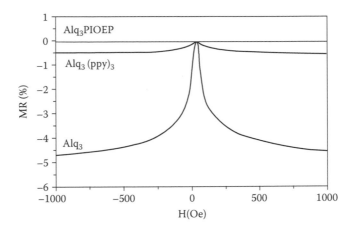

FIGURE 4.40 Magnetoresistance as a function of magnetic field for the undoped and doped Alq$_3$ devices at 300 K. (Reproduced with permission from Prigodin et al., *Synth. Met.* 156, 757 [2006]. Copyright © 2006 Elsevier.)

Within the present model we already explained the suppression of MR by introducing metal atoms into organic semiconductors (Prigodin et al., 2006). We attributed the suppression to increasing of spin-orbit coupling and shortening the spin relaxation time, T_1. As it already was pointed out, the short relaxation time washes out magnetic-field-dependent coherent spin conversion. In Figure 4.40 the changes of magnetoresistance of the Alq$_3$ LED device are shown after introducing Ir and Pt complexes.

The nonmonotonic dependence of the current on the recombination rate can lead the inversion of magnetoresistance. As it was discussed (Bergeson et al., 2008), the present model qualitatively describes the variety of magnetoresistance behavior with temperature, applied voltage, and sample thickness. Figure 4.41 schematically demonstrates that the dependence of current on the recombination mobility varies with changes of device parameters V, T, L and $\mu_{n,p}$, which determines the position of maximum μ_c. The variation of recombination mobility, μ_r, is assumed to independently be controlled with the magnetic field in some restricted window $\Delta\mu_r$.

In Figure 4.42 the experimental data for magnetoresistance of α-6T based LED devices is shown. The results are consistent with prediction of the present model. In particular, the sign of magnetoresistance can be inversed with applied voltage back and forth (Figure 4.41b). At the same time, increasing the temperature and the thickness can lead to a change of the magnetoresistance sign only once, from positive to negative.

One can mention that Desai et al. (2007) reported that the presence of a hole transport blocking layer fixes the sign of magnetoresistance to always be positive. We note that within the present model, the installation of a blocking layer is a very effective way to switch charge transport from the space charge limited regime to the recombination limited regime. It is anticipated that the position of maximum $I(\mu_r)$ is shifted to large values and the sign of magnetoresistance becomes always positive.

Niedermeier et al. (2008) have shown that the magnetoresistance increases with the introduction of recombination centers by device conditioning. The present model can be directly generalized to include these centers. We expect that the basic

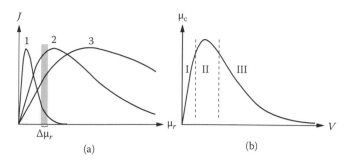

FIGURE 4.41 (a) Schematic plot of curves for current as a function of recombination mobility at low (1), intermediate (2), and high (3) values of μ_c, where μ_c is the peak of each curve. The shaded region indicates the accessible range of μ_r ($\Delta\mu_r$). The portion of a given curve within the shaded region determines whether the MR will be negative, positive, or a mixture. (b) Plot of the critical recombination mobility, as a function of applied voltage V. The different regions I, II, and III indicate where the function has a positive, maximum, or negative slope, respectively. The parameters V, T, L, and $\mu_{n,p}$ determine the curve in which the device operates. (Reproduced with permission from Bergeson et al., *Phys. Rev. Lett.* 100, 067201 [2008]. Copyright © 2008 American Physical Society. http://link.aps.org/doi/10.1103/PhysRevLett.100.067201.)

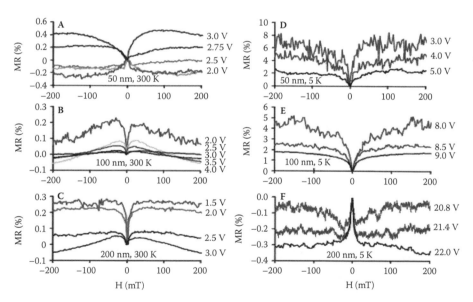

FIGURE 4.42 Magnetoresistance (MR) for devices with α-6T layers of varying thickness. Panels (a–c) show MR at 300 K for devices with α-6T layers of (a) 50 nm, (b) 100 nm, and (c) 200 nm, while panels (d–f) show MR at 5 K for layer thicknesses of (d) 50 nm, (e) 100 nm, and (f) 200 nm. Shown in (a) and (c) are examples of MR inversion with applied bias, from negative to positive, and vice versa for (c). The magnitude of the MR in (d) and (e) decreases as bias is increased. The MR in (f) is a mixture of positive and negative behavior that becomes strictly negative as bias increases. (Reproduced with permission from Bergeson et al., *Phys. Rev. Lett.* 100, 067201 [2008]. Copyright © 2008 American Physical Society. http://link.aps.org/doi/10.1103/PhysRevLett.100.067201.)

conclusions of the present model also remain valid for this case. In the present equations we use such parameters as the recombination rate $k_{S,T}$ for singlet and triplet coulomb pairs. In general, the parameter $k_{S,T}$ can include any annihilation of S,T-coulomb pair, except dissociation into free charge carriers. It could also be radiative and nonradiative recombination at recombination centers, as well as discharge of coulomb pairs at contacts. In this way, the parameter $k_{S,T}$ also can be independently modulated.

Recently the model based on bipolarons was introduced to explain the organic magnetoresistance (Bobbert et al., 2007; Wagemans et al., 2008). The concentration of bipolarons is supposed to be a factor that controls the charge transport, and in turn the concentration of bipolarons is governed by the spin dynamics of interacting polarons. In fact, the bipolaron spin dynamics is similar to the spin dynamics of the e-h bound pair. Therefore, the origin of magnetic field dependence is the same in both models. We would like to point out the differences between the models. In the bipolaron model the magnetoresistance is bulk property and takes place for unipolar transport. For the e-h bound pair model the magnetoresistance is device dependent with bipolar charge transport. Additional experimental studies are required for differentiating the models. The new evidence for the e-h recombination model was given by Majumdar et al. (2009).

In summary, the model of spin conversion for magnetoresistance in organic semiconductors was extended in a number of aspects. The conversion mechanism now includes three types of spin mixing processes: S-T_0, S-$T_{\pm 1}$, and T_0-$T_{\pm 1}$. Also, the model includes the analysis of distribution of the singlet-triplet exchange. As a result, a large variety of magnetic field dependences and even changes in the sign of magnetoresistance are anticipated. The experimental data for MR demonstrate the predicted large variety of dependences.

ACKNOWLEDGMENT

We thank Joel S. Miller and K. Z. Pokhodnya for the long-time collaborations in the study of organic/modecule-based magnets. We are thankful for funding support from AFOSR under grant No. FA9550-06-1-0175, DOE under Grant No. DE-FG02-01ER45931, No. DE-FG02-86ER45271, NSF under Grant No. DHR-0805220, and OSU Institute for Materials Research.

REFERENCES

Albrecht, U., and H. Bässler, 1995, Angevin-type charge-carrier recombination in a disordered hopping system, *Phys. Status Solidi B* 191, 455–59.

Allemand, P. M., K. C. Khemani, A. Koch, F. Wudl, K. Holczer, S. Donovan, G. Gruner, and J. D. Thompson, 1991, Organic molecular soft ferromagnetism in a fullerene-C_{60}, *Science* 253, 301–3.

de Almeida, J. R. L., and D. J. Thouless, 1978, Stability of Sherrington-Kirkpatrick solution of a spin glass model, *J. Phys. A* 11, 983–90.

Baibich, M. N., J. M. Broto, A. Fert, F. N. V. Dau, F. Petroff, P. Eitenne, G. Creuzet, A. Friederich, and J. Chazelas, 1988, Giant magnetoresistance of (001)Fe/(001)Cr magnetic superlattices, *Phys. Rev. Lett.* 61, 2472–75.

Baldo, M. A., and S. R. Forrest, 2001, Interface-limited injection in amorphous organic semi-conductors, *Phys. Rev. B* 64, 085201.

Bandyopadhyay, S., and M. Cahay, 2008, *Introduction to Spintronics*, CRC Press, Boca Raton, FL.

Barbara, B., 2005, Single-particle nanomagnetism, *Solid State Sci.* 7, 668–81.

Barbara, B., and M. Uehara, 1976, Anisotropy and coercivity in $SmCo_5$-based compounds, *IEEE Trans. Magn.* 12, 997–99.

Becker, K. W., 1982, Theory of electron-spin resonance in spin-glasses with remanence and anisotropy, *Phys. Rev. B* 26, 2394–408.

Bergeson, J. D., V. N. Prigodin, D. M. Lincoln, and A. J. Epstein, 2008, Inversion of magne-toresistance in organic semiconductors, *Phys. Rev. Lett.* 100, 067201.

Binasch, G., P. GrÄunberg, F. Saurenbach, and W. Zinn, 1989, Enhanced magnetoresistance in layered magnetic structures with antiferromagnetic interlayer exchange, *Phys. Rev. B* 39, 4828–30.

Bleuzen, A., C. Lomenech, V. Escax, F. Villain, F. Varret, C. C. D. Moulin, and M. Verdaguer, 2000, Photoinduced ferrimagnetic systems in Prussian blue analogues $C^I_x Co_4[Fe(CN)_6]_y$ (C^I = alkali cation). 1. Conditions to observe the phenomenon, *J. Am. Chem. Soc.* 122, 6648–52.

Blom, P. W. M., and M. J. M. Vissenberg, 2000, Charge transport in poly(p-phenylene vinylene), *Mater. Sci. Eng.* 27, 53–94.

Bobbert, P. A., T. D. Nguyen, F. A. van Oost, B. Koopmans, and M. Wohlgenannt, 2007, Bipolaron mechanism for organic magnetoresistance, *Phys. Rev. Lett.* 99, 216801.

Boehme, C., and K. Lips, 2003, Theory of time-domain measurement of spin-dependent recom-bination with pulsed electrically detected magnetic resonance, *Phys. Rev. B* 68, 245105.

Brillson, L. J., 1993, *Contacts to Semiconductors; Fundamentals and Technology*, Noyes, Park Ridge, NJ.

Brinckerhoff, W. B., B. G. Morin, E. J. B. J. S. Miller, and A. J. Epstein, 1996, Magnetization and dynamics of reentrant ferrimagnetic spin-glass [MnTPP]$^{.+}$[TCNE]$^{-}$·2PhMe, *J. Appl. Phys.* 79, 6147–49.

Brinckerhoff, W. B., J. Zhang, J. S. Miller, and A. J. Epstein, 1995, Magnetization of high-T_c molecule-based magnet $V/TCNE/CH_2Cl_2$, *Mol. Cryst. Liq. Cryst.* 271, A195–205.

Burroughes, J. H., D. D. C. Bradley, A. R. Brown, R. N. Marks, K. Mackey, R. H. Friend, P. L. Burns, and A. B. Holms, 1990, Light-emitting-diodes based on conjugated polymers, *Nature* 347, 539–41.

Burrows, P. E., S. R. Forrest, L. S. Sapochak, J. Schwartz, P. Fenter, T. Buma, V. S. Ban, and J. L. Forrest, 1995, Organic vapor-phase deposition—A new method for the growth of organic thin-films with large optical nonlinearities, *J. Cryst. Growth* 156, 91–98.

Butler, W. H., X.-G. Zhang, T. C. Schulthess, and J. M. MacLaren, 2001, Spin-dependent tun-neling conductance of Fe/MgO/Fe sandwiches, *Phys. Rev. B* 63, 054416.

Carlegrim, E., B. Gao, A. Kanciurzewska, M. P. de Jong, Z. Wu, Y. Luo, and M. Fahlman, 2008a, Near-edge x-ray absorption studies of Na-doped tetracyanoethylene films: A model system for the $V(TCNE)_x$ room-temperature molecular magnet, *Phys. Rev. B* 77, 054420.

Carlegrim, E., A. Kanciurzewska, P. Nordblad, and M. Fahlman, 2008b, Air-stable organic-based semiconducting room temperature thin film magnet for spintronics applications, *Appl. Phy. Lett.* 92, 163308.

de Caro, D., M. Basso-Bert, J. Sakah, H. Casellas, J. P. Legros, L. Valade, and P. Cassoux, 2000, CVD-grown thin films of molecule-based magnets, *Chem. Mater.* 12, 587–89.

Caruso, A. N., K. I. Pokhodnya, W. W. Shum, W. Y. Ching, B. Anderson, M. T. Bremer, E. Vescovo, P. Rulis, A. J. Epstein, and J. S. Miller, 2009, Direct evidence of electron spin polarization from an organic-based magnet: [Fe^{II}(TCNE)(NCMe)$_2$][Fe^{III}Cl$_4$], *Phys. Rev. B* 79, 195202.

Chiang, C. K., C. R. Fincher, Y. W. Park, A. J. Heeger, H. Shirakawa, E. J. Louis, S. C. Gau, and A. G. Macdiarmid, 1977, Electrical-conductivity in doped polyacetylene, *Phys. Rev. Lett.* 39, 1098–101.

Chitteppeddi, S., K. R. Cromack, J. S. Miller, and A. J. Epstein, 1987, Ferromagnetism in molecular decamethylferrocenium tetracyanoethenide, *Phys. Rev. Lett.* 58, 2695–98.

Chudnovsky, E. M., 1986, Disorder and spin correlations in an amorphous ferromagnet, *Phys. Rev. B* 33, 2021–23.

Chudnovsky, E. M., 1989, Dependence of the magnetization law on structural disorder in amorphous ferromagnets, *J. Magn. Magn. Mater.* 79, 127–30.

Chudnovsky, E. M., and R. A. Serota, 1982, Spin-glass and ferromagnetic states in amorphous solids, *Phys. Rev. B* 26, 2697–99.

Cinchetti, M., K. Heimer, J.-P. Wustenberg, O. Andreyev, M. Bauer, S. Lach, C. Ziegler, Y. Gao, and M. Aeschlimann, 2009, Determination of spin injection and transport in a ferromagnet/organic semiconductor heterojunction by two-photon photoemission, *Nat. Mater.* 8, 115–19.

Cole, K. S., and R. H. Cole, 1941, Dispersion and absorption in dielectrics I. Alternating current characteristics, *J. Chem. Phys.* 9, 341.

Coronado, E., J. R. Galán-Mascarós, C. J. Gómez-García, and J. M. Martínez-Agudo, 2001, Molecule-based magnets formed by bimetallic three-dimensional oxalate networks and chiral tris(bipyridyl) complex cations. The series $[Z^{II}(bpy)_3][ClO_4][(M^{II}Cr^{III}(ox)_3]$ (Z^{II} = Ru, Fe, Co, and Ni; M^{II} = Mn, Fe, Co, Cu, and Zn; ox = oxalate dianion), *Inorg. Chem.* 40, 113–20.

Coronado, E., C. J. Gómez-García, A. Nuez, F. M. Romero, E. Rusanov, and H. Stoeckli-Evans, 2002, Ferromagnetism and chirality in two-dimensional cyanide-bridged bimetallic compounds, *Inorg. Chem.* 41, 4615–17.

Culp, J. T., J. H. Park, M. W. Meisel, and D. R. Talham, 2003, Monolayer, bilayer, multi-layers: Evolving magnetic behavior in Langmuir-Blodgett films containing a two-dimensional iron-nickel cyanide square grid network, *Inorg. Chem.* 42, 2842–48.

Datta, S., and B. Das, 1990, Electronic analog of electro-optic modulator, *Appl. Phys. Lett.* 56, 665–67.

Decurtins, S., P. Gütlich, C. P. Kohler, H. Spiering, and A. Hauser, 1984, Light-induced excited spin state trapping in a transition-metal complex—The hexa-1-propyltetrazole-iron(II) tetrafluoroborate spin-crossover system, *Chem. Phys. Lett.* 105, 1–4.

Dediu, V., L. E. Hueso, I. Bergenti, A. Riminucci, F. Borgatti, P. Graziosi, C. Newby, F. Casoli, M. P. de Jong, C. Taliani, and Y. Zhan, 2008, Room-temperature spintronic effects in Alq3-based hybrid devices, *Phys. Rev. B* 78, 115203.

Dediu, V., M. Murgia, F. C. Matacotta, C. Taliani, and S. Barbanera, 2002, Room temperature spin polarized injection in organic semiconductor, *Solid State Commun.* 122, 181–84.

Delucia, J. F., T. Gustafson, Y. Wang, D. Wang, and A. Epstein, 2002, Exciplex dynamics and emission from nonbonding energy levels in electronic polymer blends and bilayers, *Phys. Rev. B* 65, 235204.

Desai, P., P. Shakya, T. Kreouzis, and W. P. Gillin, 2007, Magnetoresistance and efficiency measurements of Alq3-based OLEDs, *Phys. Rev. B* 75, 094423.

DiBenedetto, S. A., A. Facchetti, M. A. Ratner, and T. J. Marks, 2009, Molecular self-assembled monolayers and multilayers for organic and unconventional inorganic thin-film transistor applications, *Adv. Mater.* 21, 1407–33.

Dixon, D. A., and J. S. Miller, 1987, Crystal and molecular-structure of the charge-transfer salt of decamethylcobaltocene and tetracyanoethylene $(2:1):([CO(C_5Me_5)_2]^+)_2$ $[(NC)_2CC(CN)_2]^{2-}$. The electronic structure and spectra of $[TCNE]^n$ (n = 0, 1-, 2-), *J. Am. Chem. Soc.* 109, 3656–64.

Djayaprawira, D. D., K. Tsunekawa, M. Nagai, H. Maehara, S. Yamagata, N. Watanabe, S. Yuasa, Y. Suzuki, and K. Ando, 2005, 230% room-temperature magnetoresistance in CoFeB/MgO/CoFeB magnetic tunnel junctions, *Appl. Phys. Lett.* 86, 092502.

Drew, A. J., J. Hoppler, L. Schulz, F. L. Pratt, P. Desai, P. Shakya, T. Kreouzis, W. P. Gillin, A. Suter, N. A. Morley, V. K. Malik, A. Dubroka, et al., 2009, Direct measurement of the electronic spin diffusion length in a fully functional organic spin valve by low-energy muon spin rotation, *Nature Mater.* 8, 109–14.

Epstein, A. J., 2003, Organic-based magnets: Opportunities in photoinduced magnetism, spintronics, fractal magnetism, and beyond, *MRS Bull.* 28, 492–99.

Erdin, S., 2008, Ab initio studies of tetracyanoethylene-based organic magnets, *Physica B* 403, 1964–70.

Erdin, S., and M. van Veenendaal, 2006, Photoinduced magnetism caused by charge-transfer excitations in tetracyanoethylene-based organic magnets, *Phys. Rev. Lett.* 97, 247202.

Etzkorn, S. J., W. Hibbs, J. S. Miller, and A. J. Epstein, 2002, Viscous behavior in a quasi-1D fractal cluster glass, *Phys. Rev. Lett.* 89, 207201.

Fert, A., and H. Jaffrès, 2001, Conditions for efficient spin injection from a ferromagnetic metal into a semiconductor, *Phys. Rev. B* 64, 184420.

Francis, T., O. Mermer, G. Veeraraghavan, and M. Wohlgenannt, 2004, Large magnetoresistance at room temperature in semiconducting polymer sandwich devices, *New J. Phys.* 6, 185.

Frankevich, E., and E. Balabanov, 1965, New effect of increasing the photoconductivity of organic semiconductors in a weak magnetic field, *JETP Lett.* 1, 169–71.

Friedman, J. R., M. P. Sarachik, J. Tejada, and R. Ziolo, 1996, Macroscopic measurement of resonant magnetization tunneling in high-spin molecules, *Phys. Rev. Lett.* 76, 3830–33.

Fujita, W., and K. Awaga, 1996, Magnetic properties of $Cu_2(OH)_3$(alkanecarboxylate) compounds: Drastic modification with extension of the alkyl chain, *Inorg. Chem.* 35, 1915–17.

Fusco, G. C. D., L. Pisani, B. Montanari, and N. M. Harrison, 2009, Density functional study of the magnetic coupling in V(TCNE)$_2$, *Phys. Rev. B* 79, 085201.

Gabay, M., and G. Toulouse, 1981, Coexistence of spin-glass and ferromagnetic orderings, *Phys. Rev. Lett.* 47, 201–4.

Gao, E. Q., Y. F. Yue, S. Q. Bai, Z. Hea, and C. H. Yan, 2004, From achiral ligands to chiral coordination polymers: Spontaneous resolution, weak ferromagnetism, and topological ferrimagnetism, *J. Am. Chem. Soc.* 126, 1419–29.

Garcia, V., M. Bibes, A. Barthélémy, M. Bowen, E. Jacquet, J.-P. Contour, and A. Fert, 2004, Temperature dependence of the interfacial spin polarization of $La_{2/3}Sr_{1/3}MnO_3$, *Phys. Rev. B* 69, 052403.

Gatteschi, D., A. Caneschi, L. Pardi, and R. Sessoli, 1994, Large clusters of metal-ions—The transition from molecular to bulk magnets, *Science* 265, 1054–58.

Gîrţu, M. A., C. M. Wynn, W. Fujita, K. Awaga, and A. J. Epstein, 1998, Coexistence of glassiness and canted antiferromagnetism in triangular quantum Heisenberg antiferromagnets with weak Dzyaloshinskii-Moriya interaction, *Phys. Rev. B* 57, 11058–61.

Gîrţu, M. A., C. M. Wynn, W. Fujita, K. Awaga, and A. J. Epstein, 2000, Glassiness and canted antiferromagnetism in three geometrically frustrated triangular quantum Heisenberg antiferromagnets with additional Dzyaloshinskii-Moriya interaction, *Phys. Rev. B* 61, 4117–130.

Gütlich, P., and H. Goodwin, 2004, Spin crossover—An overall perspective, *Top. Curr. Chem.* 233, 1–47.

Haskel, D., Z. Islam, J. Lang, C. Kmety, G. Srajer, K. I. Pokhodnya, A. J. Epstein, and J. S. Miller, 2004, Local structural order in the disordered vanadium tetracyanoethylene room-temperature molecule-based magnet, *Phys. Rev. B* 70, 54422.

Her, J. H., P. W. Stephens, K. I. Pokhodnya, M. Bonner, and J. S. Miller, 2007, Cross-linked layered structure of magnetically ordered [Fe(TCNE)$_2$]·zCH$_2$Cl$_2$ determined by Rietveld refinement of synchrotron powder diffraction data, *Angew. Chem. Int. Ed.* 46, 1521–24.

Hibbs, W., D. K. Rittenberg, K.-I. Sugiura, B. Burkhart, B. Morin, A. Arif, A. L. R. L. Liable-Sands, M. Sundaralingam, A. J. Epstein, and J. S. Miller, 2001, Solvent dependence of the structure and magnetic ordering of ferrimagnetic manganese(III) meso-tetraphenylporphyrin tetracyanoethenide, [MnTPP]$^+$[TCNE]$^-$·x(solvent). Evidence for orientationally disordered [TCNE]$^-$, *Inorg. Chem.* 40, 1915–25.

Hozumi, T., S. Ohkoshi, and K. Hashimoto, 2005, Electrochemical synthesis, crystal structure, and photomagnetic properties of a three-dimensional cyano-bridged copper-molybdenum complex, *J. Am. Chem. Soc.* 127, 3864–69.

Ikegami, T., I. Kawayama, M. Tonouchi, S. Nakao, Y. Yamashita, and H. Tada, 2008, Planar-type spin valves based on low-molecular-weight organic materials with La$_{0.67}$Sr$_{0.33}$MnO$_3$ electrodes, *Appl. Phys. Lett.* 92, 153304.

Imai, H., K. Inoue, K. Kikuchi, Y. Yoshida, M. Ito, T. Sunahara, and S. Onaka, 2004, Three-dimensional chiral molecule-based ferrimagnet with triple-helical-strand structure, *Angew. Chem. Int. Ed.* 43, 5618–21.

Inoue, K., K. Kikuchi, M. Ohba, and H. Okawa, 2003, Structure and magnetic properties of a chiral two-dimensional ferrimagnet with T_c of 38 K, *Angew. Chem. Int. Ed.* 42, 4810–13.

Ito, F., T. Ikoma, K. Akiyama, A. Watanabe, and S. Tero-Kubota, 2003, Carrier generation process on photoconductive polymer films as studied by magnetic field effects on the charge-transfer fluorescence and photocurrent, J. *Phys. Chem.* B 109, 8707–17.

Jedema, F. J., A. T. Filip, and B. J. van Wees, 2001, Electrical spin injection and accumulation at room temperature in an all-metal mesoscopic spin valve, *Nature* 410, 345–48.

Jedema, F. J., H. B. Heersche, A. T. Filip, J. J. A. Baselmans, and B. J. van Wees, 2002, Electrical detection of spin precession in a metallic mesoscopic spin valve, *Nature* 416, 713–16.

Jiang, J. S., J. E. Pearson, and S. D. Bader, 2008, Absence of spin transport in the organic semiconductor Alq$_3$, *Phys. Rev.* B 77, 035303.

Johnson, M., 2002, Spin injection in metals and semiconductors, *Semicond. Sci. Technol.* 17, 298–309.

Johnson, M., and R. H. Silsbee, 1985, Interfacial charge-spin coupling: Injection and detection of spin magnetization in metals, *Phys. Rev. Lett.* 55, 1790–93.

Johnson, M., and R. H. Silsbee, 1987, Thermodynamic analysis of interfacial transport and of the thermomagnetoelectric system, *Phys. Rev.* B. 35, 4959–72.

Johnson, M., and R. H. Silsbee, 1988, Ferromagnet-nonferromagnet interface resistance, *Phys. Rev. Lett.* 60, 377–77.

de Jong, M. P., C. Tengstedt, A. Kanciurzewska, E. Carlegrim, W. R. Salaneck, and M. Fahlman, 2007, Chemical bonding in V(TCNE)$_x$ ($x \sim 2$) thin-film magnets grown *in situ*, *Phys. Rev.* B 75, 064407.

Joo, J., S. M. Long, J. P. Poget, E. J. Oh, A. G. MacDiarmid, and A. J. Epstein, 1998, Charge transport of the mesoscopic metallic state in partially crystalline polyanilines, *Phys. Rev.* B 57, 9567–80.

Julliere, M., 1975, Tunneling between ferromagnetic films, *Phys. Lett.* A 54, 225–26.

Jurchescu, O. D., J. Baas, and T. T. M. Palstra, 2004, Effect of impurities on the mobility of single crystal pentacene, *Appl. Phys. Lett.* 84, 3061–64.

Kahn, O., and C. J. Martinez, 1998, Spin-transition polymers: From molecular materials toward memory devices, *Science* 279, 44–48.

Kalinowski, J., M. Cocchi, D. Virgili, P. D. Marco, and V. Fattori, 2003, Magnetic field effects on emission and current in Alq$_3$-based electroluminescent diodes, *Chem. Phys. Lett.* 380, 710–15.

Kaneyoshi, T., 1992, *Introduction to Amorphous Magnets* (World Scientific, Singapore).

Khramtchenkov, D. V., H. Bässler, and V. I. Arkhipov, 1996, A model of electroluminescence in organic double-layer light-emitting diodes, *J. Appl. Phys.* 79, 9283–90.

Kinoshita, M., 1994, Ferromagnetism of organic radical crystals, *Jpn. J. Appl. Phys.* 33, 5718–33.

Kittel, C., 1949, Physical theory of ferromagnetic domains, *Rev. Mod. Phys.* 21, 541–83.

Kiveris, A., Š. Kudžmauskas, and P. Pipinys, 1976, Release of electrons from traps by an electric-field with phonon participation, *Phys. Stat. Sol.* 37, 321–27.

Kortright, J. B., D. M. Lincoln, R. S. Edelstein, and A. J. Epstein, 2008, Bonding, back-bonding, and spin-polarized molecular orbitals: Basis for magnetism and semiconducting transport in V[TCNE]$_{x-2}$, *Phys. Rev. Lett.* 100, 257204.

Koshihara, S., A. Oiwa, M. Hirasawa, S. Katsumoto, Y. Iye, C. Urano, H. Takagi, and H. Munekata, 1997, Ferromagnetic order induced by photogenerated carriers in magnetic III-V semiconductor heterostructures of (In,Mn)As/GaSb, *Phys. Rev. Lett.* 78, 4617–20.

Lampert, M. M., and P. Mark, 1970, *Current Injection in Solids*, Academic Press, New York.

Laudise, R. A., C. Kloc, P. G. Simpkins, and T. Siegrist, 1998, Physical vapor growth of organic semiconductors, *J. Cryst. Growth* 187, 449–54.

Lee, W. P., K. R. Breneman, A. D. Gudmundsdottir, M. S. Platz, P. K. Kahol, A. P. Monkman, and A. J. Epstein, 1999, Charge transport and EPR of PAN-AMSPA(DCA), *Synth. Met.* 101, 819–21.

Lis, T., 1980, Preparation, structure, and magnetic-properties of a dodecanuclear mixed-valence manganese carboxylate, *Acta Crystallogr. Sect. B* 36, 2042.

Liu, Y., S. M. Watson, T. Lee, J. M. Gorham, H. E. Katz, J. A. Borchers, H. D. Fairbrother, and D. H. Reich, 2009, Correlation between microstructure and magnetotransport in organic semiconductor spin-valve structures, *Phys. Rev. B* 79, 075312.

Lou, X., C. Adelmann, S. A. Crooker, E. S. Garlid, J. Zhang, K. S. M. Reddy, S. D. Flexner, C. J. Palmstrøm, and P. A. Crowell, 2007, Electrical detection of spin transport in lateral ferromagnet-semiconductor devices, *Nature Phys.* 3, 197–202.

Lu, Y., X. W. Li, G. Q. Gong, G. Xiao, A. Gupta, P. Lecoeur, J. Z. Sun, Y. Y. Wang, and V. P. Dravid, 1996, Large magnetotunneling effect at low magnetic fields in micrometer-scale epitaxial La$_{0.67}$Sr$_{0.33}$MnO$_3$ tunnel junctions, *Phys. Rev. B* 54, R8357–60.

Lundgren, L., P. Nordblad, and L. Sandlund, 1986, Memory behavior of the spin-glass relaxation, *Europhys. Lett.* 1, 529–34.

Ma, B. Q., S. Gao, G. Su, and G. X. Xu, 2001, Cyano-bridged 4f-3d coordination polymers with a unique two-dimensional topological architecture and unusual magnetic behavior, *Angew. Chem. Int. Ed.* 40, 434–37.

Majumdar, S., H. Huhtinen, H. S. Majumdar, R. Laiho, and R. Österbacka, 2008, Effect of La$_{0.67}$Sr$_{0.33}$MnO$_3$ electrodes on organic spin valves, *J. Appl. Phys.* 104, 033910.

Majumdar, S., R. Laiho, P. Laukkanen, I. J. Väyrynen, H. S. Majumdar, and R. Österbacka, 2006, Application of regioregular polythiophene in spintronic devices: Effect of interface, *Appl. Phys. Lett.* 89, 122114.

Majumdar, S., H. S. Majumdar, H. Aarnio, D. Vanderzande, R. Laiho, and R. Österbacka, 2009, Role of electron-hole pair formation in organic magnetoresistance, *Phys. Rev. B* 79, 201202.

Manriquez, J. M., G. T. Yee, R. S. Mclean, A. J. Epstein, and J. S. Miller, 1991, A room-temperature molecular/organic-based magnet, *Science* 252, 1415–17.

Mathieu, R., P. JÄonsson, D. N. H. Nam, and P. Nordblad, 2001, Memory and superposition in a spin glass, *Phys. Rev. B* 63, 092401.

Mathon, J., and A. Umerski, 2001, Theory of tunneling magnetoresistance of an epitaxial Fe/MgO/Fe(001) junction, *Phys. Rev. B* 63, 220403.

Matsuda, K., A. Machida, Y. Moritomo, and A. Nakamura, 1998, Photoinduced demagnetization and its dynamical behavior in a (Nd$_{0.5}$Sm$_{0.5}$)$_{0.6}$Sr$_{0.4}$MnO$_3$ thin film, *Phys. Rev. B* 58, R4203–6.

Mermer, O., G. Veeraraghavan, T. Francis, Y. Sheng, D. Nguyen, M. Wohlgenannt, A. Köhler, M. Al-Suti, and M. Khan, 2005, Large magnetoresistance in nonmagnetic π-conjugated semiconductor thin film devices, *Phys. Rev. B* 72, 205202.

Miller, J. S., 2006, Tetracyanoethylene (TCNE): The characteristic geometries and vibrational absorptions of its numerous structures, *Angew. Chem. Int. Ed.* 45, 2508–25.

Miller, J. S., 2009, Oliver Kahn lecture: Composition and structure of the V[TCNE]$_x$ (TCNE = tetracyanoethylene) room-temperature, organic-based magnet—A personal perspective, *Polyhedron* 28, 1596–605.

Miller, J. S., J. C. Calabrese, A. J. Epstein, R. W. Bigelow, J. H. Zhang, and W. M. Reiff, 1986, Ferromagnetic properties of one-dimentional decamethylferrocenium tetracyano-ethylenide (1-1)-[Fe(ETA-5-C$_5$Me$_5$)]$^+$[TCNE]$^-$, *J. Chem. Soc. Chem. Commun.* 13, 1026–28.

Miller, J. S., J. C. Calabrese, R. S. McLean, and A. J. Epstein, 1992, Meso-(tetraphenyl-porphinato)manganese(III)-tetracyanoethenide, [MnTPP]$^{\cdot+}$[TCNE]$^-$—A new structure-type linear-chain magnet with a T_c of 18 K, *Adv. Mater.* 4, 498–501.

Miller, J. S., J. C. Calabrese, H. Rommelmann, S. R. Shittipeddi, J. H. Zhang, W. M. Reiff, and A. J. Epstein, 1987, Ferromagnetic behavior of [Fe(C$_5$Me$_5$)$_2$]$^+$[TNCE]$^-$. Structural and magnetic characterization of decamethylferrocenium tetracyanoethenide, [Fe(C$_5$Me$_5$)$_2$]$^+$[TNCE]$^-$·MeCN, and decamethylferrocenium pen-tacyanopropenide, [Fe(C$_5$Me$_5$)$_2$]$^+$[C$_3$(CN)$_5$]$^-$, *J. Am. Chem. Soc.* 109, 769–81.

Miller, J. S., and A. J. Epstein, 1994, Organic organometallic molecular magnetic materials—Designer magnets, *Angew. Chem. Int. Ed. Engl.* 33, 385–415.

Miller, J. S., and A. J. Epstein, 1995, Designer magnets, *Chem. Eng. News* 73, 30–41.

Miller, J. S., and A. J. Epstein, 1998, Tetracyanoethylene-based organic magnets, *Chem. Commun.* 13, 1319–25.

Miller, J. S., A. J. Epstein, and W. M. Reiff, 1985, Linear-chain ferromagnetic compounds—Recent progress, *Mol. Cryst. Liq. Cryst.* 120, 27–34.

Miller, J. S., A. J. Epstein, and W. M. Reiff, 1988a, Ferromagnetic molecular charge-transfer complexes, *Chem. Rev.* 88, 201–20.

Miller, J. S., A. J. Epstein, and W. M. Reiff, 1988b, Molecular organic ferromagnets, *Science* 240, 40–47.

Miyazaki, T., and N. Tezuka, 1995, Giant magnetic tunneling effect in Fe/Al$_2$O$_3$/Fe junction, *J. Magn. Magn. Mater.* 139, L231–34.

Moodera, J. S., L. R. Kinder, T. M. Wong, and R. Meservey, 1995, Large magnetoresistance at room temperature in ferromagnetic thin film tunnel junctions, *Phys. Rev. Lett.* 74, 3273–76.

Morin, B. G., P. Zhou, C. Hahm, A. J. Epstein, and J. S. Miller, 1993, Complex AC suscepti-bility studies of the disordered molecular based magnets V(TCNE)$_x$—Role of spinless solvent, *J. Appl. Phys.* 73, 5648–50.

Morteani, A. C., P. Sreearunothai, L. M. Herz, R. H. Friend, and C. Silva, 2004, Exciton regen-eration at polymeric semiconductor, *Phys. Rev. Lett.* 65, 247402.

Mott, N. F., 1936a, The electrical conductivity of transitions metals, *Proc. Roy. Soc. London Ser. A* 156, 699–717.

Mott, N. F., 1936b, The resistance and thermoelectric properties of the transition metals, *Proc. Roy. Soc. London Ser. A* 156, 368–82.

Mott, N. F., 1969, Conduction in non-crystalline materials. 3. Localized states in a pseudogap and near extremities of conduction and valence bands, *Phil. Mag.* 19, 835.

Mydosh, J. A., 1993, *Spin Glasses: An Experimental Introduction* (Taylor and Francis, London).

Naber, W. J. M., S. Faez, and W. G. van der Wiel, 2007, Organic spintronics, *J. Phys. D Appl. Phys.* 40, R205–28.

Niedermeier, U., M. Vieth, R. Pötzold, W. Sarfert, and H. von Seggern, 2008, Enhancement of organic magnetoresistance by electrical conditioning, *Appl. Phys. Lett.* 92, 193309.

Nightingale, M. P., and H. W. J. Blöte, 1996, Dynamic exponent of the two-dimensional Ising model and Monte Carlo computation of the subdominant eigenvalue of the stochastic matrix, *Phys. Rev. Lett.* 76, 4548–51.

Ohkoshi, S., Y. Abe, A. Fujishima, and K. Hashimoto, 1999, Design and preparation of a novel magnet exhibiting two compensation temperatures based on molecular field theory, *Phys. Rev. Lett.* 82, 1285–88.

Ohkoshi, S., S. Yorozu, O. Sato, T. Iyoda, A. Fujishima, and K. Hashimoto, 1997, Photoinduced magnetic pole inversion in a ferro-ferrimagnet: $(Fe^{II}_{0.40}Mn^{II}_{0.60})_{1.5}Cr^{III}(CN)_6$, *Appl. Phys. Lett.* 70, 1040–42.

Ohkoshi, S. I., S. Ikeda, T. Hozumi, T. Kashiwagi, and K. Hashimoto, 2006, Photoinduced magnetization with a high curie temperature and a large coercive field in a cyano-bridged cobalt-tungstate bimetallic assembly, *J. Am. Chem. Soc.* 128, 5320–21.

Onoda, S., and M. Imada, 2001, Filling-control metal-insulator transition in the Hubbard model studied by the operator projection method, *J. Phys. Soc. Jpn.* 70, 3398–418.

Park, J.-H., E. Cizmar, M. W. Meisel, Y. D. Huh, F. Frye, S. Lane, and D. R. Talham, 2004, Anistropic photoinduced magnetism of a $Rb_jCo_k[Fe(CN)_6]_l \cdot nH_2O$ thin film, *Appl. Phys. Lett.* 85, 3797–99.

Park, J.-H., E. Vescovo, H.-J. Kim, C. Kwon, R. Ramesh, and T. Venkatesan, 1998, Magnetic properties at surface boundary of a half-metallic ferromagnet $La_{0.7}Sr_{0.3}MnO_3$, *Phys. Rev. Lett.* 81, 1953–56.

Parkin, S. S. P., C. Kaiser, A. Panchula, P. M. Rice, B. Hughes, M. Samant, and S.-H. Yang, 2004, Giant tunnelling magnetoresistance at room temperature with MgO (100) tunnel barriers, *Nature Mater.* 3, 862–67.

Parmenter, R. H., and W. Ruppel, 1959, Two-carrier space-charge-limited current in a trap-free insulator, *J. Appl. Phys.* 30, 1548–58.

Patashinskii, A. Z., and V. L. Pokrovskii, 1979, *Fluctuation Theory of Phase Transitions* (Pergamon, Oxford), p. 42-43, Table 3.

Pechan, M. J., M. B. Salamon, and I. K. Schuller, 1985, Ferromagnetic-resonance in a Ni-Mo superlattice, *J. Appl. Phys.* 57, 3678–80.

Pedersen, J., and J. Freed, 1973, Theory of chemically-induced dynamic electron polarization, *J. Chem. Phys.* 58, 2746–62.

Pejaković, D. A., C. Kitamura, J. S. Miller, and A. J. Epstein, 2002, Photoinduced magnetization in the organic-based magnet $Mn(TCNE)_x \cdot y(CH_2Cl_2)$, *Phys. Rev. Lett.* 88, 057202.

Pejaković, D. A., J. L. Manson, J. S. Miller, and A. J. Epstein, 2000, Photoinduced magnetism, dynamics, and cluster glass behavior of a molecule-based magnet, *Phys. Rev. Lett.* 85, 1994–97.

Pekker, C., A. F. H. Arts, H. W. de Wijn, A. J. van Dyneveldt, and J. A. Hydosh, 1989. Activated dynamics in a two-dimentional Ising spin glass: $Rb_2Cu_{1-x}Co_xF_4$, *Phys. Rev. B.* 40, 11243.

Petruska, M. A., B. C. Watson, M. W. Meisel, and D. R. Talham, 2002, Organic/inorganic Langmuir-Blodgett films based on metal phosphonates. 5. A magnetic manganese phosphonate film including a tetrathiafulvalene amphiphile, *Chem. Mater.* 14, 2011–19.

Pipinys, P., and A. Kiveris, 2005, Phonon-assisted tunnelling as a process determining current dependence on field and temperature in MEH-PPV diodes, *J. Phys. Condens. Mater.* 17, 4147–55.

Pivrikas, A., G. Jushka, R. Österbacka, M. Westling, M. Viliūnas, K. Arlauskas, and D. Stubb, 2005, Langevin recombination and space-charge-perturbed current transients in regiorandom poly(3-hexylthiophene), *Phys. Rev. B* 71, 125205.

Plachy, R., K. I. Pokhodnya, P. C. Taylor, J. Shi, J. S. Miller, and A. J. Epstein, 2004, Ferrimagnetic resonance in films of vanadium [tetracyanoethanide]$_x$, grown by chemical vapor deposition, *Phys. Rev. B* 70, 64411.

Podzorov, V., E. Menard, A. Borissov, V. Kiryukhin, J. A. Rogers, and M. E. Gershenson, 2004, Intrinsic charge transport on the surface of organic semiconductors, *Phys. Rev. Lett.* 93, 086602.

Pokhodnya, K. I., M. Bonner, J.-H. Her, P. W. Stephens, and J. S. Miller, 2006, Magnetic ordering (T_c = 90 K) observed for layered [Fe^{II}(TCNE)$^-$(NCMe)$_2$]$^+$[Fe^{III}Cl$_4$]$^-$ (TCNE = tetracyanoethylene), *J. Am. Chem. Soc.* 128, 15592–93.

Pokhodnya, K. I., V. Burtman, A. J. Epstein, J. W. Raebiger, and J. S. Miller, 2003, Control of coercivity in organic-based solid solution V$_x$Co$_{1-x}$[TCNE]$_2$·zCH$_2$Cl$_2$ room temperature magnets, *Adv. Mater.* 15, 1211.

Pokhodnya, K. I., A. J. Epstein, and J. S. Miller, 2000, Thin-film V[TCNE]$_x$ magnets, *Adv. Mater.* 12, 410–13.

Pokhodnya, K. I., D. A. Pejaković, A. J. Epstein, and J. S. Miller, 2001, Effect of solvent on the magnetic properties of the high-temperature V[TCNE]$_x$ molecule-based magnet, *Phys. Rev. B* 63, 174408.

Pokhodnya, K. I., E. B. Vickers, M. Bonner, A. J. Epstein, and J. S. Miller, 2004, Solid solution V$_x$Fe$_{1-x}$[TCNE]$_2$·zCH$_2$Cl$_2$ room-temperature magnets, *Chem. Mater.* 16, 3218–23.

Pope, M., and C. E. Swenberg, 1999, Electronic processes in organic crystals and polymers, Oxford University, New York, 2nd Ed.

Pramanik, S., C.-G. Stefanita, S. Patibandla, S. Bandyopadhyay, K. Garre, N. Harth, and M. Cahay, 2007, Observation of extremely long spin relaxation times in an organic nanowire spin valve, *Nat. Nanotech.* 2, 216–19.

Prigodin, V., and A. Epstein, 2008, in *Introduction to Organic Electronic and Optoelectronic Materials and Devices*, ed. S.-S. Sun and L. R. Dalton, CRC Press, Boca Raton, FL, pp. 87–127.

Prigodin, V. N., J. D. Bergeson, D. M. Lincoln, and A. J. Epstein, 2006, Anomalous room temperature magnetoresistance in organic semiconductors, *Synth. Met.* 156, 757–61.

Prigodin, V. N., F. C. Hsu, J. H. Park, O. Waldmann, and A. J. Epstein, 2008, Electron-ion interaction in doped conducting polymers, *Phys. Rev. B* 78, 035203.

Prigodin, V. N., N. P. Raju, K. I. Pokhodnya, J. S. Miller, and A. J. Epstein, 2002, Spin-driven resistance in organic-based magnetic semiconductor V[TCNE]$_x$, *Adv. Mater.* 14, 1230–33.

Raju, N. P., V. N. Prigodin, and A. J. Epstein, 2010, High field linear magnetoresistance in fully spin-polarized high-temperature organic-based ferrimagnetic semiconductor V(TCNE)$_x$ films, *Synth. Met.*, in press.

Raju, N. P., T. Savrin, V. N. Prigodin, K. I. Pokhodnya, J. S. Miller, and A. J. Epstein, 2003, Anomalous magnetoresistance in high-temperature organic-based magnetic semiconducting V(TCNE)$_x$ films, *J. Appl. Phys.* 93, 6799–801.

Rashba, E. I., 2000, Theory of electrical spin injection: Tunnel contacts as a solution of the conductivity mismatch problem, *Phys. Rev. B* 62, R16267–70.

Rothe, C., S. King, and A. Monkman, 2005, Electric-field-induced singlet and triplet exciton quenching in films of the conjugated polymer polyspirobifluorene, *Phys. Rev. B* 72, 085220.

Ruden, P. P., and D. L. Smith, 2004, Theory of spin injection into conjugated organic semiconductors, *J. Appl. Phys.* 95, 4898–904.

Santos, T. S., J. S. Lee, P. Migdal, I. C. Lekshmi, B. Satpati, and J. S. Moodera, 2007, Room-temperature tunnel magnetoresistance and spin-polarized tunneling through an organic semiconductor barrier, *Phys. Rev. Lett.* 98, 016601.

Sato, O., T. Iyoda, A. Fujishima, and K. Hashimoto, 1996a, Electrochemically tunable magnetic phase transition in a high-T_c chromium cyanide thin film, *Science* 271, 49–51.

Sato, O., T. Iyoda, A. Fujishima, and K. Hashimoto, 1996b, Photoinduced magnetization of a cobalt-iron cyanide, *Science* 272, 704–5.

Schmidt, G., D. Ferrand, L. W. Molenkamp, A. T. Filip, and B. J. van Wees, 2000, Fundamental obstacle for electrical spin injection from a ferromagnetic metal into a diffusive semiconductor, *Phys. Rev. B* 62, R4790–93.

Schulten, K., and P. Wolynes, 1977, Semiclassical description of electron spin motion in radicals including the effect of electron hopping, *J. Chem. Phys.* 68, 3292–97.

Schultz, S., E. M. Gullikson, D. R. Fredkin, and M. Tovar, 1980, Simultaneous electron-spin-resonance and magnetization measurements characterizing the spin-glass state, *Phys. Rev. Lett.* 45, 1508–12.

Seip, C. T., G. E. Granroth, M. W. Meisel, and D. R. Talham, 1997, Langmuir-Blodgett films of known layered solids: Preparation and structural properties of octadecylphosphonate bilayers with divalent metals and characterization of a magnetic Langmuir-Blodgett film, *J. Am. Chem. Soc.* 119, 7084–94.

Sessoli, R., H. L. Tsai, A. R. Schake, S. Y. Wang, J. B. Vincent, K. Folting, D. Gatteschi, G. Christou, and D. N. Hendrickson, 1993, High-spin molecules–[$Mn_{12}O_{12}(O_2CR)_{16}(H2O)_4$], *J. Am. Chem. Soc.* 115, 1804–16.

Shaw, W. W., A. J. Epstein, and J. S. Miller, 2009. Spin-polarized electronic structure for the layered two-dimensional [$Fe^{II}[TCNE](NCHe)_2$][$Fe^{III}Cl_4$] organic based magnet, *Phys. Rev. B.* 80, 064403.

Shim, J. H., K. V. Raman, Y. J. Park, T. S. Santos, G. X. Miao, B. Satpati, and J. S. Moodera, 2008, Large spin diffusion length in an amorphous organic semiconductor, *Phys. Rev. Lett.* 100, 226603.

Shirakawa, H., E. J. Louis, A. G. Macdiarmid, C. K. Chiang, and A. J. Heeger, 1977, Synthesis of electrically conducting organic polymers—Halogen derivatives of polyacetylene, $(CH)_x$, *J. Chem. Soc. Chem. Commun.* 16, 578–80.

Shklovskii, B. I., and A. L. Efros, 1984, *Electronic Properties of Doped Semiconductors*, Springer-Verlag, Berlin.

Shtein, M., H. F. Gossenberger, J. B. Benziger, and S. R. Forrest, 2001, Material transport regimes and mechanisms for growth of molecular organic thin films using low-pressure organic vapor phase deposition, *J. Appl. Phys.* 89, 1470–76.

Sirringhaus, H., P. J. Brown, R. H. Friend, M. M. Nielsen, K. B. B. M. W. Langeveld-Voss, A. J. H. Spiering, R. A. J. Janssen, E. W. Meijer, P. Herwig, and D. M. de Leeuw, 1999, Two-dimensional charge transport in self-organized, high-mobility conjugated polymers, *Nature* 401, 685–88.

Sirringhaus, H., T. Kawase, R. H. Friend, T. Shimoda, M. Inbasekaran, W. Wu, and E. P. Woo, 2000, High-resolution inkjet printing of all-polymer transistor circuits, *Science* 290, 2123–26.

Sirringhaus, H., N. Tessler, and R. H. Friend, 1998, Integrated optoelectronic devices based on conjugated polymers, *Science* 280, 1741–44.

Stryjewski, E., and N. Giordano, 1977, Metamagnetism, *Adv. Phys.* 26, 487–650.

Talham, D. R., T. Yamamoto, and M. W. Meisel, 2008, Langmuir-Blodgett films of molecular organic materials, *J. Phys. Condens. Matter* 20, 184006.

Tamura, M., Y. Nakazawa, D. Shiomi, K. Nozawa, Y. Hosokoshi, M. Ishikawa, M. Takahashi, and M. Kinoshita, 1991, Bulk ferromagnetism in the beta-phase crystal of the prar-nitrophenyl nitronyl nitroxide radical, *Chem. Phys. Lett.* 186, 401–4.

Tang, C. W., 1986, 2-layer organic photovoltaic cell, *Appl. Phys. Lett.* 48, 183–85.

Tang, C. W., and S. A. van Slyke, 1987, Organic electroluminescent diodes, *Appl. Phys. Lett.* 51, 913–15.

Tchougréeff, A. L., and R. Dronskowski, 2008, A computational study of the crystal and electronic structure of the room temperature organometallic ferromagnet $V(TCNE)_2$, *J. Comput. Chem.* 29, 2220–33.

Meservey, R., R. M. Tedrow, and P. Fulde, 1970, Magnetic field splitting of the quasiparticle States in super conducting Aluminum films, *Phys. Rev. Lett.* 25, 1270.

Tedrow, P. M., and R. Meservey, 1971b, Spin-dependent tunneling into ferromagnetic nickel, *Phys. Rev. Lett.* 26, 192–95.

Tejada, J., B. Martinez, A. Labarta, R. Gräossinger, H. Sassik, M. Vazquez, and A. Hernando, 1990, Phenomenological study of the amorphous $Fe_{80}B_{20}$ ferromagnet with small random anisotropy, *Phys. Rev. B* 42, 898–905.

Tengstedt, C., M. P. de Jong, A. Kanciurzewska, E. Carlegrim, and M. Fahlman, 2006, X-ray magnetic circular dichroism and resonant photomission of $V(TCNE)_x$ hybrid magnets, *Phys. Rev. Lett.* 96, 057209.

Tengstedt, C., M. Unge, M. P. de Jong, S. Stafstrom, W. R. Salaneck, and M. Fahlman, 2004, Coulomb interactions in rubidium-doped tetracyanoethylene: A model system for organometallic magnets, *Phys. Rev. B* 69, 165208.

Teresa, J. M. D., A. Barthélémy, A. Fert, J. P. Contour, R. Lyonnet, F. Montaigne, P. Seneor, and A. Vaurès, 1999, Inverse tunnel magnetoresistance in $Co/SrTiO_3/La_{0.7}Sr_{0.3}MnO_3$: New ideas on spin-polarized tunneling, *Phys. Lev. Lett.* 82, 4288–91.

Thomas, L., F. Lionti, R. Ballou, D. Gatteschi, R. Sessoli, and B. Barbara, 1996, Macroscopic quantum tunnelling of magnetization in a single crystal of nanomagnets, *Nature* 383, 145–47.

Thorum, M. S., K. I. Pokhodnya, and J. S. Miller, 2006, Solvent enhancement of the magnetic ordering temperature (T_c) of the room temperature $V[TCNE]_x \cdot S$ (S = solvent, TCNE = tetracyanoethylene; $x \sim 2$) magnet, *Polyhedron* 25, 1927–30.

Tokoro, H., S. Ohkoshi, and K. Hashimoto, 2003, One-shot-laser-pulse-induced demagnetization in rubidium manganese hexacyanoferrate, *Appl. Phys. Lett.* 82, 1245–47.

Tombros, N., C. Jozsa, M. Popinciuc, H. T. Jonkman, and B. J. van Wees, 2007, Electronic spin transport and spin precession in single graphene layers at room temperature, *Nature* 448, 571–74.

Vilan, A., A. Shanzer, and D. Cahen, 2000, Molecular control over Au/GaAs diodes, *Nature* 404, 166–68.

Vinzelberg, H., J. Schumann, D. Elefant, R. B. Gangineni, J. Tomas, and B. Büchner, 2008, Low temperature tunneling magnetoresistance on $(La,Sr)MnO_3/Co$ junctions with organic spacer layers, *J. Appl. Phys.* 103, 093720.

Wagemans, W., F. L. Bloom, P. A. Bobbert, M. Wohlgenann, and B. Koopmans, 2008, Correspondence of the sign change in organic magnetoresistance with the onset of bipolar charge transport, *Appl. Phys. Lett.* 93, 263302.

Wang, F. J., Z. H. Xiong, D. Wu, J. Shi, and Z. V. Vardeny, 2005, Organic spintronics: The case of $Fe/Alq_3/Co$ spin-valve devices, *Synth. Met.* 155, 172–75.

Wang, F. J., C. G. Yang, Z. V. Vardeny, and X. G. Li, 2007, Spin response in organic spin valves based on $La_{2/3}Sr_{1/3}MnO_3$ electrodes, *Phys. Rev. B* 75, 245324.

Wolf, S. A., D. D. Awschalom, R. A. Buhrman, J. M. Daughton, S. von Molnár, M. L. Roukes, A. Y. Chtchelkanova, and D. M. Treger, 2001, Spintronics: A spin-based electronics vision for the future, *Science* 294, 1488–95.

Wu, Y., and B. Hu, 2006, Metal electrode effects on spin-orbital coupling and magnetoresistance in organic semiconductor devices, *Appl. Phys. Lett.* 89, 203510.

Xia, Y., and G. M. Whitesides, 1998, Soft lithography, *Angew. Chem. Int. End. Engl.* 37, 551–75.

Xiong, Z. H., D. Wu, Z. V. Vardeny, and J. Shi, 2004, Giant magnetoresistance in organic spin-valves, *Nature* 427, 821–24.

Xu, W., G. J. Szulczewski, P. LeClair, I. Navarrete, R. Schad, G. Miao, H. Guo, and A. Gupta, 2007, Tunneling magnetoresistance observed in $La_{0.67}Sr_{0.33}MnO_3/organic$ molecule/Co junctions, *Appl. Phys. Lett.* 90, 072506.

Xu, Z., Y. Wu, and B. Hu, 2006, Dissociation processes of singlet and triplet excitons in organic photovoltaic cells, *Appl. Phys. Lett.* 89, 131116.

Yamada, H., Y. Ogawa, Y. Ishii, H. Sato, M. Kawasaki, H. Akoh, and Y. Tokura, 2004, Engineered interface of magnetic oxides, *Science* 305, 646–48.

Yee, G. T., J. M. Manriquez, D. A. Dixon, R. S. Mclean, D. M. Groski, R. B. Flippen, K. S. Narayan, A. J. Epstein, and J. S. Miller, 1991, Decamethylmanganocenium tetracyanoethenide, $[Mn(C_5Me_5)_2]^+[TNCE]^-$, a molecular ferromagnet with an 8.8 K T_c, *Adv. Mater.* 3, 309–11.

Yeomans, J. M., 1999, *Statistical Mechanics of Phase Transitions*, Oxford University, New York.

Yoo, J.-W., R. S. Edelstein, D. M. Lincoln, N. P. Raju, and A. J. Epstein, 2007, Photoinduced magnetism and random magnetic anisotropy in organic-based magnetic semiconductor $V(TCNE)_x$ films, for $x \sim 2$, *Phys. Rev. Lett.* 99, 157205.

Yoo, J.-W., R. S. Edelstein, D. M. Lincoln, N. P. Raju, C. Xia, K. I. Pokhodnya, J. S. Miller, and A. J. Epstein, 2006, Multiple photonic responses in films of organic-based magnetic semiconductor $V(TCNE)_x$, $x \sim 2$, *Phys. Rev. Lett.* 97, 247205.

Yoo, J.-W., H. W. Jang, V. N. Prigodin, C. Kao, C. B. Eom, and A. J. Epstein, 2009, Giant magnetoresistance in ferromagnet/organic semiconductor/ferromagnet heterojunctions. *Phys. Rev. B.* 80, 205207.

Yoo, J.-W., H. W. Jang, V. N. Prigodin, C. Kao, C. B. Eom, and A. J. Epstein, 2010, Tunneling vs. giant magneto-resistance in organic spin valves, *Syn. Met.* in press.

Yoo, J.-W., V. N. Prigodin, W. W. Shum, K. I. Pokhodnya, J. S. Miller, and A. J. Epstein, 2008, Magnetic bistability and nucleation of magnetic bubbles in a layered 2D organic-based magnet $[Fe(TCNE)(NCMe)_2][FeCl_4]$, *Phys. Rev. Lett.* 101, 197206.

Yu, G., J. Gao, J. C. Hummelen, F. Wudl, and A. J. Heeger, 1995, Polymer photovoltaic cells—Enhanced efficiencies via a network of internal donor-acceptor heterojunctions, *Science* 270, 1789–91.

Yuasa, S., A. Fukushima, H. Kubota, Y. Suzuki, and K. Ando, 2006, Giant tunneling magnetoresistance up to 410% at room temperature in fully epitaxial Co/MgO/Co magnetic tunnel junctions with bcc Co(001) electrodes, *Appl. Phys. Lett.* 89, 042505.

Yuasa, S., T. Nagahama, A. Fukushima, Y. Suzuki, and K. Ando, 2004, Giant room-temperature magnetoresistance in single-crystal Fe/MgO/Fe magnetic tunnel junctions, *Nature Mater.* 3, 868–71.

Zhan, Y. Q., I. Bergenti, L. E. Hueso, V. Dediu, M. P. de Jong, and Z. S. Li, 2007, Alignment of energy levels at the $Alq_3/La_{0.7}Sr_{0.3}MnO_3$ interface for organic spintronic devices, *Phys. Rev. B* 76, 045406.

Zhan, Y. Q., M. P. de Jong, F. H. Li, V. Dediu, M. Fahlman, and W. R. Salaneck, 2008, Energy level alignment and chemical interaction at Alq_3/Co interfaces for organic spintronic devices, *Phys. Rev. B* 78, 045208.

Zhang, J., J. Ensling, V. Ksenofontov, P. Gutlich, A. J. Epstein, and J. S. Miller, 1998a, $[M^{II}(tcne)_2] \cdot xCH_2Cl_2$ (M = Mn, Fe, Co, Ni) molecule-based magnets with T_c values above 100 K and coercive fields up to 6500 Oe, *Angew. Chem. Int. Ed. Engl.* 37, 657–60.

Zhang, J., L. M. Liable-Sands, A. L. Rheingold, R. E. D. Sesto, D. C. Gordon, B. M. Burkhart, and J. S. Miller, 1998b, Isolation and structural determination of octacyanobu-tanediide, $[C_4(CN)_8]^{2-}$; precursors to $M(TCNE)_x$ magnets, *Chem. Commun.* 13, 1385–86.

Zheludev, A., A. Grand, E. Ressouche, J. Schweized, B. G. Miron, A. J. Eptein, D. A. Dixon, and J. S. Miller, 1994, Experimental determination of the spin density in the tetra-cyanoethenide free radical, $[TCNE]^-$, by single-crystal polarized neutron diffraction, a view of a π^* orbital, *J. Am. Chem. Soc.* 116, 7243–49.

Zhou, P., B. G. Morin, A. J. Epstein, R. S. Mclean, and J. S. Miller, 1993a, Spin frustration and metamagnetic behavior in a molecular-based quasi-1D ferrimagnetic chain-(MnTPP)(TCNE), *J. Appl. Phys.* 73, 6569–71.

Zhou, P., B. G. Morin, J. S. Miller, and A. J. Epstein, 1993b, Magnetization and static scaling of the high-T_c disordered molecular-based magnet V(tetracyanoethylene)$_x \cdot y$(CH$_3$CN) with $x \sim 1.5$ and $y \sim 2$, *Phys. Rev. B* 48, 1325–28.

Žutíc, I., J. Fabian, and S. D. Sarma, 2004, Spintronics: Fundamentals and applications, *Rev. Mod. Phys.* 76, 323–410.

5 Magnetic Field Effects in π-Conjugated Systems

E. Ehrenfreund and Z. V. Vardeny

CONTENTS

ABSTRACT

Substantial advance has recently occurred in the field of organic magnetotransport, mainly in two directions: spintronic devices, such as organic spin
valves, where spin injection, transport, and manipulation have been demonstrated; and organic light-emitting diodes, where both conductivity and

electroluminescence have been shown to strongly depend on magnetic field. The interest in organic semiconductors has been motivated by the small spin-orbit coupling that is caused by the light elements, such as carbon and hydrogen. In this review we describe three related magnetic field effects in devices made of π-conjugated organic polymers and oligomers. These are (1) optically, electroluminescence-, and conductance-detected magnetic resonance of thin films, where spin-dependent recombination of photogenerated spin $\frac{1}{2}$ polarons is enhanced under magnetic resonance conditions; (2) magnetoelectroluminescence response and organic magnetoresistance in organic devices; and (3) giant magnetoresistance in organic spin valves, where the spin transport in the device determines its performance.

5.1   INTRODUCTION

Over the past twenty years the electron spin has transformed from an exotic subject in classroom lectures to a degree of freedom that materials scientists and engineers exploit in new electronic devices nowadays. This interest has been motivated from the prospect of using spin in addition to charge, as an information-carrying physical quantity in electronic devices, thus changing the device functionality in an entirely new paradigm, which has been dubbed spintronics.[1,2] This interest culminated by awarding the 2007 Nobel Prize in Physics to Drs. Fert and Grünberg for the discovery and application of giant magnetoresistance. More recently, the spintronics field has focused on hybrids of ferromagnetic (FM) electrodes and semiconductors, in particular spin injection and transport in the classical semiconductor gallium arsenide.[3] However, spin injection into the semiconductor has been a challenge[4]; this research has yielded much original physical insight, but no successful applications yet. In the last few years, interest has also risen in similar phenomena in organics. The organic spintronics field was launched via two articles by Dediu et al.[5] and Xiong et al.[6] that demonstrated spin-polarized currents in small organic molecules, such as T_6 and Alq_3, respectively. Organic semiconductors can absorb and emit light while transporting charge, and this leads to photovoltaic cells, organic light-emitting diodes, and organic field effect transistors. It is thus expected that adding control of the electron spin to the multifunctional characteristics of these versatile materials should yield novel magnetic devices in the near future. However, there are serious challenges to be faced in understanding the properties of spin-polarized current in organic semiconductors, and obtaining high-quality devices. Most importantly, the approach in studying spin injection and transport in organic thin films is fundamentally different from that used for inorganic crystals. In this chapter we review some of the research highlights achieved at the University of Utah in the field of organic spintronics; Section 5.4 is devoted to spin injection and transport in organic semiconducting polymers and small molecules. In addition to spin-polarized carrier injection into organic semiconductors, organic light-emitting diodes, and organic photovoltaic cells, another fascinating effect has been observed when applying a small external magnetic field (dubbed magnetic field effect[7]), in which both the current and electroluminescence increase by as much as 40% at room temperature in a relatively small magnetic field of ~100 Gauss. The magnetic field effect in organics was discovered about four decades ago; however, renewed interest in it has recently risen since it was realized that magnetic field effects in organic light-emitting diodes have a tremendous potential application in organic electronic devices. In spite of the recent tremendous research effort in organic magnetic field effects, still the underlying mechanism and basic experimental findings are hotly debated. However, it is impossible to review the field of organic spintronics done at the University of Utah without giving a glance at the research of the magnetic field effect; this is done in Section 5.3.

We devote Section 5.2 to optically detected magnetic resonance studies of thin films and organic light-emitting diodes. The reason is that this relatively mature technique and its application in organics is the basic precursor of other, more sophisticated effects, such as spin injection and magnetic field effects.[8] Optically detected magnetic resonance is actually a magnetic field effect induced by microwave absorption that mixes the spin sublevels of polaron pair species. Therefore, the understanding of optically detected magnetic resonance in organic electronic devices may shed light on the magnetic field effect phenomenon. Moreover, optically detected magnetic resonance spectroscopy is unique in that it may measure the spin relaxation time in organic semiconductors, and thus obtain information about the spin interaction strength. We thus used this technique to obtain the hyperfine interaction strength in the same materials that are used for spin injection and magnetic field effect.

5.2   OPTICALLY, ELECTROLUMINESCENCE-, AND CONDUCTANCE-DETECTED MAGNETIC RESONANCE

5.2.1   MAGNETIC RESONANCE TECHNIQUES

Resonant absorption resulting from the application of a microwave (MW) radiation with a frequency matching the energy difference of the Zeeman split levels of a spin system under magnetic field is the basis for various related magnetic resonance techniques. In conventional electron spin resonance (ESR) the resonant absorption of the MW is directly detected measuring the properties of the ground electronic state in thermal equilibrium.[9] Optically detected magnetic resonance (ODMR) techniques are extensions of the more common ESR techniques.[10] ODMR measures changes in the optical absorption or emission that occur as a result of electron spin transitions. Since the ground state of conjugated polymers is almost always spin singlet, only spin bearing electronic excited states are studied. When applied to π-conjugated semiconductor films, ODMR techniques can be used for assigning the correct spin quantum number to the optical absorption bands of long-lived photoexcitations or for studying spin-dependent reactions that may occur between these photoexcitations.[11] One major advantage of ODMR over ESR is a much higher sensitivity in the detection of resonant processes, because the energy range of the detected photons in ODMR is much higher (order of eV) than the microwave energies that monitor ESR (tens of μeV). Much better detectors with higher detectivity exist for photons in the eV range than in the μeV range. A significant increase in the sensitivity of ODMR, when compared to

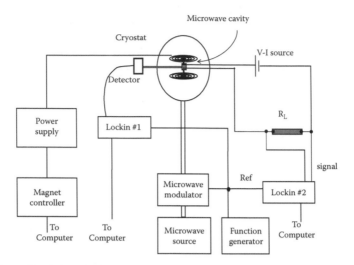

FIGURE 5.1   The 3 GHz XDMR setup.

ESR, also results from differences in the spin polarization; in ODMR experiments in π-conjugated polymers the typically obtained spin polarizations exceed the thermal equilibrium polarization by at least one order of magnitude.[11]

Related techniques are the electroluminescence- (EL),[12] conductivity (C)-,[13–15] and photoconductivity (PC)-[16]detected magnetic resonance (dubbed hereafter as XDMR, where X = EL, C, and PC, respectively). In these techniques the resonance is detected by observing the changes in the sample, or device, relevant property.

As an illustrative example, we show in Figure 5.1 the basic experimental setup for XDMR in organic light-emitting diode (OLED) devices. For CDMR measurements the OLED is biased in the forward direction at a constant voltage supplied by a DC voltage source. For PCDMR experiments, the sample is illuminated by an exciting laser under reverse (or zero) bias. The current change across the sample under magnetic resonance conditions is detected by the small load resistance, R_L, followed by a lock-in amplifier. For ELDMR experiments, the changes in the sample EL, under forward bias, are measured. The XDMR measurements described below are made using a 3 GHz (S-band) system. The empty cavity has Q ~ 1,000; however, with the OLED device inside, the metal electrodes can absorb the microwave, reducing the cavity Q by a factor of ~10.

5.2.2   The Role of Polaron Pairs

XDMR in organics is in fact a magnetic field effect (MFE) that occurs under resonance conditions with microwave radiation that induces spin sublevel mixing among the polaron pairs' (PPs) spin sublevels.[11] It is thus not surprising that models similar to those used to explain the MFE without MW radiation have been advanced to explain spin $\frac{1}{2}$ ODMR in the organics. The two ODMR versions in OLED devices, ELDMR and CDMR, may thus clarify the underlying MFE mechanism in these materials. The reason is that the spin mixing process among the PP spin sublevels in these experimental techniques is induced under controlled MW conditions, such as power and modulation frequency. Because ELDMR involves the radiative transition of singlet excitons, for explaining its response dynamics it is more convenient to treat the involved spin sublevels in terms of PP_S (singlet PP) and PP_T (triplet PP), which are precursors to intrachain singlet and triplet excitons, respectively. Thus in ELDMR the PP_S and PP_T populations continuously evolve due to carrier injection from the electrodes and subsequent PP formation, dissociation, and recombination kinetics under MW radiation. Therefore, it is expected that the spin $\frac{1}{2}$ ELDMR MW modulation frequency response would depend on both PP decay rates—γ_S and γ_T—as well as on the spin-lattice relaxation rate, γ_{SL}, of the polarons involved.

The role of PP is shown in the following experiment,[17] where $g \sim 2$ ELDMR measurements were conducted at 10 K on a well-balanced OLED (shown schematically in Figure 5.2 inset) composed of a 100 nm thick MEH-PPV active layer, sandwiched between a hole transport layer, poly(3,4-ethylenedioxythiophene) [PEDOT]-poly(styrene sulfonate) [PSS], and an electron transport layer, LiF, and capped with a transparent anode, indium tin oxide (ITO), on a glass substrate, and a cathode that is an evaporated aluminum film. The device I-V characteristics and EL-V dependence were measured and show a well-balanced OLED with relatively high EL efficiency (Figure 5.3). The current density, J, and EL emission were driven at constant forward bias voltage, and their dynamic changes, $J(f)$ and $EL(f)$, respectively, were measured while subjected to $g \sim 2$ (i.e., at magnetic field, $H \sim 0.1$ Tesla) resonance conditions at MW frequency of ~3 GHz (S-band) that was modulated at frequency f. In-phase ELDMR$_I$ and quadrature ELDMR$_Q$ components, with respect to the MW modulation phase, were measured at various MW power and current densities.

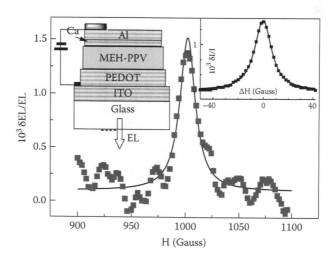

FIGURE 5.2 Spin $\frac{1}{2}$ ELDMR spectrum in an OLED based on an MEH-PPV active layer measured at 10 K at saturation MW power modulated at 200 Hz. The right inset shows the CDMR spectrum; the left inset shows the device structure composed of ITO anode, hole transport layer (PEDOT/PSS), active layer, electron transfer layer (LiF), and Al cathode. (From Yang et al., *Phys. Rev. B* 78, 205312 [2008].)

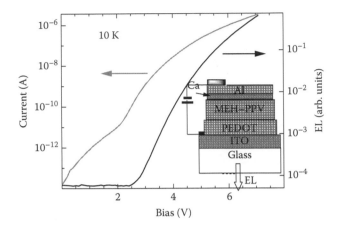

FIGURE 5.3 I-V and EL-V characteristics of the OLED. (From Wang et al., *Synth. Met.* [2009], DOI: 10.1016/j.synthmet.2009.06.014.)

Figure 5.2 shows $\Delta EL(H)/EL$ near the spin $\frac{1}{2}$ resonance field at $f = 200$ Hz. The ELDMR spectrum is composed of a single positive line at $g \sim 2$; no half-field resonance at $g \sim 4$ indicative of intrachain triplets involvement was detected. This shows that intrachain triplet excitons do not participate in organic MFE, ruling out the triplet-polaron model for explaining these phenomena. In Figure 5.2 (inset) we show CDMR at spin $\frac{1}{2}$ resonance condition measured on the same device. The $I(H)$ spectrum also consists of a single positive line of which resonance field, sign, width, and magnitude are the same as those of the spin $\frac{1}{2}$ ELDMR resonance. These indicate that the underlying mechanism for both spin $\frac{1}{2}$ ELDMR and CDMR in OLEDs is the same. This rules out the model of polaron-singlet exciton quenching for explaining ELDMR, which states that the decrease in polaron population eliminates nonradiative centers for singlet excitons. In contrast, in the experiment both spin $\frac{1}{2}$ ELDMR and CDMR are positive (Figure 5.2), showing that the free carrier density increases in the device under resonance conditions, giving rise to increased EL. We thus conclude that the two spin sublevels responsible for the spin $\frac{1}{2}$ ELDMR are loosely bound polaron pairs, rather than intrachain triplet excitons; more specifically, these are PP_S and PP_T.

The following scenario may explain the spin $\frac{1}{2}$ ELDMR and CDMR in OLEDs. The current density, J, in such devices is carried out by free charge carriers (e.g., electrons, holes, polarons, and bipolarons), but PPs may dissociate into free polarons, and thus also contribute to J. The relatively shallow PP_S and PP_T may also form intrachain singlet and triplet excitons, respectively, and also directly recombine to the ground state with combined decay rates γ_S and γ_T, respectively, for PP_S and PP_T (with $\gamma_S \neq \gamma_T$). PP_S and PP_T steady-state populations, n_S and n_T, are thus determined by the respective generation rates, $g_T = 3gs$ from free polarons, and decay rates, where $n_{S,T} = g_{S,T}/\gamma_{S,T}$. Therefore, taking $\gamma_S > \gamma_T$, then with steady state current injection conditions and MW off, $n_T > n_S$. The relatively strong magnetic field, H, forms three Zeeman split spin sublevels in the triplet manifold PP_T, namely, $m_s = 1, 0, -1,$

which are in resonance with the MW photon energy, hv; thus, transitions between $m_s = 0$ and $m_s = \pm1$ can be easily induced. The $m_s = 0$ sublevel is coupled with the singlet level, PP_S, via an intersystem conversion rate that is determined mainly by the hyperfine interaction (see Section 5.2.5) and the difference, Δg, in the individual g-factor of P^+ and P^- in the polaron pair species.[7] Thus, any population change in PP_T sublevels has a direct effect on PP_S population, and vice versa. Since $n_T > n_S$, then the $m_s = 0$ PP_T sublevel population is relatively small. Consequently, the MW transition from the $m_s = \pm1$ sublevels increases the $m_s = 0$ PP_T population, and also the PP_S population. Since $\gamma_S > \gamma_T$, then the MW transition at resonance decreases the overall PP density, similar to the well-known MW effect on the photoinduced absorption, namely, photoinduced absorption detected magnetic resonance (PADMR).[18] Since the OLED operates under the condition of constant applied bias voltage, the current density in the device adjusts itself to the new condition in the active layer. In general, the relatively immobile PP species may act as scattering centers for the mobile charges; then a reduction in the PP density increases the current, leading to positive CDMR at steady state (achieved at low f). Since the ELDMR increases with the injected current (Figure 5.3), this scenario could explain the simultaneous positive $g \sim 2$ ELDMR and CDMR resonances. Another contribution to the positive ELDMR may arise due to the resonance increase in PP_S, which is the precursor for the emitting singlet excitons.

5.2.3 THE PP MODEL FOR XDMR

In the following analysis we assume that the XDMR response dynamics exclusively occurs when all spin levels actively participate in determining the measured MFE physical quantity.[19] When PP_S effectively decays faster than PP_T, namely, $\gamma_S > \gamma_T$, then at steady state when MW is off, $n_{S,\text{off}} < n_{T,\text{off}}$.[11] Under resonant MW radiation (MW is on) a net transfer from PP_T to PP_S takes place (via the $m_s = 0$ spin sublevel in the PP_T manifold), bringing the system to a new quasi-equilibrium state, where the total PP density, $n_{PP} = n_S + n_T$, is reduced; i.e., $n_{PP,\text{on}} < n_{PP,\text{off}}$. This, in turn, leads to a current increase in the device due to free carrier mobility increase (discussed in Section 5.2.2). Under square wave MW modulation at frequency f, both $\Delta n_S(f) \equiv n_{S,\text{on}} - n_{S,\text{off}}$ and $\Delta n_T(f) \equiv n_{T,\text{on}} - n_{T,\text{off}}$ responses decrease with increasing f (for example, in the form of Lorentzians in f, if their time decays are exponentials[16]); however, since $\gamma_T < \gamma_S$, then the $n_T(f)$ response diminishes at a faster rate with f. Also, since the PP_S population change, $\Delta n_S > 0$, and PP_T population change, $\Delta n_T < 0$, the low-frequency negative, $\Delta n(f)$, signal changes sign at a frequency f_0, beyond which it stays positive; this situation is demonstrated in Figure 5.4. Therefore, any spin-dependent property that is determined by the weighted PP population (such as CDMR(f), for example) would show a sign reversal in its dynamic response.

In order to quantify the $\Delta n(f)$ response we make use of the fact that the two triplet spin sublevels ($m_s = \pm1$) should have a common dynamics. We also take the limit of strong singlet to $m_s = 0$ triplet mixing,[19] thus reducing the coupled set of four rate equations to a coupled set of two rate equations for the PP in the triplets and singlet/

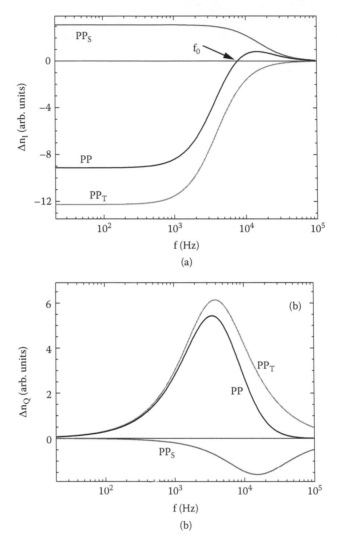

FIGURE 5.4 The in-phase (a) and quadrature (b) frequency response of the change in PP_S (singlet), PP_T (triplet), and the total PP populations based on the solution of Equation 5.1 with parameters that fit the spin $\frac{1}{2}$ ELDMR response shown in Figure 5.5. Note the zero crossing of the in-phase component at f_0 (a), which does not occur for the quadrature component (b). (From Yang et al., *Phys. Rev. B* 78, 205312 [2008].)

triplet states, respectively. These equations are written for the experimental conditions $T \gg h\nu/k_B \approx 0.14$ K as follows:

$$\frac{dn_S}{dt} = G_S - \gamma_S n_S - \frac{\gamma_{SL}}{2}(n_S - n_T) - P(n_S - n_T)$$

$$\frac{dn_T}{dt} = G_T - \gamma_T n_T - \frac{\gamma_{SL}}{2}(n_T - n_S) - P(n_T - n_S) \qquad (5.1)$$

where n_S denotes the coupled $m_s = 0$ sublevels of PP_T and PP_S and n_T denotes the $m_s = \pm 1$ in PP_T, γ_{SL} is the spin-lattice relaxation rate, P is the MW spin flip rate (assumed to be proportional to the MW power, P_{MW}), and G is the generation rate. The steady-state solutions of Equation 5.1 (i.e., $dn_S/dt = dn_T/dt = 0$) show a typical magnetic resonance saturation behavior:

$$n_{S,T} \propto \frac{P}{\Gamma_{eff} + P},$$

(5.2)

with an effective rate constant

$$\Gamma_{eff} = \frac{\gamma_{SL}}{2} + \frac{1}{\gamma_S^{-1} + \gamma_T^{-1}}.$$

(5.3)

Thus, Γ_{eff} can be directly obtained by measuring the XDMR as a function of P_{MW}, for various external parameters, such as the current density through the device in ELDMR or CDMR experiments or the light intensity in PCDMR.[17] Non-steady-state dynamic solutions were obtained by solving Equation 5.1 with square-wave-modulated MW radiation for both Δn_I and Δn_Q components as well as $(\Delta n_{S,T})_{I,Q}$ of the individual PP sublevels, as a function of the modulation frequency, f, revealing the unique properties of PP XDMR.[17,19] It was found that for $\gamma_S > \gamma_T$, $\Delta n_{S,I}(f) > 0$ and $\Delta n_{T,I}(f) < 0$ for the entire frequency range (see Figure 5.4). However, the sum $\Delta n_I(f)$ is negative at low f and positive above a certain frequency f_0. Such a sign reversal is unique to this type of models, in which all spin sublevels actively participate in the XDMR phenomenon. Further analysis of the Equation 5.1 solution shows that $f_0 \approx [(\gamma_S + \gamma_T)\Gamma_{eff}]^{1/2}/2\pi$ at low P ($P \ll \gamma_S$), increasing at first linearly with P, but tending toward saturation for $P \gg \gamma_S$. A typical $n(f)$ response based on the solution of Equation 5.1 is shown in Figure 5.4. To summarize, the unique features of the PP model are: (1) The individual sublevel components $(\Delta n_{S,T})_{I,Q}$ do not change sign as a function of f, and their absolute values decrease monotonically with f. (2) The in-phase component of the total PP population Δn_I changes sign at f_0, diminishing to zero at higher frequencies. (3) Δn_Q does not change sign within the entire f range, monotonically decreasing with f.

5.2.4 ELDMR, CDMR DYNAMICS: MEH-PPV

Referring to the experiments discussed in Section 5.2.2, Figure 5.5 shows the measured dynamics of the two ELDMR components in the MEH-PPV OLED at $J = 2$ mA/cm² and $P = 100$ mW. It is seen that the positive ELDMR$_I(f)$ reverses sign at frequency $f_0 \sim 7.5$ kHz before further decaying at higher frequencies; in contrast, ELDMR$_Q$ retains its sign throughout the measured f range. Figure 5.5 (inset) shows that f_0 increases with J; it increases more sharply at low J and tends to saturate at high J. Similar dynamics response is also typical for CDMR.[20] Figure 5.6 shows the CDMR dynamics at $J = 1$ mA/cm²; again the positive CDMR$_I$ response at low

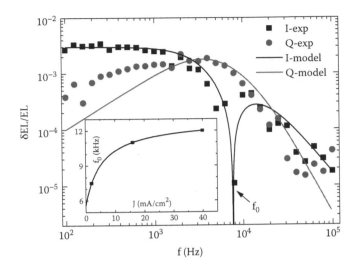

FIGURE 5.5 The spin $\frac{1}{2}$ in-phase (black squares) and quadrature (red circles) ELDMR components in the OLED device shown in Figure 5.2, plotted vs. the MW modulation frequency, f, measured at $P_{MW} = 100$ mW and T = 10 K. Zero crossing of the in-phase component occurs at $f_0 = 7.5$ kHz. The solid lines through the data points are based on a model described in the text. The inset shows the dependence of the zero-crossing frequency, f_0, vs. the device current density, J. (From Yang et al., *Phys. Rev. B* 78, 205312 [2008].)

f reverses sign at a frequency f_0, which increases with J. In addition, the ELDMR saturation behavior has also been measured. The MW power dependence of the ELDMR maximum value ([ELDMR]$_{max}$) at low f is shown in Figure 5.7 at two different current densities; [ELDMR]$_{max}$ shows a typical magnetic resonance saturation behavior, from which the recombination rates and γ_{SL} may be obtained.

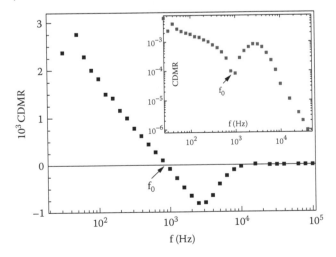

FIGURE 5.6 Same as in Figure 5.5 but for the spin $\frac{1}{2}$ CDMR$_I$ (in-phase) component measured on a different device, in both linear and logarithmical (inset) scales. (From Wang et al., *Synth. Met.* [2009], DOI: 10.1016/j.synthmet.2009.06.014.)

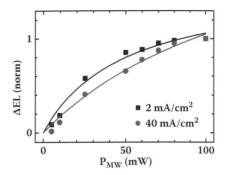

FIGURE 5.7 The dependence of spin $\frac{1}{2}$ ELDMR on the MW power, P_{MW}, measured at $f = 200$ Hz at two different current densities, $J = 2$ mA/cm^2 (black squares) and $J = 40$ mA/cm^2 (circles). Both curves show saturation behavior, where it is harder to reach saturation at larger J. (From Yang et al., *Phys. Rev. B* 78, 205312 [2008].)

Using the experimental setup value of $\alpha = 4 \times 10^3$ s^{-1}/W in the relation $P = \alpha P_{MW}$ for the loaded cavity, the fit of Equation 5.2 to the saturation curves of Figure 5.7 yields $\Gamma_{eff} = 1.9 \times 10^4$ s^{-1} for $J = 2$ mA/cm^2, whereas at $J = 40$ mA/cm^2, $\Gamma_{eff} = 5.3 \times 10^4$ s^{-1}. This increase in Γ_{eff} is related to the PP density because the density of both free polarons and PP increases as J increases. Since the dissociation of PP into free polarons and the formation of deeply bound intrachain excitons are not expected to depend strongly on the PP density, γ does not change much with J, indicating that the increase in Γ_{eff} with J is mainly caused by a γ_{SL} increase at large J. Using Γ_{eff} values obtained above from the steady-state saturation measurements, we find from the ELDMR$_I$ zero-crossing frequency f_0: $\gamma_T + \gamma_T \approx 12 \times 10^4$ s^{-1} at $J = 2$ mA/cm^2 and 11×10^4 s^{-1} at 40 mA/cm^2. The relatively small change in the $(\gamma_S + \gamma_T)$ value (i.e., ~10%) as J increases by a factor of ~20 justifies the conjecture that the main effect on the dynamics due to the current density increase is to increase γ_{SL}. Assuming that both γ_S and γ_T do not vary with J, the above extracted values of Γ_{eff} allow us to estimate an increase in γ_{SL} by at least a factor of ~5 when J increases from 2 to 40 mA/cm^2. The increase of f_0 with J (Figure 5.5, inset) is thus explained as due to an increase in γ_{SL} with the current density. This, in turn, may be caused by an increase in the spin-spin interaction rate in the active layer due to spin $\frac{1}{2}$ polaron density in the device, which increases with J. A similar effect was deduced before when spin $\frac{1}{2}$ radicals were added to MEH-PPV films.[19] A typical $\Delta n(f)$ response based on the solution of Equation 5.1 with $\gamma_S = 9.4 \times 10^4$ s^{-1}, $\gamma_T = 2.4 \times 10^4$ s^{-1}, and $\gamma_{SL} = 1 \times 10^4$ s^{-1} is shown in Figure 5.5, overlaid as a solid line on the experimental data for $J = 2$ mA/cm^2. It is apparent that (1) Δn_I changes sign at f_0, and (2) Δn_Q does not change sign within the entire f range. The good agreement obtained between the model fit and the data validates the model used.

A necessary condition for the existence of spin $\frac{1}{2}$ ELDMR and CDMR is that $\gamma_S \neq \gamma_S$.[11] The PP effective decay rates may be decomposed into three different components: $\gamma_{S,T} = d_{S,T} + k_{S,T} + r_{S,T}$, where $d_{S,T}$ is the dissociation rate to free polarons, $k_{S,T}$ is the rate at which intrachain strongly bound excitons are formed, and $r_{S,T}$ is the direct recombination rate of PP to the ground state (by direct interchain hopping). Therefore, in addition to ELDMR caused by the change in the overall effective PP recombination rate that leads to an increase in the device current density discussed above (Section 5.2.2), a more direct spin-dependent process that also leads to positive spin $\frac{1}{2}$ ELDMR should occur as well[18]; this mechanism is due to the enhanced PP_S relative population, where ELDMR $\propto k_S \Delta n_S$. However, since Δn_S alone does not change sign with f (Figure 5.4), the observed zero crossing at finite f_0 leads us to believe that this direct mechanism cannot be the dominant process of the spin $\frac{1}{2}$ ELDMR; otherwise, ELDMR$_I$ component would not reverse sign at f_0, in contrast with the data. We thus conclude that spin $\frac{1}{2}$ ELDMR in OLED devices is mainly caused by the current density increase at resonance, namely, CDMR, rather than MW-induced PP_S population increase. This scenario may explain the apparent contradiction in the literature between the similar EL and electrophosphorescence increase in OLED upon application of a strong external magnetic field,[21] as well as the MFE models based on change in $PP_T \Leftrightarrow PP_S$ interconversion rate with magnetic field.[22–24] In particular, we note that the observed positive magneto-EL with H follows the narrow positive component of the MFE in current that is due to an increase in the overall effective PP decay rate,[22] rather than the negative MFE component in current, or the overall current change, as in Wu et al.[24]

An altogether different scenario occurs in PCDMR, where the device is held under reverse bias, and no current is injected by the metallic electrodes. Here, the PP species are the intermediate stage between the photogenerated primary intrachain excitons and the subsequent mobile charges that carry the photocurrent. Therefore, the resonant decrease of the PP density directly implies a similar decrease in the photocurrent, hence to a negative PCDMR, as indeed observed experimentally (Figure 5.8). The effect of the carrier density on γ_{SL} is studied by the MW power saturation curves at various impinging light intensities (Figure 5.9). As seen, Γ_{eff} increases by about an order of magnitude when the light intensity increases from 90 to 500 mW/cm^2. It thus again strongly implies that the increase of photocurrent enhances the spin-lattice relaxation rate via the increased spin-spin interaction.

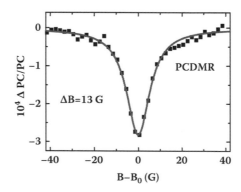

FIGURE 5.8 $g\sim 2$ PCDMR resonances vs. magnetic field, B, in an OLED based on MEH-PPV active layer, measured at 10 K and MW power modulated at $f = 200$ Hz. B_0 is the peak field. ΔB denotes the full width at half maximum. (From Wang et al., *Synth. Met.* [2009], DOI: 10.1016/j.synthmet.2009.06.014.)

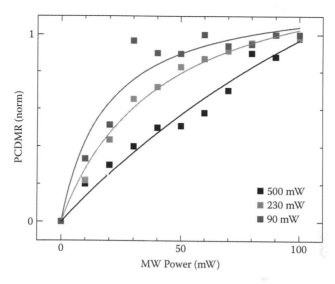

FIGURE 5.9 The dependence of spin $\frac{1}{2}$ PCDMR on the MW power, P_{MW}, measured at $f = 200$ Hz at various light intensities, as shown. Here $\Gamma_{eff}(500 \text{ mW})/\Gamma_{eff}(90 \text{ mW}) \approx 10$. (From Wang et al., *Synth. Met.* [2009], DOI: 10.1016/j.synthmet.2009.06.014.)

5.2.5 THE PP XDMR LINE SHAPE: ROLE OF THE HYPERFINE INTERACTION

The spin $\frac{1}{2}$ density of the polaron excitation in π-conjugated polymer chains is usually spread over several repeat units, which may contain N ~ 10 CH building blocks.[25] Consequently, the polaron spin $\frac{1}{2}$ XDMR (and also ESR) line shape at resonance conditions is susceptible to interactions with N protons in the immediate vicinity of the backbone intrachain carbon atoms. Thus, the XDMR line shape, similar to that in ESR, is dominated by the hyperfine interaction (HFI) strength, as well as by the wavefunction extent of the polaron on the polymer chain, which determines its spin density spread on the number of proton nuclei that are closest to the intrachain carbon atoms.[26] Having a nuclear spin $I = \frac{1}{2}$, each coupled proton splits the two electron-polaron spin levels ($m_s = \pm \frac{1}{2}$) into two electron-polaron/proton levels via the hyperfine interaction; these lines are further split by the other $N - 1$ coupled protons. Therefore, the energy levels of the ensemble system containing one electron-polaron and N protons depends on the HFI constants of the various protons a_n and the field B, and is given by the spin Hamiltonian:

$$H = H_{HF} + H_Z . \tag{5.4}$$

In Equation 5.4 H_Z is the Zeeman term:

$$H_Z = g\mu_B B S_z , \tag{5.5}$$

where g is the electron g-factor and μ_B is the Bohr magneton. Only the electronic Zeeman component is considered (we ignore the nuclear Zeeman interaction). H_{HF} is the isotropic hyperfine term:

$$H_{HF} = \sum_{n}^{N} a_n \vec{S} \cdot \vec{I}_n,$$ (5.6)

where a_n, I_n ($I_n = \frac{1}{2}$) are the HFI constant and nuclear spin operator of the nth nucleus ($n = 1, ..., N$), and S ($S = \frac{1}{2}$) is the electron-polaron spin operator. Solving the Hamiltonian Equation 5.4 in the high field limit ($g\mu_B B \gg a$) we obtain that each of the $m_s = \frac{1}{2}$ electron-polaron energy level splits into 2^N levels. As a result, there are $2^N m_s = \frac{1}{2}$ to $m_s = -\frac{1}{2}$ allowed optical transitions in the GHz range (hereafter, satellite lines) that leave the nuclear spins unchanged. In general, the satellite lines do not have equal spacing since a_n in Equation 5.6 are not equal to each other in the most general case. The satellite lines form a distribution in B of which width, Δ_{HF}, can conveniently be calculated from the square root of the second moment[9,27]:

$$\Delta_{HF} = 2\left[\sum_{n=1}^{2^N} (B_n - B_0)^2 / 2^N\right]^{1/2},$$ (5.7)

where B_0 is the resonance field in the absence of HFI, and B_n are the satellite lines. For a uniform spin distribution we take $a_n = a/N$, and consequently there are $N + 1$ nondegenerate electron/proton energy levels, which, when using Equation 5.7, yield $\Delta_{HF} \sim a/\sqrt{N}$. Thus, for example, for the polaron excitation in polymer chains with $N = 10$, we find $\Delta_{HF} \approx 0.3a$. Similar analysis can be readily extended to the HF coupling of the polaron with N nuclei having $I \neq \frac{1}{2}$.

In amorphous solids such as π-conjugated polymer films, the spin $\frac{1}{2}$ resonance line is inhomogeneously broadened. In this case the HFI split resonance is composed of 2^N bands (or $N + 1$ nondegenerate bands for a_n constant and $I = \frac{1}{2}$) centered at B_n, each having a line shape, $L(B - B_n)$, with full width at half maximum (FWHM), Δ_L, that is determined by the inhomogeneity in the sample. When the inhomogeneous broadening Δ_L is large enough such that the HFI split resonance appears as a single, nonsplit continuous band (i.e., $\Delta_L > \Delta_{Hf}$), then the composite resonance FWHM is approximately given by the sum $\Delta_{HF} + \Delta_L$.

Figure 5.10a shows the spin $\frac{1}{2}$ photoluminescence-detected magnetic resonance (PLDMR) of polarons in poly [2,5-dioctyloxy p-phenylene-vinylene] [DOO-PPV] films taken with an S-band resonance system at ~3 GHz and ~6 K, where the microwave (MW) power, P_{MW}, was modulated at $f \sim 400$ Hz.[19] In the experiment, the PL intensity increases due to enhancement of photogenerated PP recombination in the polymer film upon MW absorption at resonance conditions; therefore, the PLDMR signal appears at $B \sim 100$ mT, which corresponds to spin $\frac{1}{2}$ species with $g \sim 2$. At small P_{MW} the resonance FWHM is $\Delta B_H \sim 1.2$ mT, increasing gradually with P_{MW} (see also Figure 5.10a). The blue solid line in Figure 5.10a is a model calculation for the spin $\frac{1}{2}$ resonance, in which the polaron interacts with $N = 10$ protons, assuming uniform spin distribution. The excellent agreement with the data was obtained using $a(H) = 3.5$ mT, where each of the resulting eleven satellite lines was broadened by $\delta B_{inh} + \delta B_0 = 0.65$ mT.

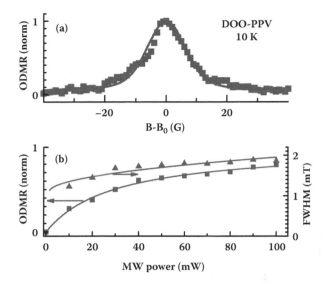

FIGURE 5.10 (a) The normalized spin $\frac{1}{2}$ PLDMR response in the DOO-PPV (blue squares) polymer; the solid line though the data points is based on a model fit taking into account the HFI splitting and the polaron spin density spread on the polymer chain, as well as inhomogeneous and natural line-broadening contributions to the resonance width (Equation 5.9). (b) The microwave power dependence of the integrated PLDMR signal intensity for DOO-PPV (squares, left scale) and line width (triangles, right scale); the lines through the data points are model fits (see text).

In Figure 5.10b the dependence of the PLDMR intensity, ΔPL, on P_{MW} is displayed showing a saturation behavior,[19] namely,

$$\Delta PL \propto \frac{P_{MW}}{(P_{MW} + P_S)}, \tag{5.8}$$

where $P_S = \alpha \Gamma_{eff}$, with α being an experimentally determined constant, and the rate Γ_{eff} is given in Equation 5.3. Γ_{eff} is determined by both $\gamma_{S,T}$ and γ_{SL}. It is possible to obtain an estimate for $\gamma_{S,T}$ from the PP recombination dynamics in the two polymers using the photomodulation technique,[19] but this is not a straightforward procedure.[17] Alternatively, a different method for estimating γ_{SL} can be employed. Figure 5.10b shows that the PLDMR line width, δB, increases nonlinearly with P_{MW}. The homogeneous line width, δB_o increases with P_{MW} according to the relation[27]

$$\delta B_0(P_{MW}) = \delta B_0(0)\left[1 + \frac{\beta}{\gamma_{SL}}P_{MW}\right], \tag{5.9}$$

where $b = (T_2 \delta B_0 \alpha)^{-1}$, $P_{MW} = \alpha B_1^2$, and B_1 is the field strength of the MW radiation. The fit through the data using Equation 5.9 is excellent (Figure 5.10b), and consequently, γ_{SL} can be estimated.

5.3 MAGNETOELECTROLUMINESCENCE AND MAGNETORESISTANCE IN ORGANIC DEVICES

5.3.1 EXPERIMENTAL DETAILS

Magnetoelectroluminescence (MEL) and magnetoresistance measurements on organic devices are conducted generally on devices with a typical area of ~5 mm². For MEL, the light-emitting diodes are made from an electroluminescing polymer (e.g., DOO-PPV polymer) layer sandwiched between a hole transport layer (e.g., poly(3,4-ethylenedioxythiophene) [PEDOT]–poly(styrene sulfonate) [PSS]), and capped with a transparent anode (e.g., indium tin oxide), and a cathode (e.g., calcium protected by aluminum film). The OLED structure is thus in the form of ITO/PEDOT:PSS/DOO-PPV/Ca/Al. Such devices show sizable electroluminescence, which for biasing voltage above the built-in potential $V > V_{bi}$ ($V_{bi} \sim 2$ V) approximately follows the device I-V characteristic. The devices are placed in a variable-temperature optical cryostat placed in between the pole pieces of an electromagnet producing magnetic field, B, up to 300 mT with 0.1 mT resolution. The devices are driven at constant V, and the current, I, and EL intensity are simultaneously measured while sweeping the magnetic field. The magnetoconductance (MC), magnetoresistance (MR), and MEL are defined, respectively, via:

$$MC(B) \equiv \frac{\Delta I(B)}{I(0)} = \frac{I(B) - I(0)}{I(0)} \, ,$$

$$MR(B) \equiv \frac{\Delta R(B)}{R(0)} = \frac{R(B) - R(0)}{R(0)} \, ,$$

$$MEL(B) \equiv \frac{\Delta EL(B)}{EL(0)} = \frac{EL(B) - EL(0)}{EL(0)} \, , \tag{5.10}$$

where ΔI, ΔR, and ΔEL are the field-induced changes in the current, resistance, and EL.

5.3.2 MAGNETIC FIELD EFFECTS IN ORGANICS: SPIN PAIR MECHANISM

Various models have been put forward to explain the MFE in devices based on π-conjugated organic solids. These models are based on the role of the hyperfine interaction (HFI) between the spin $\frac{1}{2}$ of the injected carriers and the proton nuclear spins in the active layer. The general understanding is that the HFI mixing of spin configurations of bound pairs of carriers becomes less effective as the magnetic field increases, thereby causing the MFE. Polaron pairs are comprised of negative (P^-) and positive (P^+) polarons (these are PP), and same-charge pairs comprised of either pairs of P^- or P^+ (these may be termed as loosely bound negative or positive bipolarons, respectively) are examples of such bound pairs species.

In organic light-emitting devices based on π-conjugated polymers some of the injected free carriers form loosely bound polaron pairs (PP), which comprise P^- and

P^+ polaron excitation separated by a distance of a few nanometers, each having spin $\frac{1}{2}$. The free carriers and PP excitations are in dynamic equilibrium in the device active layer, which is dominated by the processes of creation/dissociation, and recombination of PP via intrachain excitons. A PP excitation can be in either a spin singlet (PP_S) or spin triplet state (PP_T), depending on its mutual polarons' spin configuration. The steady-state PP density depends on the PP_S and PP_T effective rate constant, γ, which is the sum of the formation, dissociation, and recombination rate constants, as well as the triplet-singlet (T-S) mixing via the intersystem coupling (ISC) interaction. If the effective rates, γ_S for PP_S and γ_T for PP_T, are not identical to each other, then any disturbance of the T-S mixing rate, such as by the application of an external magnetic field, would perturb the dynamic steady-state equilibrium, resulting in MEL as well as MC. In the absence of an external magnetic field, T-S mixing is caused by both the exchange interaction between the unpaired spins of each polaron belonging to the PP excitation, and the hyperfine interaction (HFI) between each polaron spin and the adjacent nuclei. For PP excitation in π-conjugated polymer chains, where the polarons are separated by a distance, $R > \sim 1 - 2$ nm, the HFI with protons is expected to be dominant.

The MFE in organic devices is analogous to the MFE observed in chemical and biochemical reactions, where it is explained via the radical pair (RP) mechanism.[28,29] In this model the HFI, Zeeman, and exchange interactions are taken into account. It is assumed that the RPs are immobile, and hence RP diffusion is ignored, but the overall rate, k, of RP decay is taken into account. The steady-state singlet fraction of the RP population (singlet yield, ϕ_S) is then calculated from the coherent time evolution of RP wavefunctions subject to the above interactions. It is clear that when $\hbar k$ is much larger than either the HFI or exchange interactions, the MFE should be negligibly small, since the pairs disappear before any spin exchange can occur. For relatively small $\hbar k$, the MFE becomes substantially larger and (for negligible exchange interaction) the singlet yield behaves as $\phi_S \propto B^2/(B_0^2 + B^2)$ (upside-down Lorentzian), with $B_0 \sim a_{HF}/g\mu_B$, where a_{HF} is an average HFI constant. The half width at half maximum (HWHM) is then $B_{1/2} = B_0$. In organic devices, PPs are playing the role of radical pairs, and the calculated MC (and MEL) response may then be expressed as a weighted average of the singlet and triplet (ϕ_T) PP yields in an external magnetic field, B.

5.3.3 MAGNETOELECTROLUMINESCENCE IN DOO-PPV: THE ROLE OF THE HYPERFINE INTERACTION

Electroluminescence in OLED devices results from eventual recombination of polaron pairs in the spin singlet configuration, PP_S, that are formed from the injected carriers via the metal electrodes into the active layer; the PP_S species may also dissociate with a rate constant, d_S. The PPs in the spin triplet configuration, PP_T, do not recombine radiatively, and dissociate with a different rate constant, d_T.[16] It was shown that MEL is in fact directly related to the MC in the device, which subsequently increases the PP population.[30] The PPs in the two spin configurations exchange spins via the intersystem crossing (ISC), and their steady-state populations are also determined by their spin exchange interaction, as well as by the individual recombination and dissociation rates.[17] It is believed that the MEL effect is due to field-induced changes of the PP spin sublevels that determine the ISC rate,[31] but the underlying mechanism for the ISC rate change with B is still debated.[32,33] In this section we bring experimental evidence that shows that the HFI and its consequent spin sublevel mixing are the main cause of MEL.

MEL in OLEDs may be considered an example of a much broader research field that deals with magnetic field effects (MFEs) in physics,[34] chemistry, and biology,[28,29] such as phosphorescence, delayed fluorescence, biochemical reactions, etc.[7] In some MFE examples it was empirically realized[35] that $B_{1/2}$ scales with the HFI constant a_{HF} of the involved radical ions in the active material. Actually, an empirical law was written that, for positive and negative radicals with the same a_{HF}, reads[35]:

$$B_{1/2} \cong 2a_{HF}[I(I+1)]^{1/2} , \tag{5.11}$$

where I is the nuclear spin quantum number. If a_{HF} is the same for positive and negative polarons in OLEDs, then from Equation 5.11 we get $B_{1/2} \cong \sqrt{3}a_{HF}/g\mu_B$ with $I=\frac{1}{2}$. Figure 5.11 shows the room temperature MEL of DOO-PPV. The experimentally measured HWHM is $B_{1/2} \cong 5$ mT; thus, from Equation 5.11 a value of $a_{HF}/g\mu_B \cong 3$ mT is obtained. It is generally estimated that the exchange interaction is much smaller, and therefore it is concluded that the MEL response in OLED is mainly due to the HFI; this settles the discussion regarding the MFE origin in OLED.[16,32,33] A popular scenario used for explaining the MEL is that the two PP_T spin sublevels with $m_s = \mp 1$ split linearly with B due to the Zeeman interaction,[7,36–38] so that at $B > \sim a/g\mu_B$ the intersystem crossing (ISC) between PP_S and these two PP_T sublevels becomes increasingly less effective, and consequently this increases the steady-state PP_S population that leads to $MEL > 0$.

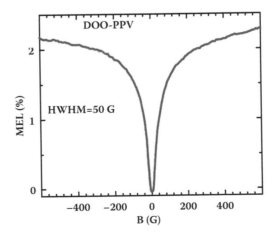

FIGURE 5.11 Room temperature MEL response of DOO-PPV measured at bias voltage $V = 2.5$ V, plotted as a function of the magnetic field in the range –600 to 600 Gauss. The half width at half maximum is HWHM~50 G.

5.3.4 MAGNETORESISTANCE IN ORGANIC BLENDS

Sizable organic magnetoresistance (OMAR) has been recently observed in a variety of light-emitting diodes of π-conjugated polymers and small molecules with non-magnetic electrodes.[22,23,30,31,36,39–43] In fact, OMAR has been studied for several years in the past[44,45]; however, renewed interest has recently increased due to the high OMAR value obtained at room temperature (up to ~10%), induced by a relatively small magnetic field of ~50 mT.[30] OMAR is the highest known magnetoresistance response in the class of semiconductor materials, and therefore has the potential to be used in magnetically controlled optoelectronic devices and magnetic sensors. It is thus one of the most unusual phenomena of plastic electronics, and more specifically, organic spintronics,[46] which makes this field attractive for both basic research and applications. In spite of the latest surge of interest in OMAR, its underlying mechanism is still hotly debated. Due to the weak magnetic field involved, it is largely believed that OMAR originates from spin sublevel mixing via the hyperfine interaction (HFI)[22,23,30,31,36,39–45] (but see Lupton and Boehme[32] for another view), which is relatively small in π-conjugated organic semiconductors.[45] However, two competing basic models based on HFI have been proposed for explaining the spin mixing process that causes OMAR. These are the excitonic model, in which the magnetic field modulates the singlet-to-triplet interconversion rate,[22] or triplet-exciton polaron quenching,[23] and the bipolaron model.[41] The exciton model is based on the spin-dependent electrostatically bound polaron pairs (P^+P^-) formed from the oppositely injected P^+ and P^- currents in the device,[22] whereas the bipolaron model relies on the spin-dependent formation of doubly charged excitations.[41] Therefore, the exciton model is based on the existence of both P^+ and P^- in the device active layer, whereas the bipolaron model is viable also when only one type of charge carrier is injected into the device. OMAR has been shown to be positive or negative, depending on the applied bias voltage, V, and temperature[30,36,42]; it can be tuned between positive and negative values by changing the device architecture,[31] and is enhanced by device conditioning at high current densities.[43] In recent years the field of organic photovoltaic (OPV) has also dramatically advanced, and power conversion efficiency of organic solar cells approaching 6% has been reached in bulk heterojunction-type devices made of polymer/fullerene blends.[47] The polymer and fullerene constituents in such blends were shown to form separate phases, in which the photogenerated P^+ and P^- move rather independently from each other in the polymer and fullerene phases, respectively, for reaching the opposite electrodes.[48] In agreement with this ansatz, the bimolecular recombination kinetics in the blends was shown to be much weaker than in polymer films.[48] It was thus tempting to study OMAR in OPV devices made of polymer/fullerene blends, because the formation of PP species in such devices would be suppressed due to the phase separation of the constituents, and this might unravel the correct OMAR underlying mechanism.

In a recent work,[49] two magnetic field effects, namely, MC and MEL in OPV devices, made of blends of poly(phenylene-vinylene) (PPV) derivatives/fullerene molecules with various concentrations, c, ranging from $c = 0\%$ to 50% (optimum blending) to 100% in weight, were studied. The devices used for the MC measurements were 5 mm^2 diodes made from a blend of PPV derivative,

2-methoxy-5-(2'-ethylhexyloxy) (MEH-PPV), with 1-(3-methoxycarbonyl) propyl-1-phenyl-(6,6)-methanofullerene (PCBM) at various concentrations, ranging from $c = 0\%$ (unblended MEH-PPV) to $c = 50\%$ (optimal blending) to $c = 0\%$ (unblended PCBM), which was sandwiched between a hole transport layer, poly(3,4-ethyl-enedioxythiophene) (PEDOT)–poly(styrene sulfonate) (PSS), and capped with an ITO transparent anode and a cathode, calcium (protected by aluminum film). The unblended $c = 0$ (or $c = 100\%$) device was in the form of ITO/PEDOT:PSS/MEH-PPV (or PCBM)/Ca/Al. The $c = 0$ hole-unipolar device was ITO/PEDOT-PSS/MEH-PPV/ Au, whereas the electron-unipolar device was Al/MEH-PPV/Ca/Al. These devices did not show any measurable EL, whereas the blended devices showed very weak EL. The devices were transferred to an optical cryostat with variable temperature that was placed in between the pole pieces of an electromagnet-producing magnetic field, B, up to 300 mT. The devices were driven at constant V using a Keithley 236 apparatus to obtain $\Delta I/I$ and $\Delta EL/EL$ as defined in Equation 5.10. In Figure 5.12 we show the room temperature MC response in MEH-PPV/PCBM diodes at various PCBM concentrations ranging from $c = 0\%$ to 100%.[49] Several observations are noteworthy: (1) The positive MC response decreases by approximately two orders of magnitude in the blends, changing from a few percent at $c = 0$ to ~0.1% at $c = 50\%$; at the same time it also substantially broadens. (2) The MC response changes with c and V. At $c = 0$ (panel (a)) MC changes sign from negative to positive at large V[42]; this

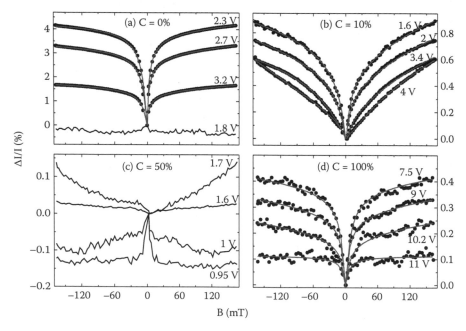

FIGURE 5.12 Magnetoconductance (MC) response, $\Delta I/I$, vs. field, B, in unblended diodes ($c = 0$ (a) and $c = 100\%$ (d)), and OPV devices ((b) and (c)) based on blends of MEH-PPV/PCBM with various PCBM concentrations, c (in weight), and biasing voltage, V, at room temperature. The lines through the data points in (a), (b), and (d) are fits using a two-mechanism model for the spin mixing process (see text). (From Wang et al., *Phys. Rev. Lett.* 101, 236805 [2008].)

is more pronounced for $c = 50\%$ (panel (c)). (3) The MC response broadens at large V; in fact, the positive low-field (LF) MC component decreases even at $c = 0$, and a much broader MC component, namely, a high-field (HF) component, is unraveled. This is more pronounced for $c = 10\%$ (panel (b)).

The LF MC response in the unblended devices (both $c = 0$ [panel (a)] and $c = 100\%$ [panel (d)]) can be approximately fit by the empirical relation[30] $f(B, B_0) \approx B^2/(|B| + B_0)^2$ at both positive and negative MC values. However, for $c = 10\%$ the MC response is less strongly field dependent at all V's; it is thus clear that the positive HF component dominates the MC response in the blends. We note that the HF component also exists in the unblended devices, but it is difficult to be resolved due to the strong LF component. (4) Finally, it was noted[49] that the positive LF component completely disappears from the MC response in the $c = 50\%$ device (Figure 5.12c). Instead, a very small negative MC at low V followed by a small positive HF response at large V was found. Figure 5.13a summarizes the maximum MC value, defined as $(\Delta I/I)_{max} = [I(160) - I(0)]/I(0)$ (see Figure 5.12) vs. V in four devices with various c. (Note: The MC response vs. V for the $c = 100\%$ device is not shown in Figure 5.13a; nevertheless, the decrease in MC with V can be inferred from Figure 5.12d.) For $c = 0$, 1, and 10% the MC first increases, saturates, and then decreases with V. This response is in contradiction with the bipolaron model,[41] since at large V we expect larger P^+ and P^- concentrations, and consequently, an increase of bipolaron density is anticipated but not observed. Also, as seen in Figures 5.12 and 5.13, the MC is very small in the $c = 50\%$ device, where phase separation of the polymer and fullerene constituents is known to exist.[48] This shows that the positive MC is due to both P^+ and P^- together in the active layer, permitting PP formation.[22] We also note that the bias voltage onset, $V_0 (= 1.7 \text{ V})$ for the positive LF MC component in the $c = 0$ device coincides with the appearance of EL, showing exciton formation from PPs; this also favors the exciton model.[22] Clear evidence in support of this interpretation is that in all devices the MEL response basically follows the LF MC component, rather than the negative, or HF, components. This was demonstrated in Figure 5.14, where the MEL and MC responses in devices with $c = 1\%$ and 50%, respectively, are compared.[49] It is seen that the MEL response is positive at all c, and is much narrower than the MC response; it follows the positive LF component even in the $c = 50\%$ device, where the EL is very small anyhow. This characteristic MEL behavior attests that the main MC response in the unblended devices is caused by spin mixing of tightly bound PPs via the HFI.[22,23]

For understanding the positive HF MC response that is revealed in the blends, it was noted that this component also decreases with blending; therefore, it is also related to PPs in the active layer. However, it does not correlate well with the MEL response (Figure 5.14), and therefore is due to a different PP type—most probably a charge transfer (CT) pair across the polymer/fullerene interface, with P^+ in the polymer chain and P^- in the fullerene molecule.[50] Such a CT pair does not luminesce in the visible spectral range, and hence cannot contribute to visible MEL. However, since P^+ and P^- in the CT pair are each influenced by a different environment, they possess different gyromagnetic factors, namely, $g^+ \neq g^-$ (i.e., $\Delta g \neq 0$)[51]; therefore, another mechanism for spin mixing with B becomes available. This process was

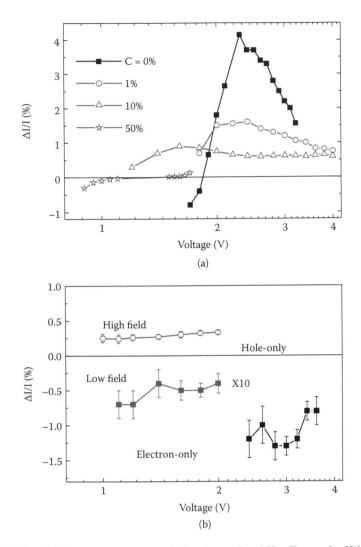

FIGURE 5.13 (a) The room temperature MC value at $B = 160$ mT, namely, $[I(160 \text{ mT}) - I(0)]/I(0)$, as a function of the biasing voltage, V, for MEH-PPV devices at four PCBM concentrations, c. (b) Same as in (a) but for unipolar MEH-PPV devices at 6 K; electron-only (black squares) and hole-only (circles for HF, and squares for the LF components [multiplied by a factor of 10]). (From Wang et al., *Phys. Rev. Lett.* 101, 236805 [2008].)

dubbed in the MFE literature as Δg-mechanism,[45,52] which shows a response given by $g(B) = [B/B^*]^{1/2}$,[7] where B^* is of the order of 5 Tesla.[7,52] Therefore, the MC response was fitted in the blends by the two spin mixing mechanisms, the HFI-mechanism for the LF component and the Δg-mechanism for the HF component,[7] respectively, using the relation $MC(B) = af(B, B_0) + bg(B)$, where a, b, and B_0 are free parameters (see Figures 5.12 and 5.14). In the unblended devices for $c = 0$ (Figure 5.12a) it was found that $B_0 = 3.5$ mT, whereas a/b ratio was about 3 at $V = 2.3$ V, and continuously

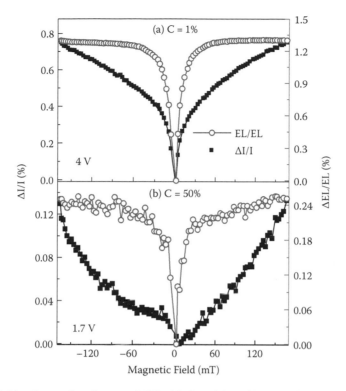

FIGURE 5.14 Comparison between MEL (circles; right axis) and MC (black squares; left axis) responses in MEH-PPV/PCBM OPV devices at two PCBM concentrations and biasing voltages: (a) $c = 1\%$, $V = 4$ V; (b) $c = 50\%$ $V = 1.7$ V. The lines through the data points in (a) are fits using the two-mechanism model as in Figure 5.12. (From Wang et al., *Phys. Rev. Lett.* 101, 236805 [2008].)

decreased with V in agreement with the decrease of the LF component with V seen in Figures 5.12 and 5.13, whereas for $c = 100\%$ (Figure 5.12d) it was found that $B_0 = 2.5$ mT, and there was a similar decrease in the a/b ratio with V. (We note here that the decrease of B_0 in the PCBM device [$c = 100\%$] is significant and might be due to a smaller HFI of the carbon atoms in the fullerene; the hydrogen-based moiety of this molecule [having proton-related HFI] resides outside the C_{61} complex.) For $c = 1\%$ (Figure 5.14) and $c = 10\%$ (Figure 5.12b), however, the a/b ratio decreased by an order of magnitude, showing the dominance of the HF component due to the g-mechanism in the MC response, whereas B_0 increases somewhat to ~5 mT. For the MEL fit (Figure 5.14) the a/b ratio remains large (~4), showing that this MFE still originates from PPs in the polymer phase of the blends. For understanding the negative MC response seen in most devices at low V (Figure 5.12), the MFE was studied in hole- and electron-unipolar $c = 0$ devices.[49] In both devices was observed (Figure 5.15) a negative LF MC that did not depend much on V (Figure 5.13b). Therefore, this response is in agreement with spin mixing in hole- and electron-bipolaron species, rather than with neutral PPs. It was also noted that the MC response is much larger

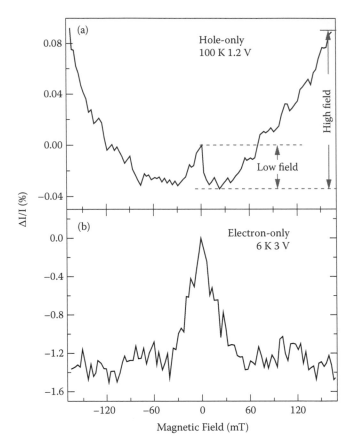

FIGURE 5.15 MC response of unipolar MEH-PPV devices ($c = 0$) at 6 K. (a) Hole-only diode at $V = 1.2$ V, where the low-field and high-field components are defined. (b) Electron-only diode at $V = 3$ V. (From Wang et al., *Phys. Rev. Lett.* 101, 236805 [2008].)

for the electron-unipolar device than that for the hole-unipolar device. This is in contrast with the MC response in Alq_3-unipolar devices,[53] where the MC value was found to be much larger in hole-unipolar devices than in electron-unipolar devices. In general, bipolaron formation is not favorable in organic semiconductors, and requires large relaxation energy to compensate the gain in energy due to the coulomb repulsion. It may thus be that deep traps are involved in bipolaron formation, such as trions,[54] for example. Deep traps are more abundant for minority carriers, which are therefore less mobile.[43] This explains the larger MC negative response in electron MEH-PPV-based unipolar devices, in which electrons are the minority carriers, as opposed to the larger MC response in hole Alq_3-based unipolar devices, where holes are the minority carriers.[53] The bipolaron model also explains the MC response of OPV devices obtained at optimum blending, namely, $c = 50\%$. Since the donor and acceptor phases are separated in this blend, then MC is governed by P^-P^- and P^+P^+ pairs in each constituent. P^-P^- pairs occur in the fullerene phase, where MC is quite weak (Figure 5.12d); thus, the MC response in this blend is dominated by the P^+P^+

response in the MEH-PPV phase. This MC response was found to be negative and very weak in hole-unipolar devices (Figure 5.15a), consistent with the weak negative MC response for OPV with $c = 50\%$ (Figure 5.12c).

5.4 ORGANIC SPIN VALVES

5.4.1 INTRODUCTION

The electronic spin sense was basically ignored in charge-based electronics, until few decades ago. The technology of spintronics (or spin-based electronics), where the electron spin is used as the information carrier in addition to the charge, offers opportunities for a new generation of electronic devices combining standard micro-electronics with spin-dependent effects that arise from the interaction between the carrier spin and externally applied magnetic fields. Adding the spin degree of free-dom to conventional semiconductor charge-based electronics increases substan-tially the functionality and performance of electronic products. The advantages of these new devices are increased data processing speed, decreased electric power consumption, and increased integration densities compared to conventional semi-conductor electronic devices, which are nearly at their physical limits nowadays. The discovery of the giant magnetoresistive (GMR) effect in 1988 is considered the beginning of the new generation of spin-based electronics[55]; this discovery led to the Nobel Prize in Physics (2008). Since then, the role of the electron spin in solid-state devices and possible technology that specifically exploits spin rather than charge properties have been studied extensively.[56] A good example of rapid transi-tion from discovery to commercialization for spintronics is the application of GMR and tunneling magnetoresistance (TMR)[57] in magnetic information storage. Since the first laboratory demonstration of GMR in 1988, the first GMR device as a mag-netic field sensor was commercialized in 1994, and "read-heads" for magnetic hard disk drives were announced in 1997 by IBM. Major challenges in the field of spin-tronics are the optimization of electron spin lifetimes, detection of spin coherence in nanoscale structures, transport of spin-polarized carriers across relevant length scales and heterointerfaces, and manipulation of both electron and nuclear spins on sufficiently fast timescales.[2] The success of these ventures depends on a deeper understanding of fundamental spin interactions in solid-state materials as well as the roles of dimensionality, defects, and semiconductor band structure in modify-ing the spin properties. With proper understanding and control of the spin degrees of freedom in semiconductors and heterostructures, the potential for realization of high-performance spintronic devices is excellent. The research in this field so far has led to the understanding that the future of spintronics relies mainly on suc-cessful spin injection into multilayer devices and optimization of spin lifetimes in these structures. Hence, for obtaining multifunctional spintronic devices operating at room temperature, different materials suitable for efficient spin injection and spin transport have to be studied thoroughly.

In recent years spintronics has benefited from the class of emerging materi-als, mainly semiconductors. The III-V and II-VI systems and also magnetic atom-doped III-V and II-VI systems (dilute magnetic semiconductors) have been studied extensively as either promising spin transport materials or as spin-injecting electrodes.[2,58] Little attention has been paid so far to the use of organic semiconductors such as small molecules or π-conjugated polymers (PCP) as spin transporting materials. The conducting properties of the PCPs were discovered in the late seventies, and later on the semiconducting properties of the organic semiconductors (OSECs) and PCPs gave birth to a completely new field of electronics: organic, or plastic, electronics. OSECs and PCPs are mainly composed of light atoms such as carbon and hydrogen, which leads to large spin correlation length due to weak spin-orbit coupling. This makes the small molecules and PCPs more promising materials for spin transport than their inorganic counterparts.[59,60] The ability to manipulate the electron spin in organic molecules offers an alternative route to spintronics.

Key requirements for success in engineering spintronics devices using spin injection via ferromagnetic (FM) electrodes in a two-terminal device such as diode (or junction) include the following processes[2,4]: (1) efficient injection of spin-polarized (SP) charge carriers through one device terminal (i.e., FM electrode) into the semiconductor interlayer, (2) efficient transport and sufficiently long spin relaxation time within the semiconductor spacer, (3) effective control and manipulation of the SP carriers in the structure, and (4) effective detection of the SP carriers at a second device terminal (i.e., another FM electrode). Usually FM metals have been used as injectors of SP charge carriers into semiconductors, and they can also serve to detect a SP current. However, more recently magnetic semiconductors have been sought to serve as the spin injector because of the conductivity mismatch between the metallic FM and the semiconductor interlayer (see below). The conductivity mismatch was thought to be less severe using OSEC as the medium in which spin-aligned carriers are injected from the FM electrodes, since carriers are injected into the OSEC mainly by tunneling, and the tunnel barrier may be magnetic field dependent and defuse the conductivity mismatch.[4] Spin relaxation lifetimes in conventional, inorganic semiconductors are primarily limited by the spin-orbit interaction.[1,2] However, OSECs are composed of light elements such as carbon and hydrogen that have weak spin-orbit interaction for the relevant electronic states that participate in the electrical conductance process, and consequently are thought to possess long spin relaxation times.[5,61] Therefore, OSECs appear to offer significant potential applications for spintronic devices.[46,61] The inset in Figure 5.16a schematically shows the canonical example of a spintronic device, the spin valve (SV).[6,62] Two FM electrodes (in this example $La_{2/3}Sr_{1/3}MnO_3$ [LSMO] and Co, respectively[6]) used as spin injector and spin detector, respectively, are separated by a nonmagnetic spacer (which in organic SV is an OSEC layer). By engineering the two FM electrodes so that they have different coercive fields (H_c), we may switch their relative magnetization directions from parallel (P) to antiparallel (AP) alignment (and vice versa) upon sweeping the external magnetic field, H (see Figure 5.16b). The FM electrode capability for injecting SP carriers depends on its interfacial spin polarization value, P, which is defined in terms of the density, n, of carriers close to the FM metal Fermi level with spin-up, $n\uparrow$, and spin-down, $n\downarrow$, and is given by the relation $P = [n\uparrow - n\downarrow]/[n\uparrow + n\downarrow]$. The spacer decouples the FM electrodes, while allowing spin transport from one contact to the other. The device electrical resistance depends on the relative orientation of the magnetization in the two FM-injecting electrodes; this has

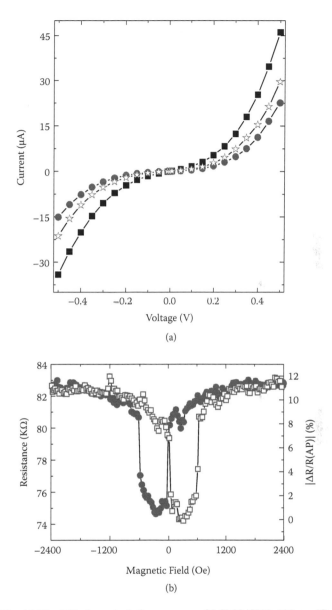

(a)

(b)

FIGURE 5.16 (a) The I-V characteristic response of LSMO/CVB 102 nm/Co/Al diode at three temperatures: 14 K (dark circles), 140 K (dark stars), and 200 K (black squares). (b) The magnetoresistance response of the spin valve device shown in (a) for an applied biasing voltage of 10 mV at 14 K; the empty squares (filled circles) are for H swept in the forward (backward) direction. The relative magnetization directions of the FM electrodes are parallel for |H| > 1200 Oe, and antiparallel for |H| < 600 Oe. (From Wang et al., *Phys. Rev. B* 75, 245324 [2007].)

been dubbed magnetoresistance (MR). The electrical resistance is usually higher for the AP magnetization orientation, an effect referred to as GMR,[1] which is due to spin injection and transport through the spacer interlayer. The spacer usually consists of a nonmagnetic metal, semiconductor, or a thin insulating layer (in the case of a magnetic tunnel junction). The magnetoresistance effect in the latter case is referred to as tunnel magnetoresistance, and does not necessarily show spin injection into the spacer interlayer as in the case of GMR response, but rather, the SP carriers are injected through the nonmagnetic overlayer. Semiconductor spintronics is very promising because it allows for electrical control of the spin dynamics, and due to the relatively long spin relaxation time, multiple operations on the spins can be performed when they are out of equilibrium (i.e., transport via SP carriers occurs).[3] This type of spin valves (for example, based on GaAs as a spacer[63]) may have other interesting optical properties, such as circular polarized emission, that may be controlled by an external magnetic field.[64] Significant spin injection from FM metals into nonmagnetic semiconductors is challenging, though, because at thermal equilibrium conditions the carrier densities with spin-up and spin-down are equal, and no spin polarization exists in the semiconductor layer. Therefore, in order to achieve SP carriers the semiconductor needs be driven far out of equilibrium and into a situation characterized by different quasi-Fermi levels for spin-up and spin-down charge carriers. Early calculations of spin injection from an FM metal into a semiconductor showed[65–67] that the large difference in conductivity of the two materials inhibits the creation of such a nonequilibrium situation, and this makes efficient spin injection from metallic FM into semiconductors difficult; this has been known in the literature as the conductivity mismatch hurdle. However, a tunnel barrier contact between the FM metal and the semiconductor may effectively achieve significant spin injection.[68] The tunnel barrier contact can be formed, for example, by adding a thin insulating layer between the FM metal and the semiconductor.[69] Tunneling through a potential barrier from an FM contact is spin selective because the barrier transmission probability, which dominates the carrier injection process into the semiconductor spacer, depends on the wavefunctions of the tunneling electron in the contact regions.[4] In FM materials the wavefunctions are different for spin-up and spin-down electrons at the Fermi surface, which are referred to as majority and minority carriers, respectively, and this contributes to their spin injection capability through a tunneling barrier layer.

5.4.2 ORGANIC SPIN VALVES: RECENT EXPERIMENTS

Experimental evidence of large MR in organic SV (OSV) devices has been reported during the last few years.[6,68–78] In the experimentally explored structures, some of the FM contacts were made from half-metallic materials, such as LSMO that has a spin polarization $P \sim 95\%$[79] (see Figure 5.16b); such devices showed a substantial GMR response up to 40% at low temperatures,[6] even though tunnel barriers were not used (this occurs because the half-metal FM electrode has large P, see below). However, the GMR response was found to substantially decrease at high temperature and large biasing voltage.[6,75] Other OSV devices were fabricated from more conventional FM, such as Co and Fe, which have not shown as large an MR response as the former OSV devices,[68,76,78,80] or not at all.[81] In other reports it was demonstrated that the MR response in polymer OSV[71,77] and small molecules[69,73] survives up to room temperature. In addition, it was also shown that organic diodes based on one FM electrode (namely, LSMO) possess another MR response at high fields, which is intimately related to the LSMO electrode magnetic properties.[6,24] This response, however, was shown to have no relation with spin-polarized carrier injection,[24] and thus it casts doubt on the original report in the organic spintronics field,[5] which claimed evidence for spin-polarized current in OSEC using a lateral device with T6 molecules in between two identical LSMO overlayers.

Figure 5.16b shows[75] a typical magnetoresistance loop at low H obtained with an LSMO/CVB/Co SV device at $T = 14$ K and bias voltage of V ~10 mV applied to the LSMO electrode, CVB, or 4,4'-bis-(ethyl-3-carbazovinylene)-1,1'-biphenyl is a light emissive oligomer. Figure 5.16b shows that R(AP) < R(P), where R(AP) [R(P)] is the device electrical resistance in the antiparallel [parallel] electrode's magnetization orientation. This is opposite to many other metallic spin valves, and may be caused by the different P signs of the two FM electrodes in these devices,[6,82] where negative P shows that minority electrons with spin-down have a larger density of states at the Fermi level of the FM electrode than do majority carriers with spin-up. In agreement with this hypothesis, it is noteworthy that a positive MR response was reported for OSV based on Co/Alq$_3$/Py tunnel junction[76] (here Alq$_3$ is 8-hydroxy-quinoline aluminum), in which P of both FM electrodes have the same sign (in this case, negative). However, this point has been extensively debated in the literature.[6,69,82] The obtained spin-valve-related MR value (MR$_{SV}$) in the device shown in Figure 5.16, defined as MR$_{SV} = max[R(P) - R(AP)]/[(R(AP)]$ at low H (< 1 kG), was inferred from the MR response to be about 11%. It was also verified that the OSV device switched resistance values at H_{c1} ~ 5 mT and H_{c2} ~ 50 mT, in agreement with the respective coercive fields of the LSMO and Co electrodes measured using the magnetic optical Kerr effect.[6,74]

5.4.2.1 Magnetoresistance Bias Voltage Dependence

The MR$_{SV}$ bias voltage dependence in many OSVs was found to decrease at large bias,[6,75,78] as seen in Figure 5.17 for the OSV device shown in Figure 5.16. It is seen that MR$_{SV}$ monotonically decreases with V. However, it decreases less at negative bias, V, where electrons are injected from the LSMO electrode into the OSEC interlayer; this apparent asymmetry is reduced when the MR is plotted vs. the current density in the device.[75] The MR$_{SV}$ dependence on V is seen to be the same at the two measured temperatures, in spite of the apparent current increase obtained at high temperatures (Figure 5.16a). A similar MR decrease with V, including the polarity asymmetry, has been measured in numerous LSMO/Co-based OSVs,[6,83,84] as well as in inorganic magnetic tunneling junctions based on the same two FM electrodes.[85-87] The TMR steep decrease with V has been explained in the latter devices as due to changes in P(Co) upon sweeping V,[85] or by the increase of the electron-magnon scattering in the LSMO electrode upon current increase.[88,89]

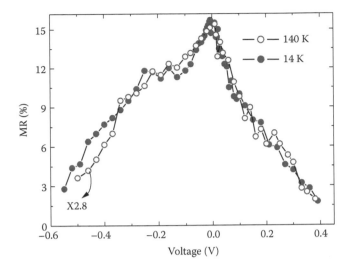

FIGURE 5.17 The spin-valve-related MR value of the diode shown in Figure 5.16 vs. the applied biasing voltage, *V*, at 14 K (filled circles) and 140 K (empty circles); the 140 K data were multiplied by a factor of 2.8 for normalization purposes. (From Wang et al., *Phys. Rev. B* 75, 245324 [2007].)

5.4.2.2   Magnetoresistance Temperature Dependence

The temperature dependence of the MR response in OSV made from a variety of organic small molecules and polymer spacers was measured by several groups.[6,69,71,75] It was found that MR_{SV} dramatically decreases with *T*. Figure 5.18a shows the MR_{SV} temperature dependence obtained in three OSV devices based on different OSEC interlayer molecules.[75] These molecules are Alq_3, which is most commonly used in organic light-emitting diodes with green emission; CVB, discussed above; and N,N'-bis (1-naphtalenyl)-N-N'-bis (phenyl) benzidiane (α-NPD), which is commonly used as a hole transport layer in organic optoelectronic devices. It is seen that MR_{SV} monotonically decreases with *T*, and vanishes (within the noise level) at *T* ~220 K independent of the specific OSEC interlayer.

There are two possible explanations for the MR_{SV} decrease with temperature: (1) the FM electrode's spin polarization degree, *P*, is temperature dependent, and (2) the spin-aligned transport through the organic spacer diminishes at high *T* due to an increase of the spin-lattice relaxation rate for the charge polaron excitations injected into the OSEC layer. The latter explanation was recently refuted,[75] since the spin-lattice relaxation rate in a typical small OSEC molecule (namely, Alq_3) was measured by the technique of optically detected magnetic resonance, and found to be temperature independent. However, recent experiments using the technique of low-energy muon spin rotation claim that the spin-lattice relaxation rate increases with temperature in Alq_3.[90] Also, the similarity of the obtained MR_{SV} response for the different OSECs used indicates that the FM electrode's response, rather than

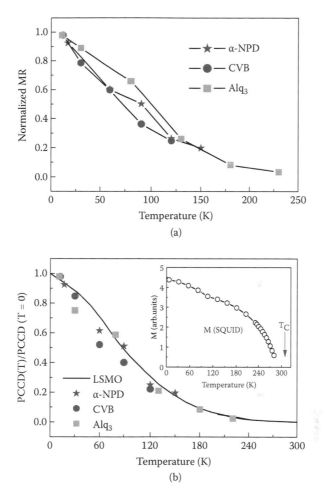

FIGURE 5.18 (a) The MR values of three different LSMO/OSEC/Co spin valve devices vs. temperature, *T*, normalized at *T* = 14 K. The OSEC interlayers in these devices are Alq₃ (squares), CVB (circles), and NPD (stars); their chemical formulae are given in the text. (b) The calculated LSMO spin polarization, P(LSMO), using the modified Jullière model[6] (Equation 5.12) and the data in (a); the symbols and colors are the same as in (a). The line through the data points is the calculated polarized charge carrier density (PCCD) of the LSMO electrode normalized at *T* = 0 using the model given in Drew et al.,[90] with T_c = 325 K (see inset). The inset shows the magnetization, *M* (empty circles), of the LSMO film used as the bottom FM electrode for the OSV shown in (a); the Curie temperature, T_c = 325 K, is assigned. (From Wang et al., *Phys. Rev. B* 75, 245324 [2007].)

the OSEC interlayer response, dominates the MR_{SV} decrease with temperature. The Co magnetic properties do not change much with *T*; however, the LSMO magnetic properties strongly depend on the temperature (Figure 5.18b, inset),[91,92] and it was thus concluded that its particular response is the underlying mechanism for the MR_{SV} temperature dependence.[75]

5.4.2.3   Spin Polarization Properties of the LSMO Electrode

The MR response of various OSVs has been interpreted using a modified Jullière model,[93] in which $MR_{SV} = [\Delta R/R]_{max}$ is given by[6]

$$[\Delta R/R]_{max} = \frac{2P_1 P_2 D}{(1 + P_1 P_2 D)},\tag{5.12}$$

where P_1 and P_2 are the spin polarizations of the two FM electrodes, respectively. In Equation 5.12, $D = exp[-(d - d_0)/\lambda]_s$, where λ_s is the spin diffusion length in the OSEC, d is the OSEC thickness, and d_0 (G60 nm) is an ill-defined OSEC layer thickness,[6] where inclusions of the upper FM metal may be abundantly found. Equation 5.12 is in fact written in the spirit of a more rigorous analysis of MR responses in inorganic semiconductor spin valves,[94,95] where the processes of spin injection, spin accumulation, and spin transport through a semiconductor are all explicitly taken into account. As shown in Figure 5.18a, the normalized spin polarization value $P_1(T)$ (= P(LSMO)) for the LSMO electrode vs. temperature can be calculated (Figure 5.18b) from the MR_{SV} temperature dependence using Equation 5.12, and the temperature independent parameters $P_2(Co)$ and D.[75] Based on this ansatz, it is seen that P(LSMO) steeply decreases with T, indicating that the surface spin polarization of the LSMO electrode strongly depends on the temperature. For comparison with the FM bulk properties, the LSMO magnetization moment, M, vs. temperature was also measured (Figure 5.18b inset).[75] In contrast to P(LSMO) vs. T obtained from the above-mentioned analysis of the OSV temperature response, $M(T)$ is less temperature dependent; in particular, M vanishes at a Curie temperature, $T_c = 325$ K, rather than at $T = 220$K, where P(LSMO) diminishes (Figure 5.18b). It is thus apparent that the surface LSMO spin polarization, rather than the bulk LSMO magnetization, is responsible for the steep decrease of P(LSMO) with T. In fact, the obtained P(LSMO) temperature response agrees very well with the surface spin polarization of LSMO films as probed by spin-polarized photoemission spectroscopy,[91] which measures the polarized charge carrier density (PCCD) at the surface boundary within 0.5 nm depth. Since P is proportional to PCCD—which in turn is given by $DOS(E_F) \times M$, where $DOS(E_F)$ is the density of electronic states at the metal Fermi level—P at the surface is not determined by M alone. Actually in LSMO for $T < T_c$, $DOS(E_F)$ varies strongly with T, and consequently, $P(T)$ does not follow $M(T)$. PCCD(T/T_c) for LSMO was calculated by Park et al.[91] from the spin-polarized photoemission spectrum, and this functional dependence was used with $T_c = 325$ K to calculate PCCD(T) for the LSMO electrode used in our OSV, as seen in Figure 5.18b. The agreement between the T dependencies of the calculated PCCD and P(LSMO) obtained directly from the three OSV temperature responses was found to be excellent[75]; this verifies that the LSMO interface spin polarization is the mechanism responsible for explaining the steep MR temperature dependence of the fabricated OSV. However, this behavior is not universal[69] since the LSMO surface properties may be enhanced by proper surface techniques, and this may change the PCCD(T) functional dependence.

5.4.3 ORGANIC SPIN VALVES BASED ON CONVENTIONAL FERROMAGNETIC ELECTRODES

In addition to OSV based on LSMO, which has an SP-injecting capability P of ~95%, other OSV devices based on more conventional FM electrodes having smaller P but less steep temperature dependencies, such as Fe, Co, and Ni, have been also studied.[68,76,78,80,81] Originally it was reported[68] that OSV based on an Alq$_3$ interlayer sandwiched between Fe and Co FM electrodes showed MR$_{SV}$ ~ 3% at low temperature; recently this value was measured to be ~7% by another group.[80] However, the original data were challenged[81]; it was claimed that when carefully fabricated, namely, deposition in a chamber of high vacuum and without breaking the vacuum in between the OSEC and electrode's deposition, such an OSV does not show a spin valve MR response.[81] This claim has cast some doubts on the obtained MR response in OSV in general, since the response might have been due to artifacts such as FM inclusions in the OSEC film, as claimed in Xu et al.,[72] although numerous laboratories around the world have repeated the original OSV response data.[6] These doubts, however, were refuted when the spin diffusion length in an amorphous OSEC film of rubrene ($C_{42}H_{28}$) of SP electrons injected from conventional FM electrodes was directly obtained by spin-polarized tunneling into an Al superconductor film at ultra-low temperatures.[78] A spin diffusion length of ~13 nm was measured at low temperature with a relatively small decrease up to room temperature; the authors predicted a spin diffusion length of a few millimeters in rubrene single crystal.[78] Moreover, an MR$_{SV}$ of ~15% was measured for an OSV composed of Co/Al$_2$O$_3$/rubrene/Fe magnetic tunneling junction, in which a tunnel barrier layer was introduced between the Co and OSEC interlayer[78]; this is in direct contradiction with the claims in Jiang et al.,[81] which argued a null MR response for OSV made of conventional FM electrodes, even when a tunnel barrier was introduced between one of the FM electrodes and the OSEC layer. In addition, recently Liu et al.[80] reported measurements on LSMO/Alq$_3$/Co OSV. They found that a correlation exists between the MR and the FM/Alq$_3$ interface microstructure. These results demonstrate that it is possible to realize room temperature spin injection from transition metals to OSEC by careful interface modification.

A recent excellent theoretical review has dealt with the MR response of OSV under different growing conditions.[4] In this theoretical contribution the SP injection current was calculated and compared to the charge current in diverse organic diodes based on (1) conventional FM electrodes without a tunnel barrier, (2) conventional FM with a tunnel barrier (insulating buffer layer) between the FM electrode and OSEC interlayer, and (3) electrode composed of half-metallic FM materials with low conductivity, such as LSMO. It was shown[4] that spin injection is indeed difficult to achieve for case 1 since the "conductivity mismatch"[65,66,96] acts for OSEC similarly as for inorganic semiconductors. This may explain the reason that MR was not achieved in OSV devices based on two conventional FM electrodes without introducing a tunnel barrier in between the FM electrode and OSEC layer.[81] However, according to the model used, the SP current dramatically increases in case 2 when a tunnel barrier is introduced, and thus spin-dependent tunneling is the limiting process for carrier injection. This explains the earlier result,[68] where a tunnel barrier was inadvertently introduced between the OSEC and the capped FM layer (Co), due to the fabrication process; in that case, the vacuum in the evaporation chamber was broken before evaporating the upper FM electrode. It also explains the more recent finding,[80] where a dead magnetic region was seen near the Fe electrode, which resulted in an excellent MR value for the Fe/Alq$_3$/Co spin valve. Moreover, it was also shown in the theoretical work[4] that SP current is substantial in case 3 for half-metallic FM, because its conductivity is low (thus reducing the conductivity mismatch with the OSEC interlayer) and P is high. This explains the high MR response in OSV based on LSMO as an SP-injecting electrode, such as in Xiong et al.,[6] without the need to use tunnel barriers.

5.4.4 ROOM TEMPERATURE LSMO-BASED ORGANIC SPIN VALVE OPERATION

There is no real obstacle for obtaining LSMO-based OSV operation at room temperature, provided that the signal/noise (S/N) ratio of the MR loop measurement is improved to observe the weak, anticipated MR signal at ~300 K. To achieve this task, the LSMO-based OSV needs be very stable in order to improve the S/N ratio in the MR measurements. We recently found that OSV devices based on the C$_{60}$ spacer layer indeed possess such a stable operation. This is probably caused by a superior interface between the LSMO and C$_{60}$ molecular layer that may be formed due to the ability of the fullerene molecule to diffuse at the deposition temperature, thus filling the LSMO rough surface.[97] Another advantage of the C$_{60}$ spacer is the weak hyperfine interaction (HFI) of this molecule, which is based only on carbon atoms. The carbon nucleus ^{12}C isotope has spin zero, and thus does not count for the HFI. Although ^{13}C isotope nucleus has spin $\frac{1}{2}$, its natural abundance is <2%, and thus the overall HFI of the natural C$_{60}$ molecule may be approximately two orders smaller than that of the hydrogen atom. We therefore conjecture that the weak HFI of the C$_{60}$ molecule may increase the spin diffusion length in OSV based on fullerene molecules so that the corresponding MR would be only limited by the LSMO ability of injecting spin-aligned carriers into the OSEC.

Figure 5.19 shows the MR loop (symbols) of an LSMO/C$_{60}$/Co OSV at room temperature (RT). The excellent S/N ratio achieved in these measurements reveals an MR$_{SV}$ of $\cong$ 0.16% at 200 mV bias; the MR$_{SV}$ increases to $\cong$ 0.3% at low bias voltage, namely, $V < 50$ mV.[97] The solid lines are simulations using the diffusion model of Bobbert et al.[98] with a very small HFI constant ($a_{HF} \cong 0.15$ mT), indicating long spin diffusion lengths. The obtained MR$_{SV}$ value is similar to that measured using OSVs with a polymer spacer,[71] or superior LSMO surface.[69,73] This shows that the MR$_{SV}$ RT value is independent of the OSEC used, or the quality of the LSMO surface; instead, it depends on the intrinsic properties of the spin-polarized injection capability of the LSMO substrate at RT. It is also noteworthy that the coercive fields of the LSMO and Co FM electrodes at RT are both substantially smaller than the corresponding fields at low temperature[97]; this makes the OSV devices very attractive for RT applications.

5.4.5 CONCLUSIONS AND FUTURE OUTLOOK

In this contribution we reviewed some of the latest achievements in organic spintronics, with emphasize on OSV research at the University of Utah. The organic

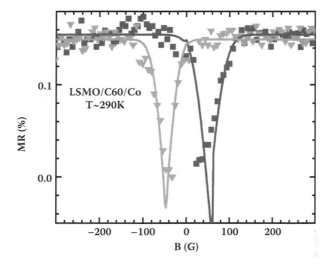

FIGURE 5.19　The MR experimental response (symbols) of an OSV based on a 40-nm thick C_{60} spacer layer in between LSMO and Co FM electrodes at room temperature and bias voltage $V = 200$ mV. The solid lines are simulations using the diffusion model of Bobbert et al.[98] The boxes (triangles) denote up (down) magnetic field sweep.

spintronics field is in its debut; much more work has to be accomplished before the field will mature. At the present time, controversies regarding the exact operation of OSV still exist, especially related to the MR sign in these devices, as well as the interface properties between the FM electrode and the organic overlayer. Recently, SP carrier injection into an OSEC has been directly proven using low-energy muon spin rotation,[90] so the doubts raised at the beginning of the organic spintronics field may be safely defused. At the present time spin injection achieved in OSV is of the same order of magnitude as in spin valves based on inorganic semiconductor spacers. In particular, both organic and inorganic semiconductor interlayers suffer from the same conductivity mismatch problem, and show only little MR at room temperature.

　　The organics do not possess the polarized emission properties of inorganic direct gap semiconductors such as GaAs, which can be used to directly detect SP current in spin valve devices.[99,100] The reason for this is that the emission in the organics results from tightly bound singlet excitons,[101] rather than pairs of electrons and holes as in inorganic semiconductors. Also, the method of spin-induced magnetic Kerr effect, which was used successfully to image SP carrier injection into inorganic semiconductors,[102] is not useful for OSV because of the small current involved and the small spin-orbit coupling in the organic layer. Direct imaging of SP current injection into the organic semiconductor is highly needed because MR alone may be prone to artifacts.[103] Therefore, other detection methods should be used to more directly measure depth-resolved information on the SP charge carriers within the buried layers of organic spin devices. Such a method was recently reported[90,104] using low-energy muon spin rotation; however, this method is not tabletop. Other methods, such as two-photon photoemission, have also been recently advanced for probing spin injection from the FM electrode into the organic overlayer.[105] The field of organic spintronics has very much benefited from such direct measurements of spin injection.[104]

At the present time, MR of few fractions of 1% has been achieved with OSV based on LSMO at room temperature[71,73]; this is too low for generating industrial interest. For this field to take off, FM spin injectors other than LSMO need be discovered, of which high SP injection capabilities would survive at room temperature. We predict that it should be possible to reach sizable room temperature MR_{SV} values in OSV based on FM electrodes with large P, but with a milder dependence on temperature than the LSMO. Such an FM electrode, for example, might be the half-metallic CrO_2 with $T_c \sim 395$ K,[106] although no spin valve devices based on it have been successful so far; or the double-perovskite oxides with $T_c > 400$ K[107]; or the recently discovered EuO.[108] It is also noteworthy that MR of a few percent has been recently measured in Fe/Alq$_3$/Co OSV at room temperature.[80] Another possibility would be to use organic FM as spin-injecting electrodes[109]; since the conductivity of these materials matches that of OSEC, the conductivity mismatch problem would be naturally solved. We also note that sizable SP carrier injection was recently achieved with graphene,[110] which is, after all, also a (two-dimensional) OSEC material.

ACKNOWLEDGMENTS

This work benefitted from the contributions of many students and postdocs at the University of Utah, as well as other collaborators during the years 2003–2009. We thank Profs. X-G. Li and J. Shi; Drs. T. D. Nguyen, F. Wang, D. Wu, C. G. Yang, and Z. H. Xiong; as well as L. Wojcik and G. Hukic-Markosian, who contributed immensely to all parts of the work reviewed here. We also acknowledge useful discussion with Profs. C. Grissom and M. Wohlgenannt. This work was supported in part by the NSF-DMR Grant 08-03172 and DOE Grant 04-ER 46109 at the University of Utah, and by the Israel Science Foundation (745/08).

REFERENCES

1. S. A. Wolf, D. D. Awschalom, R. A. Buhrman, J. M. Daughton, S. von Molnar, M. L. Roukes, A. Y. Chtchelkanova, and D. M. Treger, *Science* 294, 1488 (2001).
2. I. Zutic, J. Fabian, and S. Das-Sarma, *Rev. Modern Phys.* 76, 323 (2004), and references therein.
3. D. D. Awschalom and M. E. Flateé, *Nature Phys.* 3, 153 (2007).
4. M. Yunus, P. P. Ruden, and D. L. Smith, *J. Appl. Phys.* 103, 103714 (2008).
5. V. Dediu, M. Murgia, F. C. Matacotta, C. Taliani, and S. Barbanera, *Solid State Commun.* 122, 181 (2002).
6. Z. H. Xiong, D. Wu, Z. V. Vardeny, and J. Shi, *Nature* (London) 427, 821 (2004).
7. H. Hayashi, *World Scientific Lecture and Course Notes in Chemistry: Introduction to Dynamic Spin Chemistry; Magnetic Field Effects on Chemical and Biochemical Reactions*, Vol. 8, World Scientific Publishing Co., Singapore, (2004).
8. M. Wohlgenannt, E. Ehrenfreund, and Z. V. Vardeny, Spectroscopy of long-lived photoexcitations in π-conjugated systems, in *Photophysics of Molecular Materials*, ed. G. Lanzani, 183–261, Wiley-VCH Verlag, Weinheim, 2006.
9. A. Carrington and A. D. McLachlan, *Introduction to Magnetic Resonance*, Harper and Row, New York, 1967.

10. B. C. Cavenet, *Adv. Phys.* 30, 475 (1981).
11. Z. V. Vardeny and X. Wei, *Handbook of Conducting Polymers II*, chap. 22, Marcel Dekker, New York, 1997.
12. G. Li, C. H. Kim, P. A. Lane, and J. Shinar, *Phys. Rev. B* 69, 165311 (2004).
13. I. Solomon, D. Biegelsen, and J. C. Knights, *Solid State Commun.* 22, 505 (1977).
14. A. Swanson, J. Shinar, A. Brown, D. Bradley, R. H. Friend, P. Burn, A. Kraft, and B. Holmes, *Phys. Rev. B* 46, 15072 (1992).
15. T. Eickelkamp, S. Roth, and M. Mehring, *Mol. Physics* 95, 967 (1998).
16. D. R. McCamey, H. A. Seipel, S. Y. Paik, M. J. Walter, N. J. Borys, J. M. Lupton, and C. Boehme, *Nature Mater.* 7, 723 (2008).
17. C. G. Yang, F. Wang, T. Drori, E. Ehrenfreund, and Z. V. Vardeny, *Phys. Rev. B* 78, 205312 (2008).
18. M. Wohlgenannt, C. Yang, and Z. V. Vardeny, *Phys. Rev. B* 66, 241201 (2002).
19. C. G. Yang, E. Ehrenfreund, and Z. V. Vardeny, *Phys. Rev. Lett.* 99, 157401 (2007).
20. F. J. Wang, C. G. Yang, E. Ehrenfreund, and Z. V. Vardeny, *Synth. Met.* (2009), DOI: 10.1016/j.synthmet.2009.06.014.
21. M. Reufer, M. J. Walter, G. Lagoudakis, A. B. Hummel, J. S. Kolb, H. G. Rosko, U. Scherf, and J. M. Lupton, *Nature Mater.* 4, 340 (2005).
22. V. Prigodin, J. Bergeson, D. Lincoln, and A. Epstein, *Synth. Met.* 156, 757 (2006).
23. P. Desai, P. Shakya, T. Kreouzis, W. P. Gillin, N. A. Morley, and M. R. J. Gibbs, *Phys. Rev. B* 75, 094423 (2007).
24. D. Wu, Z. H. Xiong, X. G. Li, and Z. V. Vardeny, *Phys. Rev. Lett.* 95, 016802 (2005).
25. K. Fesser, A. Bishop, and D. K. Campbell, *Phys. Rev. B* 27, 4804 (1983).
26. B. R. Weinberger, E. Ehrenfreund, A. Pron, A. J. Heeger, and A. MacDiarmid, *J. Chem. Phys.* 72, 4749 (1980).
27. A. Abragam, *The Principles of Nuclear Magnetism*, Oxford University Press, London, 1961.
28. B. Brocklehurst and K. A. McLauchlan, *Int. J. Radiat. Biol.* 69, 3 (1996).
29. C. R. Timmel, U. Till, B. Brocklehurst, K. A. McLauchlan, and P. J. Hore, *Mol. Physics* 95, 71 (1998).
30. O. Mermer, G. Veeraraghavan, T. L. Francis, Y. Sheng, D. T. Nguyen, M. Wohlgenannt, A. Köhler, M. K. Al-Suti, and M. S. Khan, *Phys. Rev. B* 72, 205202 (2005).
31. B. Hu and Y. Wu, *Nature Mater.* 6, 985 (2007).
32. J. M. Lupton and C. Boehme, *Nature Mater.* 7, 598 (2008).
33. B. Hu and Y. Wu, *Nature Mater.* 7, 600 (2008).
34. R. P. Groff, A. Suna, P. Avakian, and R. E. Merrifield, *Phys. Rev. B* 9, 2655 (1974).
35. A. Weller, F. Nolting, and H. A. Staerk, *Chem. Phys. Lett.* 96, 24 (1983).
36. J. D. Bergeson, V. N. Prigodin, D. M. Lincoln, and A. J. Epstein, *Phys. Rev. Lett.* 100, 067201 (2008).
37. Y. Sheng, T. D. Nguyen, G. Veeraraghavan, O. Mermer, M. Wohlgenannt, S. Qiu, and U. Scherf, *Phys. Rev. B* 74, 045213 (2006).
38. P. Desai, P. Shakya, T. Kreouzis, and W. P. Gillin, *Phys. Rev. B* 76, 235202 (2007).
39. Y. Iwasaki, T. Osasa, M. Asahi, M. Matsumura, Y. Sakaguchi, and T. Suzuki, *Phys. Rev. B* 74, 195209 (2006).
40. Y. Wu, Z. Xu, B. Hu, and J. Howe, *Phys. Rev. B* 75, 035214 (2007).
41. P. A. Bobbert, T. D. Nguyen, F. W. A. van Oost, B. Koopmans, and M. Wohlgenannt, *Phys. Rev. Lett.* 99, 216801 (2007).
42. F. L. Bloom, W. Wagemans, M. Kemerink, and B. Koopmans, *Phys. Rev. Lett.* 99, 257201 (2007).
43. U. Niedermeir, M. Vieth, R. Pätzold, W. Sarfert, and H. von Seggern, *Appl. Phys. Lett.* 92, 193309 (2008).
44. E. L. Frankevich, A. A. Lymarev, I. Sokolik, F. E. Karasz, S. Blumstengel, R. H. Baughman, and H. H. Hörhold, *Phys. Rev. B* 46, 9320 (1992).

45. M. Pope and C. E. Swenberg, *Organic Electronic Processes in Crystals and Polymers*, Oxford University Press, New York (1999).
46. W. J. M. Naber, S. Faez, and W. G. van der Wiel, *J. Phys. Appl. Phys.* 40, R205 (2007).
47. M. A. Green, K. Emery, Y. Hishikawa, and W. Warta, *Prog. Photovolt. Res. Appl.* 16, 61 (2008).
48. A. Pivrikas, N. S. Sariciftci, G. Juška, and R. Österbacka, *Prog. Photovolt. Res. Appl.* 15, 677 (2007).
49. F. Wang, H. Bässler, and Z. V. Vardeny, *Phys. Rev. Lett.* 101, 236805 (2008).
50. T. Drori, C. X. Sheng, A. Ndobe, S. Singh, J. Holt, and Z. V. Vardeny, *Phys. Rev. Lett.* 101, 037401 (2008).
51. M. C. Scharber, N. A. Schultz, and N. S. Sariciftci, *Phys. Rev. B* 67, 085202 (2003).
52. U. E. Steiner and T. Ulrich, *Chem. Rev.* 89, 51 (1989).
53. T. D. Nguyen, Y. Sheng, J. Rybicki, and M. Wohlgenannt, *Phys. Rev. B* 77, 235209 (2008).
54. A. Kadashchuk, V. I. Arkhipov, C. H. Kim, J. Shinar, D. W. Lee, Y. R. Hong, and J. I. Jin, *Phys. Rev. B* 76, 235205 (2007).
55. M. Baibich, J. M. Broto, A. Fert, F. NguyenVanDan, F. Petroff, P. Etienne, G. Creuzert, A. Friederich, and J. Chazelas, *Phys. Rev. Lett.* 61, 2472 (1988).
56. G. A. Prinz, *Science*, 282, 1660 (1998).
57. J. S. Moodera, L. R. Kinder, T. M. Wong, and R. Meservey, *Phys. Rev. Lett.* 74, 3273 (1995).
58. H. X. Tang, et al., in *Semiconductor Spintronics and Quantum Computation,* ed. D. Awschalom, D. Loss, and N. Samarth, pp. 31–92, Springer, New York, 2002.
59. P. P. Ruden and D. L. Smith, *J. Appl. Phys.* 95, 4898 (2004).
60. A. R. Rocha, V. M. Garcia-Suárez, S. W. Baily, C. J. Lambert, J. Ferrer, and S. Sanvito, *Nature Mater.*, 4, 335 (2005).
61. S. Sanvito and A. R. Rocha, *J. Comput. Theor. Nanosci.* 3, 624 (2006).
62. B. T. Jonker, *Proc. IEEE* 91, 727 (2003).
63. X. Lou, C. Adelmann, S. A. Crooker, E. S. Garlid, J. Zhang, S. M. Reddy, S. D. Flexner, C. J. Palmstrøm, and P. A. Crowell, *Nature Phys.* 3, 197 (2007).
64. S. Sanvito, *Nature Mater.* 6, 803 (2007).
65. G. Schmidt, D. Ferrand, L. W. Molenkamp, A. T. Filip, and B. J. van Wees, *Phys. Rev. B* 62, R4790 (2000).
66. D. L. Smith and R. N. Silver, *Phys. Rev. B* 64, 045323 (2001).
67. J. D. Albrecht and D. L. Smith, *Phys. Rev. B* 66, 113303 (2002).
68. F. J. Wang, Z. H. Xiong, D. Wu, J. Shi, and Z. V. Vardeny, *Synth. Met.* 155, 172 (2005).
69. V. Dediu, L. E. Hueso, I. Bergenti, A. Riminucci, F. Borgatti, P. Graziosi, C. Newby, F. Casoli, M. P. D. Jong, C. Taliani, and Y. Zhan, *Phys. Rev. B* 78, 115203 (2008).
70. T. X. Wang, H. X. Wei, Z. M. Zeng, X. F. Han, Z. M. Hong, and G. Q. Shi, *Appl. Phys. Lett.* 88, 242505 (2006).
71. S. Majumdar, H. S. Majumdar, P. Laukkanen, I. J. Väyrynen, R. Laiho, and R. Österbacka, *Appl. Phys. Lett.* 89, 122114 (2006).
72. W. Xu, G. J. Szulczewski, P. LeClair, I. Navarrete, R. Schad, G. Miao, H. Guo, and A. Gupta, *Appl. Phys. Lett.* 90, 072506 (2007).
73. L. E. Hueso, I. Bergenti, A. Riminucci, Y. Zhan, and V. Dediu, *Adv. Mater.* 19, 2639 (2007).
74. S. Pramanik, C. G. Stefanita, S. Padibandla, S. Bandyopadhyay, K. Garre, N. Harth, and M. Cahay, *Nature Nanotech.* 2, 216 (2007).
75. F. J. Wang, C. G. Yang, Z. V. Vardeny, and X. G. Li, *Phys. Rev. B* 75, 245324 (2007).
76. T. S. Santos, J. S. Lee, P. Migdal, I. C. Lekshmi, and J. S. Moodera, *Phys. Rev. Lett.* 98, 016601 (2007).
77. N. A. Morley, A. Rao, D. Dhandapani, M. R. J. Gibbs, M. Grell, and T. Richardson, *J. Appl. Phys.* 103, 07F306 (2008).
78. J. H. Shim, K. V. Raman, Y. J. Park, T. S. Santos, G. X. Miao, B. Satpati, and J. S. Moodera, *Phys. Rev. Lett.* 100, 226603 (2008).

79. M. Bowen, M. Bibes, A. Barthelemy, J. P. Contour, A. Anane, Y. Lemaitre, and A. Fert, *Appl. Phys. Lett.* 82, 233 (2003).
80. Y. Liu, S. M. Watson, T. Lee, J. M. Gorham, H. E. Katz, J. A. Borchers, and H. D. Fairbrother, *Phys. Rev. B* 79, 075312 (2009).
81. J. S. Jiang, J. E. Pearson, and S. D. Bader, *Phys. Rev. B* 77, 035303 (2008).
82. Y. C. Zhang, M. P. deJong, F. H. Li, V. Dediu, M. Fahlman, and W. R. Salaneck, *Phys. Rev. B* 78, 045208 (2008).
83. A. Riminuci, I. Bergenti, L. E. Hueso, M. Muriga, C. Taliani, Y. Zhan, F. Casoli, M. P. de Jong, and V. Dediu, cond-mat/0701603 (2007).
84. L. E. Hueso, J. M. Pruneda, V. Ferrari, G. Burnell, J. P. Valdés-Herrera, B. D. Simons, P. B. Littlewood, E. Artacho, A. Fert, and N. D. Mathur, *Nature* 445, 410 (2007).
85. J. M. De-Teresa, A. Barthélémy, A. Fert, J. P. Contour, F. Montaigne, and P. Seneor, *Science* 286, 507 (1999).
86. J. S. Moodera and G. Mathon, *J. Magnetism Magnetic Mater.* 200, 248 (1999).
87. J. Hayakawa, K. Ito, S. Kokado, M. Ichimura, A. Sakuma, M. Sugiyama, H. Asano, and M. Matsui, *J. Appl. Phys.* 91, 8792 (2002).
88. S. Zhang, P. M. Levy, A. C. Marley, and S. S. P. Parkin, *Phys. Rev. Lett.* 79, 3744 (1997).
89. A. M. Bratkovsky, *Phys. Rev. B* 56, 2344 (1997).
90. A. J. Drew, J. Hoppler, L. Schulz, F. L. Pratt, P. Desai, P. Shakya, T. Kreouzis, W. P. Gillin, A. Suter, N. A. Morley, V. K. Malik, A. Dubroka, K. W. Kim, H. Bouyanfif, F. Bourqui, C. Bernhard, R. Scheuermann, G. J. Nieuwenhuys, T. Prokscha, and E. Morenzoni, *Nature Mater.* 8, 109 (1982).
91. J. H. Park, E. Vescovo, H. J. Kim, C. Kwon, R. Ramesh, and T. Venkaresan, *Phys. Rev. Lett.* 81, 1953 (1998).
92. V. Garcia, M. Bibes, A. Barthélémy, M. Bowen, E. Jacquet, J. P. Contour, and A. Fert, *Phys. Rev. B* 69, 052403 (1999).
93. M. Julliere, *Phys. Rev. A* 54, 225 (1975).
94. A. Fert and H. Jaffrès, *Phys. Rev. B* 64, 184420 (2001).
95. A. Fert, J. M. George, H. Jaffrès, and R. Mattana, cond-mat/0612495 (2006).
96. E. I. Rashba, *Phys. Rev. B* 62, R16267 (2000).
97. F. J. Wang, PhD thesis, University of Utah (2009).
98. P. A. Bobbert, W. Wagemans, F. W. A. van Oost, B. Koopmans, and M. Wohlgenannt, *Phys. Rev. Lett.* 102, 156604 (2009).
99. R. Flederling, M. Kelm, G. Rauscher, W. Ossau, G. Schmidt, A. Waag, and L. W. Molenkamp, *Nature* 402, 787 (1999).
100. Y. Ohno, D. K. Young, B. Beschoten, F. Matsukura, H. Ohno, and D. D. Awschalom, *Nature* 402, 790 (1999).
101. S. Singh, T. Drori, and Z. V. Vardeny, *Phys. Rev. B* 77, 195304 (2008).
102. S. A. Crooker, M. Furis, X. Lou, C. Adelmann, D. L. Smith, C. J. Palmstrøm, and P. A. Crowell, *Science* 309, 2191 (2005).
103. J. M. Coey, *J. Appl. Phys.* 85, 5576 (1999).
104. Z. Vardeny, *Nat. Mater.* 8, 91 (2009).
105. M. Cinchetti, K. Heimer, J.-P. Wüstenberg, O. Andreyev, M. Bauer, S. Lach, C. Ziegler, Y. Gao, and M. Aeschlimann, *Nature Mater.* 8, 115 (2009).
106. R. S. Keizer, S. T. B. Goennenwein, T. M. Klapwijk, G. Miao, G. Xiao, and A. Gupta, *Nature* 439, 825 (2006), and references therein.
107. D. Serrate, J. M. De-Teresa, and M. R. Ibarra, *J. Phys. Condens. Matter* 19, 023201 (2007).
108. J. F. Gregg, *Nature Mater.* 6, 798 (2007).
109. E. Carlegrim, A. Canciurzewska, P. Nordblad, and M. Fahlman, *Appl. Phys. Lett.* 92, 163308 (2008).
110. N. Tombros, C. Jozsa, M. Popinciu, H. T. Jonkman, and B. J. van Wees, *Nature* 448, 571 (2007).

6 Investigating Spin-Dependent Processes in Organic Semiconductors

Christoph Boehme and Dane R. McCamey

CONTENTS

6.1 INTRODUCTION

Today, organic semiconductors are used commercially for a variety of organic electronic devices, including organic light-emitting diodes (OLEDs),[1,2] organic electronic displays,[3–6] and organic solar cells.[7–11] Some of the technological advantages organic electronics provide include excellent energy efficiency, physical flexibility, and extraordinary economical production. Using organic materials for solar cells provides a number of important advantages over conventional inorganic approaches: the material is flexible and lightweight, allowing for novel installation and applications; it can be printed onto a flexible substrate in a reel-to-reel process, enabling large-scale production; organic solar cells are reasonably energy efficient and, while they still trail the efficiencies of inorganic solar cells,[12] their significantly lower cost makes them highly competitive.

Given the large range of technological and commercial applications of organic electronics resulting from decades of research on these materials,[13] it is remarkable that there are still unanswered fundamental questions about the most basic and most crucial physical processes in organic semiconductor materials. Many of these questions concern the influence of the spin degree of freedom of electrons and atomic nuclei on electronic and optical properties of these materials. Since organic materials are carbon based, the atomic order number of most atoms in organic semiconductors is low, leading to very weak spin-orbit interactions. This imposes spin conservation, and therefore spin selection rules on many electronic processes affecting macroscopic materials properties. Examples of these processes include charge carrier recombination and charge carrier transport, both of which are known to be important for device efficiencies in solar cells[14] and OLEDs,[15] as well as magnetoresistance and therefore sensor applications.[16–18]

6.1.1 Observing Spins in Semiconductors

The experimental investigation of spin-dependent transitions requires the observation of spins and their interaction with energy and charge transport. Traditional ways to observe spins, magnetic resonance techniques such as electron paramagnetic resonance (EPR) or nuclear magnetic resonance (NMR), have been used for decades in materials research.[19–23] NMR and EPR are able to detect nuclear[20] and electron[19,22] spins, respectively, and both methods also allow the observation of some interactions between spins, spins and their environment, spin propagation, and spin relaxation. For the application to organic semiconductors, these methods are limited: First, conventional magnetic resonance methods are purely volume-sensitive techniques. Their detection limits in organic semiconductors and commercially available spectrometers are more than approximately 10^{14} spins for NMR, while EPR requires more than approximately 10^{10} spins.[22] This is not sufficient since many organic semiconductors in real devices are manufacturable only as thin films. Moreover, conventional magnetic resonance spectroscopy reveals little information about electronic transitions between paramagnetic states, and as a result, the insight that it provides when spins are detectable is often limited. Finally, as NMR and EPR are sensitive to all (nuclear or electronic) spins within a given sample, one can rarely distinguish the spins that directly impact materials properties from those that do not.

6.1.2 Electrically and Optically Detected Magnetic Resonance

In order to overcome the limitations of EPR and NMR for the investigation of spin-dependent processes, experiments to observe the direct influence of spins on electronic and optical properties have been explored since the late 1950s.[24] A variety of implementations for such experiments have been demonstrated.[23–30] Most of these different approaches follow a common principle to observe spin-dependent processes: manipulate the involved spin with magnetic resonance so that the macroscopic observables (optical or electrical properties) change due to the change of spin. The experimental methods resulting from this approach are called optically detected magnetic resonance (ODMR) when spin-controlled radiative emission or absorption is measured, and electrically detected magnetic resonance (EDMR) when the conductivity of a sample or device is measured.

The first ODMR experiments were carried out in 1959[24,31] on chromium ions in an aluminum oxide host matrix. These experiments were the first direct experimental proof of the existence of spin selection rules and triggered a series of additional studies that eventually led to the first demonstration of EDMR in 1966 by Honig,[25] Maxwell and Honig,[26] and Schmidt and Solomon.[32] Since then EDMR and

ODMR spectroscopy have been continuously developed and their range of application extended to a large variety of organic[33–40] and inorganic[27,41–48] materials. EDMR and ODMR provide solutions to a number of the drawbacks of conventional EPR experiments: They overcome sample volume limitations since they are not reliant on the observation of microwave photons, but on the measurement of currents and photons (in the IR to UV range). Both EDMR and ODMR are significantly more sensitive than EPR, and since the sensitivity of EDMR and ODMR scales with the size of the sample,[49] extremely small samples can be investigated. Single spin detection with both EDMR[50] and ODMR[51] has been reported. EDMR and ODMR provide the additional benefit that only spins that directly impact spin-dependent rates are observed. For example, spin effects from geminate charge carrier pairs, which have little impact on the photocurrent, are not seen in EDMR experiments, but are observed in ODMR experiments. In contrast, nongeminate recombination is observed with EDMR and, depending on whether it influences radiative processes, may or may not be observed with ODMR.[52,138]

EDMR and ODMR do not generally rely on spin polarization but on spin selection rules, such as permutation symmetries of pairs of paramagnetic centers between which transitions occur. Because of this, EDMR and ODMR are well suited to experimental conditions like low magnetic fields or high temperatures, where conventional EPR is inapplicable.[19] Recently, EDMR has been observed in silicon with magnetic fields lower than 1 mT[53]—similar experiments in organic materials could allow spin-dependent processes to be investigated in the regime where the hyperfine interaction is comparable to the externally applied magnetic fields. This would allow the investigation of currently discussed organic magnetoresistance (OMAR) models. EDMR and ODMR can also be performed with a broad excitation bandwidth, since a fixed frequency due to a resonant microwave detection circuit is not needed.[54]

6.1.3 PULSED ELECTRICALLY DETECTED MAGNETIC RESONANCE

EPR, ODMR, and EDMR have, for the most part, been performed as continuous wave (cw) experiments where a magnetic field, B_0, is swept adiabatically and microwave radiation at a fixed frequency is continuously applied.[55] At the same time, observables such as microwave radiation (EPR), light (ODMR), or conductivity (EDMR) are monitored. Such experiments are experimentally simple, but the information available from cw spectra is limited to Landé (g) factors and strongly convoluted spin relaxation, coupling, and electronic lifetime parameters that determine the line shapes and line widths of the resonances in a way that is not easily experimentally accessible. Because of this, the development of coherent, pulsed (p) EPR[21] and pODMR spectroscopy[56] has been aggressively pursued. These methods allow the coherent propagation of spins during a short pulsed resonant excitation to be accessed, and by doing so, are a strong experimental tool for obtaining both qualitative and quantitative parameters of the spin being investigated.[21]

PEPR[21] and pODMR[56] have been utilized since the 1970s. This followed the development in the 1950s of pulsed NMR, a technique that is now performed almost exclusively in the coherent time domain. PEDMR, the direct electrical detection of coherent spin propagation, was first demonstrated in 2002.[29] The implementation of pEDMR experiments was prevented mainly by technological challenges, including (1) generating strong homogeneous electromagnetic pulses (which provide the B_1 field) around conductive semiconductor samples and devices,[57] (2) preventing the conducting samples from absorbing the B_1 field (antenna effect) and thus producing strong perturbing currents, and (3) measuring the (usually) subtle spin-dependent current on top of strong spin-independent current offsets at a time resolution that is appropriate for the observation of coherent spin motion.[58] Measuring the subtle currents presented a particularly difficult challenge until an indirect spin pump-current probe measurement scheme for pEDMR was demonstrated and developed in 2002,[29,57] as discussed in detail in Section 6.3.3.2. Since that first demonstration, many pulse sequences developed for pEPR have been demonstrated, either identically or with slight modification, using pEDMR. These include rotary echoes[59] (see Figure 6.1a) and spin Rabi nutation experiments[60] (see Figure 6.1b), which have been used to demonstrate that exchange (J),[52,61] dipolar (D),[52] and hyperfine (A)[30] coupling can influence pEDMR data. Electrical detection of Hahn echoes[62–64] as well as inversion-recovery-like experiments[64] have also been demonstrated.

So far, most pEDMR experiments have been performed on inorganic semiconductors. Considering both the significant development of pEDMR and the equally extensive application of cw EDMR to carbon-based semiconductors, which have revealed the existence of many spin-dependent processes that affect fluorescence and conductivity,[34,37,40,65,66] it is clear that applying pEDMR to organic materials presents a significant opportunity to obtain new insights and discoveries. As a

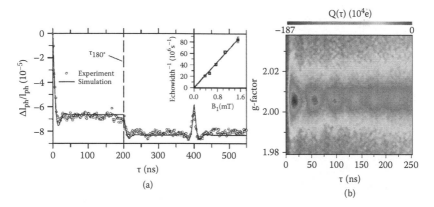

FIGURE 6.1   (a) An electrically detected spin echo produced by silicon dangling bond states in hydrogenated microcrystalline silicon.[29] The measurement is conducted as a rotary echo experiment where a high-intensity spin resonant microwave field induces a fast dephasing Rabi oscillation. When a sudden (subnanosecond range) 180° phase jump takes place at a time $\tau_{180°} = 200$ ns, a temporary rephasing (the echo effect) occurs at a time $2\tau_{180°} = 400$ ns.[29,57] (From Boehme and Lips, *Phys. Rev. Lett.* **91**, 246603 [2003]. Copyright © 2003 by the American Physical Society. With permission.) (b) The plot displays a pEDMR signal as a function of the applied magnetic field, B_0, and the length of the excitation pulse. The plot reveals an electrically detected spin Rabi nutation caused by spin-dependent transport of charge carriers through a silicon solar cell device. (From Herring, PhD thesis, University of Utah, 2008. With permission.)

result, the initial work to apply pEDMR to carbon-based materials has recently begun.[61,67-69]

In the following, a review of spin-dependent processes that are important for the properties of organic semiconductors is given. Various physical phenomena that impact these processes, such as spin coupling and spin relaxation, are discussed, and a detailed general model for spin-dependent transitions via exclusive pairs of paramagnetic states is developed. This model is then used to explain various pEDMR experiments and to discuss various pulse sequences and other experimental modes, along with the information about spin-dependent transitions that can be accessed through these experiments. We stress that many of the statements made in the following apply equally to pODMR and pEDMR experiments, as both methods are based on the measurement of coherently altered spin-dependent transition rates. However, since pEDMR experiments are technically more challenging, we will focus our discussion on this technique.

6.2 SPIN-DEPENDENT PROCESSES IN ORGANIC SEMICONDUCTORS

Spin-dependent transitions in organic semiconductors influence dark[70] and photoconductivity,[68] photo- and electroluminescence,[71] as well as photoabsorption.[71,72] Spin states, and thus spin-dependent rates, are changed not only by magnetic resonance but also by static magnetic fields.[73,74] Thus, the influence of spins on these properties causes magnetoresistance[75] and magnetoluminescence[76] in the materials mentioned above. In the following, some of these phenomena are presented, and we show that, in spite of the different physical nature of these effects, the underlying models are very similar, as they are all based on spin-dependent transitions between paramagnetic electron states. This realization allows us to derive a generalized pair model for spin-dependent transitions, similar to previous models used for the description of charge carrier recombination and transport in inorganic semiconductors.[27,28,57,77-79]

6.2.1 EXCITON GENERATION

One of the most important spin-dependent processes in organic semiconductors is the formation of excitonic states. Excitons in organic materials are very strongly exchange-coupled pairs of electrons and holes with coupling strength of the order of hundreds of meV (see Knupfer[80] and references therein). They are generated when positive and negative charge carriers encounter each other, and they can decay into the uncharged ground state via radiative recombination. It is this radiative decay that is used for light generation in OLED and display applications. Since excitons form from electron and hole polarons, which both have spin $s = \frac{1}{2}$, they provide a spin $s = 1$ system that can exist in four possible spin eigenstates, namely, one singlet state and three triplet states.[81] Light generation with organic materials is usually completely based on singlet exciton decay, which causes fluorescence. Radiative triplet exciton decay (usually called phosphorescence) is negligible in most materials because (1) it is slow due to the spin-forbidden triplet/singlet decay, (2) it is often dominated by non-radiative processes such as triplet-triplet annihilation,[82,83] and (3) the phosphorescent

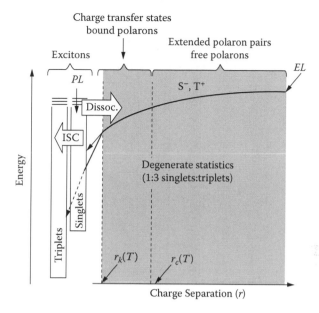

FIGURE 6.2 Binding energy of electrons and holes as a function of spatial separation. At a large separation, the two spin states of electrons and holes injected into the organic semiconductor are degenerate. The two characteristic length scales, r_c and r_x, describe the distances for which the coulombic and exchange interactions become relevant, respectively. Once electron and hole are bound coulombically, a charge transfer state is formed that can ultimately lead to the population of a molecular excitation. (Reprinted with permission from Segal et al., *Phys. Rev. B* 68, 075211 [2003]. Copyright © 2003 by the American Physical Society.)

light has a significantly longer wavelength due to the strong exchange interaction within excitonic states.[76] Hence, since exciton formation and decay conserves electronic spin, it is clear that the spin of charge carriers (polarons) in organic materials is very important for the internal quantum efficiencies.

The formation of excitons from polarons is illustrated in the sketch of Figure 6.2, which plots the binding energy of a charge carrier pair as a function of its distance. The pair formation is a gradual, multistep process where the randomly approaching, originally noninteracting localized charge carriers first experience a non-spin-dependent Coulomb attraction that causes them to move increasingly closer before strongly spin-exchange-coupled exciton states are formed.[84] During the approach, on length scales of several nanometers, the pair will gradually lose energy due to the increased electrostatic coupling. The spin coupling of the pair, however, remains negligible as both spin-exchange and spin-dipolar coupling are significantly weaker than the Coulomb energy. During this phase of the exciton formation, the two charge carriers are likely on different polymer chains. Since they are constrained to move along these chains, they will eventually reach a minimal distance point between the two polymer chains. The polaron pair will remain in this minimal distance state until an abrupt intermolecular charge transfer takes place, completing the exciton formation.

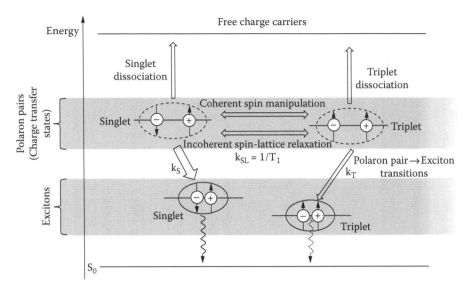

FIGURE 6.3 Interplay between exchange splitting and molecular level population. Singlets are populated from the charge transfer (CT) states (the bound electron-hole pairs) at a rate of k_S, whereas triplets are created at a rate of k_T. A conversion of the spin configuration can occur in the charge transfer state. If such a conversion occurs due to spin-lattice relaxation (k_{SL}) (also referred to as intersystem crossing within the CT state) with associated conservation of energy through absorption or emission of phonons, then the ultimate quantum efficiency of the device can exceed the spin statistical limit of 25%, assuming that singlet formation occurs faster than triplet formation.[75]

Due to the spin conservation, this final transition, from a strongly coulomb-coupled but weakly spin-coupled charge transfer state into a strongly exchange-coupled excitonic state, determines whether the resulting exciton is in a singlet or triplet manifold. The spin dynamics of the charge transfer state before the transition occurs is therefore fundamentally important in determining whether radiative or nonradiative recombination occurs.

Figure 6.3 displays a rate diagram for excitonic recombination in organic semiconductors simplifying both (1) the exciton generation model into a generation process of long-lived, weakly spin-coupled polaron pairs from previously uncoupled charge carriers, and (2) the polaron pair decay into transitions to singlet and triplet excitons. The rate picture allows the possibility of spin mixing (also referred to as intersystem crossing[85–88]), and dissociation of the polaron pairs. Note that the recombination rate in this model is fully determined by the polaron pair to exciton transitions since exciton decay times are negligibly short.[75]

The polaron pair formation and recombination model described here is central to a number of electrical and optical phenomena in organic semiconductors. Depending on the values of the various rate coefficients, as well as their dependence on other parameters, different behaviors for conductivity, quantum efficiencies of radiative decay, and magnetoresistance can be expected. Similarly, this model also describes how the manipulation of polaron pair spins with magnetic resonance leads to changes of conductivity and fluorescence, and thus, it reveals the physical parameters that are accessed with EDMR and ODMR experiments.

6.2.2 Bipolaron Pair Generation and Dissociation

The polaron pair model[73,84,89] describes a bipolar spin-dependent process that involves both positive and negative charge carriers. In contrast, the bipolaron pair process, the pair formation of either electron-only or hole-only pairs, is an example of a spin-dependent process involving only unipolar charge carriers. (Note that *bipolaron* does not refer to the polarity of the charge carrier but to a pair of identical unipolar polarons.) Such pair formation is possible due to the ability of localized states to become doubly occupied in a singlet state. Thus, pair formation is spin dependent, and since the ability of two charge carriers to form a pair increases hopping mobility, spin-dependent conductivity is obtained. Note that bipolaron pairs mainly impact conductivity and thus contribute to magnetoresistance effects. Radiative processes such as electroluminescence will be influenced only indirectly by this process, such as when the luminescence rates change due to current changes caused by the spin-controlled charge carrier mobility.

6.2.3 A Generalized Pair Model for Spin-Dependent Processes

The two examples for spin-dependent electronic processes in organic semiconductors described in Sections 6.2.1 and 6.2.2 are physically fundamentally different. However, they can show similar qualitative behavior, as they are both based on the formation of spin pairs that can then either dissociate or undergo a spin-dependent transition. Spin selection rules have been studied in many materials, especially weakly spin-coupled inorganic materials such as silicon. In the course of those studies, many different spin-dependent processes have been identified, including spin-dependent scattering,[46] spin trapping,[63,89] spin polarization,[27] and intermediate pair processes.[28] The nature of most of these processes has been investigated experimentally with ODMR and EDMR. These measurements revealed that the EDMR/ODMR signals of all but the intermediate pair processes are strongly spin polarization dependent. As a result, most spin-dependent processes, other than the intermediate pair processes, are difficult to observe under the typical operation conditions of electronic devices (i.e., low magnetic fields and room temperature). In contrast, EDMR and ODMR signals from intermediate pair processes, as first described by Kaplan, Solomon, and Mott,[28] are completely polarization independent and have therefore been extensively studied at many temperatures and magnetic fields (down to excitation frequencies of 20 MHz[53]) where conventional magnetic resonance detection is nearly impossible. Following the picture of Kaplan, Solomon, and Mott, intermediate pair processes are based on the formation of exclusive pairs of unipolar or bipolar charge carriers that can transition into a singlet state. An exclusive pair consists of two charge carriers that may only interact with each other unless dissociation occurs. For weakly spin-orbital-coupled systems, this situation implies spin dependency, and remarkably, while the encounter of two charge carriers without the formation of intermediate pairs results in magnetic-resonance-induced spin-dependent rate changes with

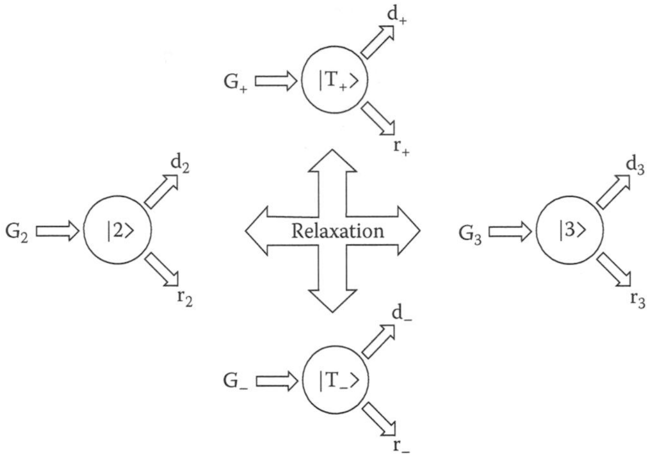

FIGURE 6.4 Illustration of the intermediate pair picture for spin-dependent transitions as proposed by Kaplan, Solomon, and Mott.[28] The rate picture illustrates a generalized model for spin-dependent pair processes that includes the polaron pair model or the bipolaron pair model that describes spin-dependent recombination and transport in organic semiconductors. The physical nature of a given pair system is encoded quantitatively into the model by choice of model parameters (rate coefficients, pair Hamiltonian, etc.).

strong dependence on polarization,[27] the intermediate pairs lead to strong polarization *independent* rate changes.[28] This occurs as long as polarization is weak enough that the polarization-dependent signals do not dominate.

Figure 6.4 illustrates the generalized intermediate pair picture that applies to many spin-dependent processes in organic and inorganic semiconductors. Crucial to this model is the formation of exclusive pairs of spin $s = \frac{1}{2}$ charge carriers prior to a spin-dependent transition. Note that the actual physical nature of the pairs can vary largely and includes systems of weakly exchange-coupled degenerate electronic states,[52] coulomb-coupled pairs of charge carriers,[68,84] and pairs of weakly interacting charge carriers in which exclusive pairs arise through nearest-neighbor distances in disordered semiconductors.[30] The polaron pair and the bipolaron pairs described in Sections 6.2.1 and 6.2.2 are important examples of intermediate pair systems in organic semiconductors. The precise physical nature of an intermediate pair system is encoded into the model illustrated in Figure 6.4 via the quantitative parameters such as the rate coefficients, the values of the spin pair Hamiltonian parameters, and the relaxation matrix (the Redfield matrix). The details of this description are outlined in the following. It is important to realize that, for any given ensemble of intermediate pairs, the spin-dependent rate will always depend on the permutation symmetry of the pairs and not the polarization.[90] As a consequence, changes of polarization due to magnetic field or temperature variations will not directly change the spin-dependent transition rate, but may do so indirectly (e.g., by influencing spin mixing processes[73] or the spin eigenstates of the pairs).

6.2.4   THE SPIN-HAMILTONIAN OF SPIN ($s = 1/2$) PAIRS

In the following, a generalized Hamiltonian for a two spin $s = \frac{1}{2}$ pair in the presence of a magnetic field is discussed in detail. A deep understanding of this Hamiltonian is necessary, as it describes both the energy structure and the relevant spin-spin couplings present in the system. The Hamiltonian contains terms that act on the individual spin pair partners, as well as terms to describe the interaction of the two spin pair partners with each other and also with any other spins surrounding the pair. The time-independent pair Hamiltonian (in absence of a magnetic resonant excitation) is given by a set of contributions

$$\hat{\mathcal{H}} = \hat{\mathcal{H}}_{\text{Zeeman}} + \hat{\mathcal{H}}_{\text{exchange}} + \hat{\mathcal{H}}_{\text{hyperfine}} + \hat{\mathcal{H}}_{\text{dipolar}} + \hat{\mathcal{H}}_{\text{orbit}} \tag{6.1}$$

which may or may not be relevant (depending of the given nature of the pair). These terms represent the general influence of the Zeeman interaction, exchange coupling between the two spins within the pair, the isotropic and anisotropic hyperfine interaction of each spin to the surrounding nuclear bath, as well as the dipolar coupling and the spin-orbit coupling-induced zero-field splittings.

6.2.4.1   Zeeman Interaction

The Zeeman interaction is due to the interaction of each of the spins with the applied magnetic field. It is described by

$$\hat{\mathcal{H}}_{\text{Zeeman}} = \frac{1}{2} \mu_B B_0 \left[g_a \hat{\sigma}_z^a + g_b \hat{\sigma}_z^b \right] \tag{6.2}$$

where μ_B is the Bohr magneton, g_i the Landé g-factor of pair partner i, and B_0 the magnetic field strength to which the pair is exposed. As the direction of the B_0-field is defined as the $\hat{z}$-direction, the Zeeman Hamiltonian simplifies into a term proportional to the Pauli spinors, $\hat{\sigma}_z^i$, of the spin i. As B_0 is an experimental parameter, the Zeeman term can be adjusted experimentally such that it dominates all other contributions to the Hamiltonian.

6.2.4.2   Exchange Interaction

For organic semiconductors only isotropic exchange is relevant, as anisotropic exchange requires the presence of significant spin-orbit interaction.[91,92] Thus, the exchange interaction Hamiltonian becomes

$$\hat{\mathcal{H}}_{\text{exchange}} = -J \hat{\mathbf{S}}_a \cdot \hat{\mathbf{S}}_b \tag{6.3}$$

where J represents the exchange integral between the two spin pair wavefunctions.[93] The exchange interaction arises due to the Pauli principle, and is one of the most important factors that determines the nature of spin-dependent transitions. With very weak exchange, the spin-dependent transition becomes increasingly unlikely (the rate coefficient vanishes) and the spin-dependent transition does not influence the properties of a material. When J becomes very large (larger than the difference in Larmor frequency $(g_a - g_b)\mu_B B_0 \ll J$), spin pairs become strongly coupled (like excitons) and neither magnetic resonance nor strong magnetic fields can influence the rate of spin-dependent transitions (singlet–triplet transitions are forbidden for magnetic resonance). In contrast to the low J regime, spin-dependent rate coefficients do not vanish for large J but become constant as a function of all other system parameters. Only in the intermediate range, where $(g_a - g_b)\mu_B B_0 \approx J$,[57] do spin-dependent rates become nonnegligible and strongly dependent on the magnetic field and polarization. Pairs with the exchange interaction in this regime are often referred to as distant pairs.[94] Distant pair states are involved in many spin-dependent transitions in both organic and inorganic semiconductors, and they play a crucial role for device efficiencies and degradation effects.[95] For organic semiconductors there is strong evidence that polaron pairs are distant pairs, as they are usually separated by several nanometers on polymer chains.[84]

6.2.4.3 Isotropic and Anisotropic Hyperfine Interaction

The interaction of nuclear and electron spins is called the hyperfine interaction. Hyperfine interactions between charge carrier pairs and the surrounding nuclear spin bath in carbon materials are of importance for spin mixing[19,96,97] and magnetoresistance.[73–75] Because of this, the hyperfine interactions have become, in recent years, one of the most intensely investigated phenomena in organic semiconductors. The hyperfine interaction $\hat{\mathcal{H}}_{\text{hyperfine}} = \hat{\mathcal{H}}_{\text{isoHF}} + \hat{\mathcal{H}}_{\text{anisoHF}}$ consists of two physically different mechanisms:

1. The isotropic hyperfine interaction:

$$\hat{\mathcal{H}}_{isoHF} = \sum_{j\in\{a,b\}} \sum_{i=1}^{n} A_{ij} \hat{\mathbf{S}}_j \cdot \hat{\mathbf{I}}_i \tag{6.4}$$

with $\hat{\mathbf{I}}_i$ representing the nuclear spin operator of the ith nucleus in the environment of the charge carrier pair partner j. The isotropic hyperfine interaction is an exchange term between an electron and a nucleus that can be represented by

$$A_{ij} = \frac{2\mu_0}{3\hbar^2} g_j g_i \mu_B \mu_N \mid \Psi_j(\mathbf{r}_i) \mid^2 \tag{6.5}$$

with g_i and g_j representing the Landé factors of the ith nucleus and pair partner j, μ_0 the vacuum permeability, μ_B the Bohr magneton, μ_N the nuclear magneton, and $|\Psi_j(r_i)|^2$ the probability of finding pair partner j at the position of nucleus r_i. Note that the isotropic hyperfine coupling vanishes for electron orbitals with angular momentum $l \neq 0$.[19] For this reason, isotropic hyperfine coupling does not play a role for many polaronic states in organic semiconductors.

2. The anisotropic hyperfine interaction is represented by the term

$$\hat{\mathcal{H}}_{anisoHF} = \sum_{j\in\{a,b\}} \sum_{i=1}^{n} \hat{\mathbf{S}}_j^\dagger \tilde{\mathbf{D}} \hat{\mathbf{I}}_i \tag{6.6}$$

describing the interaction of the magnetic dipolar fields of electrons and nuclei. Equation 6.6 represents the dipolar interaction in the form of a zero-field splitting matrix, $\tilde{\mathbf{D}}$, that assumes the form

$$(\tilde{\mathbf{D}})_{kl} = \frac{g_i g_j \mu_0 \mu_B \mu_N}{4\pi r^5 \hbar^2} \left(r^2 \delta_{kl} - 3 r_k r_l \right) \tag{6.7}$$

with $r := |\vec{r}|$ being the magnitude of the vector $r = (r_x, r_y, r_z) = r_i - r_j$, which connects an electron spin in the pair and a nuclear spin in the environment of the pair, and $k, l \in \{x, y, z\}$. There have been a number of studies treating the influence of anisotropic hyperfine interaction on electron spins[96] and even electron spin pairs[97,98] (especially on electron spin relaxation). As the quantitative analytical description for large nuclear spin baths rapidly becomes too complex, some of these studies[96,97] have considered only small numbers or even individual nuclear spins in the environment of electron spin or spin pairs. The reliability of the predictions for magnetoresistance or spin mixing derived from this approach remains unknown, and thus the precise mechanisms by which hyperfine coupling influences macroscopic properties of organic semiconductors remain disputed.[73,74,76,99] However, given the large number of nuclear spins (mostly protons) in organic semiconductors, it is clear that anisotropic hyperfine interaction plays a profound role for spin relaxation, and thus spin mixing and magnetoresistance effects.

6.2.4.4 Dipolar Interaction-Induced Zero-Field Splitting

Similar to the anisotropic hyperfine interaction of electron and nuclear magnetic moments, the two charge carriers within an intermediate spin pair each feel the magnetic fields produced by their respective pair partner's magnetic dipolar moments. Following Equations 6.6 and 6.7 for the dipolar interaction of electron and nuclear spins, the intrapair dipolar interaction can be described by a zero-field splitting term:

$$\hat{\mathcal{H}}_{dipolar} = \hat{\mathbf{S}}_a^\dagger \tilde{\mathbf{D}}_d \hat{\mathbf{S}}_b \tag{6.8}$$

with the zero-field splitting matrix, $\tilde{\mathbf{D}}_d$, representing

$$(\tilde{\mathbf{D}}_d)_{ij} = \frac{g_a g_b \mu_0 \mu_B^2}{4\pi r^5 \hbar^2} \left(r^2 \delta_{ij} - 3 r_i r_j \right) \tag{6.9}$$

with $r := |\vec{r}|$ being the magnitude of the vector $\vec{r}$, connecting the electron spins of the two pair partners. Intrapair dipolar coupling is always present but rarely relevant. It is strongly dependent on the pair size (= distance of the two pair partners), and for distant pairs, often too weak in comparison to other coupling effects (e.g., hyperfine coupling).[19] At short distance its strength grows cubically while the strength of exchange interaction grows exponentially. Thus, for strongly exchange-coupled pairs (such as excitons) it is often irrelevant, too. An example of pairs whose dipolar coupling is not negligible is electron–hole pairs in the band tail states of hydrogenated amorphous silicon.[52,100] PODMR measurements can resolve the presence of two symmetric Pake-like-shaped spectral features,[101] which evolve from an ensemble of spin pairs whose coupling is dominantly dipolar, whose pair axis (the vector $\vec{r}$), is randomly oriented with regard to the external magnetic field, and whose size (the distance, r, between the pair partners) is randomly distributed.[101]

6.2.4.5 Zero-Field Splitting and Spin-Orbit Coupling

In addition to the zero-field splitting contribution due to dipolar coupling of electron spins with other electron or nuclear spins, a similar term, $\hat{\mathcal{H}}_{orbit} = \mathbf{S}_a^\dagger \tilde{D} \mathbf{S}_b$, can arise when significant spin-orbit coupling is present.[19] In carbon-based materials this is mostly insignificant and can be neglected in the description of most organic semiconductors.

6.2.5 THE LIOUVILLE EQUATION FOR THE PROPAGATION OF SPIN PAIR ENSEMBLES

With the spin pair Hamiltonian given in Section 6.2.4, the dynamics of an ensemble of many identical spin pairs can be calculated using a statistical Liouville equation. This approach has been used intensively for the prediction of spin ensemble propagation as required for the simulation of magnetic resonance experiments. Its application to spin-dependent transitions was first suggested by Haberkorn and Dietz[77] and has since been widely used for the calculation of pEDMR and pODMR experiments.[57,102–105]

6.2.5.1 Calculation of the Dynamics of a Spin Pair Ensemble

The basis of a statistical treatment of an ensemble of identical quantum systems such as charge carrier pairs in organic semiconductors is the Liouville operator, $\hat{\rho}$, also called a density operator.[106] Based on the Hamiltonian described in Section 6.2.4 and the rate picture shown in Figure 6.4, the statistical Liouville equation assumes the form

$$\partial_t \hat{\rho} = \frac{i}{\hbar}\left[\hat{\rho}, \hat{H}\right]^- + S\left[\hat{\rho}\right] + \mathcal{R}\{\hat{\rho} - \hat{\rho}_0\} \tag{6.10}$$

in which the stochastic operator, S, represents the external changes of the ensemble due to generation, spin-dependent recombination, and dissociation of spin pairs.[57,77,102] The operator R in the last term describes the influence of spin relaxation, as described by Redfield.[19,107] R is a fourth-rank tensor whose matrix representation has 256 entries.[102,108] This complexity indicates that the description of spin relaxation within pairs of charge carriers in organic semiconductor by a single spin mixing constant, as often found in the literature,[40,73,109,110] is only phenomenological, and that the conclusion drawn from such descriptions has finite validity for coherently propagating spin ensembles.

Finding the time-dependent solution $\hat{\rho}(t)$ of Equation 6.10 is analytically possible only for very simple systems where both the intrapair interaction and the stochasticity are weak.[57] For more complex situations, where most of the statistical terms as well as constituents of the Hamiltonian are not negligible, numerical simulations are necessary, as has been demonstrated for the simulation of pEDMR and pODMR experiments.[103–105]

6.2.5.2   Linking the Ensemble Dynamics with Spin-Dependent Rates

Once $\hat{\rho}(t)$ is found, the spin-dependent electronic transition rates can be obtained using the rate picture in Figure 6.4. For recombination processes such as luminescence rate, we obtain

$$R(t) = \sum_{i \in \{|S\rangle,|T_+\rangle,|T_0\rangle,|T_-\rangle\}} r_i \mathrm{Tr}[|\,i\rangle\langle i\,|\,\hat{\rho}(t)] \qquad (6.11)$$

while conductivity controlled by spin-dependent pair transitions derives from dissociation rates through

$$\sigma_{\mathrm{spin}}(t) = e\mu_e \tau \sum_{i \in \{|S\rangle,|T_+\rangle,|T_0\rangle,|T_-\rangle\}} d_i Tr[|\,i\rangle\langle i\,|\,\hat{\rho}(t)] \qquad (6.12)$$

when the spin-dependent process involves either unipolar or bipolar charge carrier pairs.[57,90]

6.2.6   Simplifying the Pair Model for the Description of Incoherent Pair Propagation

The intermediate pair model illustrated in Figure 6.4 explicitly considers all four electron spin eigenstates of the two spin $s = \frac{1}{2}$ pairs. The description of the dynamics of pair systems with these four eigenstates (as given in Sections 6.2.4 and 6.2.5) can be drastically simplified to a statistical rate model when coherent spin propagation is neglected. The resulting simplified picture is displayed in Figure 6.5. The most notable change is the unification of the three triplet states into one triplet density. Obviously, the two densities, n_S and n_T, do not represent densities of energy eigenstates, but densities of permutation-antisymmetry and permutation-symmetry. While this picture does not reveal the effect of coherent spin motion when observed with pEDMR and pODMR experiments, including T_2 spin relaxation effects, it does very accurately describe the random rates between singlet and triplet densities. Note that intrapair spin interactions that change the basis of energy eigenstates have no impact on this model even when these interactions lead to mixtures between the

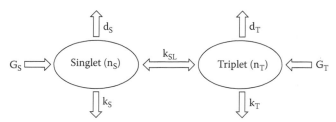

FIGURE 6.5   Illustration of the intermediate pair picture for the simplified case when coherent spin propagation is neglected. Instead of four densities corresponding to the four energy eigenstates of the system, only two densities for singlet and triplet states are considered, which influences the transition rates in a different way. For details, see text.

singlet and triplet states. They do not impact the accuracy of this model because any change of the singlet-to-triplet ratio due to spin interaction leaves the overall singlet and triplet projection of all eigenstates constant at one-quarter and three-quarters, respectively.[99] Hence, the application of this rate picture to the description of magnetoresistance and efficiencies of organic devices as well as continuous wave (cw) EDMR and ODMR experiments[39] is justified.

6.2.7 MACROSCOPIC PHENOMENA DETERMINED BY SPIN-DEPENDENT PAIR TRANSITIONS

The generalized pair model outlined above can be used to describe a number of macroscopic properties and effects in organic semiconductors. By using exclusive pairs in order to model spin-dependent transitions through either unipolar or bipolar pairs, fundamental properties such as the efficiency limits of organic solar cells can be understood directly. For other properties, such as magnetoresistance, this is not as straightforward since different pair systems modeled with the exclusive pair picture can produce similar results. In the following these processes are briefly discussed and the current state of their exploration is summarized.

6.2.7.1 Fundamental Efficiency Limitations of OLEDs

Based on the polaron pair model explained in Section 6.2.1, the fundamental limit for the quantum efficiency of OLEDs is directly dependent on the ability of exciton precursor spin pairs to convert between singlet and triplet manifolds. In the absence of this process (spin mixing), the random spin statistics that applies when charge carriers form will lead to the creation of singlet excitons out of one-fourth of all pairs (independent of the pair eigenbase or the intrapair spin coupling). In the presence of spin mixing, triplet pairs can convert into singlet pairs, and while the opposite process will also be possible, the net spin mixing rate will increase the singlet densities due to the energy-level spacing, which makes singlet precursor–to–singlet exciton transitions much more likely than triplet precursor–to–triplet exciton transitions. Hence, spin mixing increases the quantum efficiency of OLEDs, and consequently, being able to observe spin mixing[68] is crucial for the investigation of fundamental efficiency limitations in OLEDs.

6.2.7.2 Spin Mixing

Spin mixing has been introduced as the random spontaneous process that changes singlet into triplet spin states, and vice versa. Spin mixing is rarely discussed in the literature in connection to conventional spin relaxation theory as developed by Redfield,[107] which comprehensively describes all transverse (T_2) and longitudinal (T_1) relaxation processes to which spins are subjected. As pointed out in Section 6.2.6, the spin mixing process described by the rate picture in Figure 6.5 is a composite rate coefficient that depends on different spin relaxation contributions (i.e., it is a function of Redfield matrix elements). As it has rarely been discussed in the literature on spin mixing processes, it is not clear at this point what the analytical form of this functional dependence is. However, due to the importance of spin mixing for the discussion of magnetoresistance phenomena in organic semiconductors,[73,83,111–114] investigation of this relationship is of great importance, especially in order to elucidate what role hyperfine interactions play for spin mixing, and how this changes when the strength of an applied magnetic field is changed.

6.2.7.3 Magnetoresistance

There are currently at least three models[73,115,116] describing spin-dependent processes in organic semiconductors that predict organic magnetoresistance (OMAR) effects. There is general agreement that OMAR is related to spin-dependent electronic transitions that change conductivity as spin ensemble states are changed. The hypothesis is that hyperfine coupling influences these spins due to a change (quenching) of spin mixing rates that occur when an externally applied magnetic field exceeds the hyperfine field created by the ensemble of nuclear spins.[73,115] Currently, the spin-dependent process responsible for this effect is disputed, as is the question of whether the magnitude of the spin-dependent rates, as well as the spin mixing rates, is high enough to be varied by the change of the external magnetic field/hyperfine field ratio.[73,76,99,115] Given the experimental evidence for strong OMAR in bipolar devices and weaker OMAR in unipolar devices, it is conceivable that both bipolaron pairs and electron-hole polaron pairs can contribute to OMAR.[112,117] In any case, a comprehensive microscopic understanding of these processes requires direct experimental access of spin mixing rates at various magnetic fields (above and below the hyperfine fields). For high magnetic fields ($\approx$340 mT) such access has been demonstrated using pEDMR.[68]

6.3 THE INVESTIGATION OF SPIN-DEPENDENT TRANSITIONS

Experimental access to spin-dependent transitions can be achieved with EDMR and ODMR experiments where spin-dependent rates are observed during or after magnetic resonant excitation. The pair models described in Section 6.2 can be used for the description of these experiments by extending the Hamiltonian $\hat{\mathcal{H}}$ in Equation 6.1 by an addend containing the time-dependent electromagnetic perturbation $\hat{\mathcal{H}}_1(t) = \frac{1}{2}\mu_B B_1 \left[g_a \hat{\sigma}_+^a + g_b \hat{\sigma}_+^b \right] e^{-i\omega t}$. In this term, B_1 represents the amplitude and ω represents the angular frequency of the exciting radiation. The solution of this perturbation is usually found by transformation into a rotating frame where the B_1 field is not time dependent.[19] As this procedure complicates the already complex Hamiltonian, and therefore the Liouville equation, the simulation of EDMR and ODMR (both for cw and pulsed experiments) typically requires numerical methods.[103–105]

In this section, the experimental implementation of both cw and pulsed EDMR and ODMR will be discussed. This includes the discussion of the technical requirements for the sample implementation and a variety of experimental modes (such as modulation frequency dependence measurements, pulse sequences, etc.) that allow access to various parameters of the spin-dependent transitions under investigation.

6.3.1 Technical Requirements

EDMR and ODMR spectrometers are, in many regards, technically much more simple than EPR spectrometers, since detection of radio- and microwaves necessary for

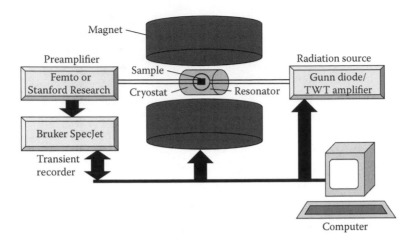

FIGURE 6.6 Illustration of an EDMR spectrometer. The setup consists of a magnet and radiation source that is able to manipulate spins via EPR excitation. The detection is done through measurement of electrical currents. An ODMR spectrometer would differ from the sketched setup only in that an IR to UV light detector able to receive radiation coming from the sample would substitute the input of the preamplifier.

conventional magnetic resonance is not needed. A typical EDMR setup is illustrated in Figure 6.6. Similar to a conventional EPR spectrometer, it consists of magnetic field coils and a radiation source in order to establish magnetic resonance. However, in contrast to EPR spectrometers, it does not require complex radio- or microwave-compatible interferometric devices (mixers, circulators, splitters, etc.) in order to obtain sensitive quadrature detection of the EPR signals. For EDMR, the detection is based solely on the sensitive measurement of sample currents, which requires a current amplifier and appropriate contact structures for the sample. Similarly, for ODMR, which is not displayed in Figure 6.6, a sufficiently fast radiation detector for IR to UV wavelengths is required.

6.3.1.1 Magnetic Resonance Excitation

For most EDMR and ODMR experiments described in the literature, commercial EPR spectrometers were used to conduct the experiments. EPR spectrometers are always built for the operation in one (or very few) very narrow frequency range (e.g., X-, K-, Q-, W-bands, etc.). As signal detection in EDMR or ODMR measurements does not require restricted frequency ranges,[53,54] both EDMR and ODMR can be performed over a broad frequency range. Aside from this technical versatility, EDMR and ODMR allow high sensitivity and, due to miniaturization, rather strong B_1 excitation fields at very low excitation powers[49] and very low frequencies,[53] since EDMR and ODMR signals are neither polarization dependent (see discussion in Section 6.2 and references 28, 53, 55, and 77) nor volume sensitive.[49,54,89] As mentioned above, a recent continuous wave (cw) EDMR study demonstrated electron spin resonance detection at 20 MHz.[53] This huge frequency versatility for magnetic resonance spectroscopy provides a significant advantage for the investigation of coupling strengths and zero-field parameters.[19] In addition, for the exploration of spin-dependent processes in organic materials, a multifrequency/multimagnetic field EDMR system[54] can also provide insights into the organic semiconductor spin mixing processes mentioned in Section 6.2.7.3.

6.3.1.2 Detection of Spin-Dependent Rates

The greatest technical challenge for the detection of spin-dependent rates is to observe at sufficiently high time resolution, often very small relative rate changes that occur on top of large, constant offset rates. For current detection, the required sensitivity, dynamic range, and time resolution present opposing requirements. When one of these is improved, the others become worse due to the inherent physical nature of electronic noise. This challenge is particularly tricky for EDMR experiments, as these are oftentimes limited not only by noise but also by signal currents induced by nonmagnetic resonant effects of the microwave radiation (antenna effects). A significant amount of work has therefore been done on detection electronics for fast spin-dependent currents after a pulsed magnetic resonant excitation.[29,90,118,119]

6.3.1.3 Contact Challenges for EDMR Experiments on Organic Semiconductors

As with EDMR with inorganic semiconductor materials, organic materials must be contacted in a way that will prevent the contacts from distorting, and thus inhomogenizing, the radiation modes needed for the spin manipulation. For most inorganic semiconductors that are prepared with clean-room techniques, this problem can be solved by the use of appropriate lithographically prepared structures for contact geometries.[30,89,120] Applying this approach to organic semiconductors is challenging since these often require preparation in inert atmospheres, usually obtained with a glove box. The application of photolithography in this environment is extremely difficult, and overcoming this problem with shadow masks is also technically demanding, as it provides less structural resolution, and is not always feasible with the vertical sample structures (layer stacks) needed for low-mobility materials. An additional challenge for the EDMR device preparation is that samples with great reliability and reproducibility require fast preparation times and little time between the preparation and the measurement. Due to the extreme volatility of many organic semiconductors in the presence of ambient air, the measurement device structures must provide sufficient encapsulation for the measured materials. Thus, the materials preparation for EDMR measurement devices should be quick and easy. This also helps prevent sample preparation becoming a bottleneck for the entire measurement process. Finally, the device preparation process for organic semiconductors must be versatile enough so that different materials and combinations thereof can be measured without needing a new device design for each measurement. EDMR spectroscopy is most insightful when performed on materials under operating conditions, preferably even in devices such as light-emitting diodes or solar cells.[121] Thus, sample design must allow the quick preparation of organic solar cells or OLEDs.

6.3.1.4 Meeting the Challenges of EDMR Experiments on Organic Semiconductors Using Microwave-Compatible Thin-Film Templates

The requirements for EDMR measurements on carbon materials described above can be met using photolithographically prepared templates for the measurement device preparation. The idea behind this approach is to carry out all lithography steps first using materials that do not need to be confined to a glove box before the sensitive organic semiconductor materials are prepared in a glove box. Figure 6.7 displays a sketch of a sample structure that uses EDMR device templates made for a dielectric cylindrical microwave resonator (the commercially available Bruker Flexline pulse EPR resonators[122]). The device is fabricated on a long match-like substrate. One end of this substrate has two contact pads that are positioned outside the resonator. These pads are connected through thin-film metal layers with the measurement device on the other end of the substrate, which is designed to be located at the center of the resonator where the microwave radiation is homogeneous and maximal during the measurement. The connecting thin-film Al wires are less than 100 nm thick, which is about an order of magnitude smaller than the microwave penetration depth for the Al wires at 10 GHz and $T = 5K$. This is thin enough to keep the radiation mostly unperturbed. As the templates must allow vertical current paths, one sample contact is established through a metal back layer that is separated from the front contact by an insulating silicon nitride layer that covers the entire device except for four windows: two windows at the contact pads, one at the device area, and one for a via that is located close to the middle of the substrate. The device templates shown in Figure 6.7 can be made quickly in large numbers with straightforward clean-room procedures. They allow the implementation of a large variety of cw- and

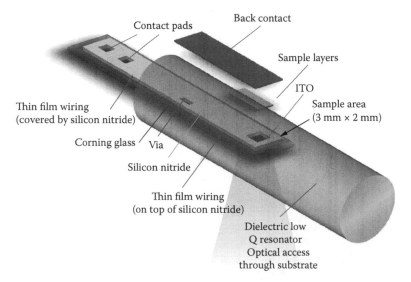

FIGURE 6.7 Schematics of an organic semiconductor measurement device prepared on an EDMR-compatible template that is placed within a dielectric low-Q resonator. The picture is not to scale. (Adapted from Boehme et al., *Phys. Stat. Sol.* b, 246, 2750 [2009].)

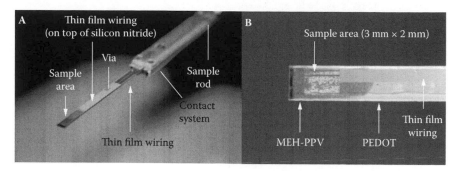

FIGURE 6.8 (a) Photo of OLED device based on an Al/MEH-PPV/PEDOT/ITO stack fabricated on an EDMR template as sketched in Figure 6.7. (b) When a voltage is applied to the OLED, the electroluminescence becomes visible. (Adapted from Boehme et al., *Phys. Stat. Sol.* b, 246, 2750 [2009].)

pulsed-EDMR-compatible measurement devices for many organic materials (at least all those that are deposited through spin coating) without using lithography during or after the deposition of the organic layers. For the spin coating procedure of the templates, a special template holder is needed that fixes the long substrate in a position where the rotation axis is close to its center. Note that while spin coating cannot produce laterally well-defined layers, the active area of the measurement device will still be well defined by the sample window through the silicon nitride layer. Stacks of material can be formed through additional spin coating steps, and when a desired device stack is made, the back contact of the sample is deposited through metal evaporation, which covers the entire substrate area except for the region at the contact pads. This metal cover connects the back contact of the device with the via and at the same time provides a simple encapsulation of the sample that preserves the sample well enough to allow a transfer from the glove box to the EDMR setup.

Figure 6.8a shows an OLED device based on an indium tin oxide (ITO) front contact, a poly(3,4-ethylenedioxythiophene) (PEDOT) hole injection layer, and a poly[2-methoxy-5-(20-ethyl-hexyloxy)-1,4-phenylene vinylene] (MEH-PPV) active layer. The OLED is held by a pulsed EDMR sample rod system that was developed for the particular sample template design. This mechanical device holds the sample in the correct position within the resonator during the measurement and also establishes electrical contact from the sample to the outside of the cryostat into which the entire experimental setup (including the resonator) is placed. Figure 6.8b displays the same OLED device as Figure 6.8a under operating conditions. The orange light, due to the current-induced electroluminescence, becomes visible when a voltage is applied.

6.3.2 CONTINUOUS WAVE EDMR AND ODMR

Most EDMR and ODMR studies performed on organic semiconductors in the past have used incoherent, cw EPR, EDMR, and ODMR.[33–40,123–125] In these experiments, the sample is placed into an external magnetic field and microwaves at a fixed power are applied continuously. The electrical conductivity (in the case of EDMR) or an optical property (usually luminescence or absorption for ODMR) is continuously

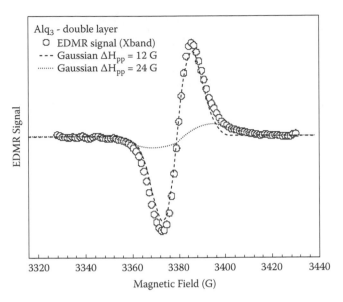

FIGURE 6.9 Cw EDMR spectrum showing two resonance lines detected in an Alq3-based OLED at room temperature. (Adapted from Castro et al., *J. Non-Cryst. Solids* 662, 328–340 [2004].)

monitored as the external field is slowly (adiabatically) varied. When the frequency of the microwaves and the magnitude of the magnetic field satisfy the resonance condition, the spins are incoherently driven between different states and the observed spin-dependent rates change. While the approach is simple, the information obtained from the cw spectra is limited to Landé (*g*) factors and strongly convoluted spin relaxation, coupling, and electronic lifetime parameters that determine the line shapes and line widths of the resonances.

Figure 6.9 displays an example for an EDMR spectrum recorded on an indium tin oxide/tris(8-Hydroxyquinolinolato) aluminum (Alq3) and N,N-Di(naphthalene-1-yl)-N,N-diphenyl-benzidine (NPD) diode.[35] Similar to most cw EPR experiments described in the literature, this experiment was conducted using lock-in detection of the modulated B_0 field. The experimental data were fit with the derivative of two Gaussian functions with almost the same g-factor but significantly different line width.

Most cw EDMR and ODMR experiments found in the literature are performed using lock-in detection of the modulated B_0 field or the microwave radiation. Using this approach has demonstrated extraordinary sensitivities[126]; however, even at low modulation frequencies (several kHz), this approach does not allow the interpretation of the data as an adiabatic field sweep experiment. This circumstance is often neglected in the literature.[35] In order to correct for this problem, the dynamics of spin-dependent transitions during a given experiment has to be taken into account correctly. As cw EDMR and ODMR experiments conduct measurements in an incoherent transient regime, this can be done with the simplified rate model discussed in Section 6.2.6 and displayed in Figure 6.5.[127,128] While this model appears to be simple and depends on only a few spin-dependent generation, recombination, dissociation, and relaxation rate coefficients, its mathematical description becomes rather complex for the general case where no simplifying assumptions are made.[127] To our knowledge, no analytical solution for the general dynamics of this model has been presented in the literature, and as a result, the interpretation of EDMR and ODMR experiments and their dependence on the lock-in modulation frequency is often calculated using numerical simulation[127] or simplifying assumptions about the rate coefficients.

We note that the interpretation of cw EDMR and ODMR data based on the latter approach can easily lead to incorrect results, as there is great ambiguity in the choice of the simplifying conditions, based on which the fit of the experimental data can provide strong agreement with the data. A systematic exploration of various qualitative behaviors of cw EDMR and ODMR signals based on fundamentally different quantitative choices for the rate coefficients has shown[129] that completely different spin-dependent rate dynamics can lead to almost identical cw EDMR and ODMR signals, resulting in similar dependency of the signal to the modulation frequency. We conclude that, for the interpretation of cw EDMR and ODMR data, a spin-dependent rate model that explains cw EDMR or ODMR data does not prove the model; it only does not refute the model. For a true confirmation, all other models that would lead to these cw EDMR or ODMR data need to be excluded by other experimental means.

6.3.3 Transient EDMR and ODMR

Because of the limitations of cw EDMR and ODMR methods, much effort has been invested in recent years into the development of transient EDMR[30,57,63,130,131] and ODMR spectroscopy.[52,132–134] The idea behind this work is to enable a similar drastic improvement to the experimental parameters available for these measurement techniques as was previously accomplished for transient EPR and NMR techniques by introducing another dimension (the time domain) into the available experimental data space.

Transient EDMR and ODMR measurements can be classified into two qualitatively different groups, the transient measurement with magnetic resonant excitation in (1) the incoherent time domain and (2) the coherent time domain. Technologically, the two time domains correspond to weak and strong excitation (B_1) fields, respectively. Transient EDMR and ODMR in the incoherent time domain are the time domain equivalents of cw spectroscopy, which is why this is referred to as time domain cw EDMR in the following. In fact, microwave-modulated cw EDMR and ODMR spectra can be simulated by first calculating the spin-dependent rate transients under application of chopped microwave irradiation with weak intensity and a duty cycle of 0.5, and subsequently applying a Fourier transform. In contrast, the physics of transient EDMR and ODMR experiments in the coherent time domain is fundamentally different. In these pulsed (p) experiments, strong B_1 fields (typically in the 0.1 to 1 mT range) are applied for very short times (typically nanoseconds) to the spin ensemble in order to prepare a coherent superposition of eigenstates.[21] The observation of these coherent spin states during or after the excitation pulses then reveals information about the observed system.

6.3.3.1 Time Domain cw EDMR and ODMR and the Measurement of Spin Mixing Rates

Transient EDMR in the incoherent time domain was first applied to organic materials by Hiromitsu et al.,[131] who investigated the transient behavior of spin-dependent rates in phthalocyanine/C_{60}-devices. Figure 6.10 displays the typical transient behavior of the spin-dependent rate transients during such an experiment. Characteristic for these measurements is that the spin-dependent rate approaches steady-state values after abrupt changes take place from the on-resonance to the off-resonance state (or vice versa). The steady-state values may be different or equal (depending on the rate parameters of the given spin-dependent transitions[129]); however, the induced change is always observed to be a double exponential decay function. This double exponential behavior can be explained by the rate picture in Figure 6.5, where discrimination between only the two permutation symmetries (singlet and triplet), and not the four energy eigenstates, is made (in contrast to Figure 6.4). For each of the two permutation symmetries, a different single exponential decay appears, and thus the net transient becomes biexponential.[68,102,131]

The standard description of this behavior for spin-dependent rates during or after a resonant radiation pulse is obtained by solving a set of inhomogeneous ordinary differential equations (ODEs). The analytical solution is complex, but simplifies drastically when spin mixing is considered to be extremely weak or extremely strong ($k_{SL} >$ or $k_{SL} <$ all other rate coefficients). For the case of weak spin mixing, this solutions assumes the form

$$\Delta\sigma(t) = 2\mu e\tau \, \Delta n \left[d_{PP_S} \, e^{(-\Gamma_s t)} - d_{PP_T} \, e^{(-\Gamma_t t)} \right] \tag{6.13}$$

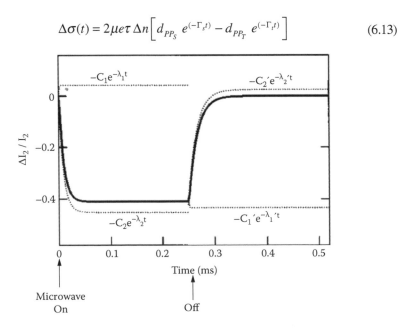

FIGURE 6.10 Transient cw EDMR measurement of spin-dependent transport and recombination in phthalocyanine/C_{60} devices. The measurement shows the current through the device as a function of time during one excitation cycle. The current follows a double exponential approach after the microwave radiation is turned on and off. (From Hiromitsu et al., *Phys. Rev. B* 59, 2151 [1999]. Copyright © 1999 by the American Physical Society. With permission.)

where τ is the free polaron lifetime, d_{PP_S} and d_{PP_T} the singlet and triplet dissociation rates, and Γ_s and Γ_t the sums of the dissociation and recombination rates for the singlets and triplets, respectively.[135] Clearly, in this case, a significant amount of information regarding the system is obtained, particularly from the decay rates of the exponential functions, which are directly related to the recombination and dissociation rates of the system. Care, however, needs to be taken that spin mixing does not take place on a relevant timescale, since, in the limit of fast spin mixing, the transient is given by

$$\Delta \upsilon(t) = 2\mu e \tau \Delta n [(d_{PP_S} - d_{PP_T})e^{-k_{SL}t} + d_{PP_S}\rho(1-\rho)(e^{-k_{SL}t} - e^{-((1-\rho)\Gamma_s + \rho\Gamma_t)t})] \qquad (6.14)$$

where ρ describes the Fermi distribution between the singlet and triplet states.[135] As can be seen here, the faster of the two times is given solely by the spin mixing rate, and the other by a weighted combination of the other times constants. Hence, similar to modulation frequency domain EDMR, it is difficult to determine precise rate coefficients from the data. However, one important and unambiguous insight that these consideration give is that, as long as the double exponential decay of a spin-dependent rate is observed with EDMR or ODMR, the spin mixing cannot be faster than the fastest of the observed time constants.

Figure 6.11 displays a plot of the current change, ΔI, to the forward current of an MEH-PPV device (identically prepared to those shown in Figure 6.8) at low temperature ($T = 5$ K). The measurement of the displayed data occurred after the application of a microwave pulse that ended at the time $t = 0$. The plot shows ΔI as a function of the time and the magnetic field, B_0. The plot clearly shows a resonantly induced current response that occurs only around a magnetic field of $B_0 \approx 237.5$ mT corresponding to the known spin resonance of polarons in MEH–PPV ($g \approx 2.003$). The current transient shows, directly after the pulse, a short, intense enhancement, which is followed by a weaker, more slowly decaying quenching of the current. Both signals can be fit excellently with exponential decay functions, and the fast, strong signal reveals a decay rate of about 8 μs. The analysis of the magnetic field, B_0, dependence of the two signals (not shown here) reveals both an identical line shape and line width and an identical g-factor. Therefore, both signals are with high probability caused by the same process, as it is extremely unlikely that two uncorrelated single exponential signals belonging to different physical processes will show, except for an arbitrary proportionality factor, exactly the same B_0 dependence. Therefore, using the conclusion drawn from Equations 6.13 and 6.14, the data shown in Figure 6.11 demonstrates that spin mixing in MEH-PPV under the given experimental conditions must be significantly slower than previously assumed.[75]

In contrast to the data measured on the phthalocyanine/C_{60} devices shown in Figure 6.10, the two exponential functions in Figure 6.11 have different signs. This is caused by the short (pulsed) excitation that was used for the experiment in Figure 6.11. The magnetic resonant pulse ending at $t = 0$ is very strong ($B_1 = 0.15$ mT) and only $\tau = 48$ ns long. On the microsecond timescale of the incoherent exponential decays this is instantaneous, and the effect of this coherent excitation on the spin ensemble during the pulse is not a double exponential approach to an

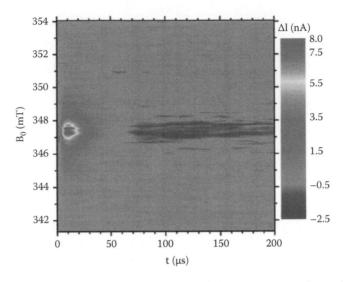

FIGURE 6.11 Plot of the current change ΔI in an MEH-PPV OLED after a short micro-wave pulse as a function of time and the magnetic field, B_0. At resonance, $B_0 \approx 347.5$ mT, the data shows an initial strong enhancement followed by a slowly decaying weak quenching as expected for spin-dependent pair transitions. (Adapted from Boehme et al., *Phys. Stat. Sol. b*, 246, 2750 [2009].)

on-resonant steady state, but the inversion of the singlet and triplet densities. Since there is a high triplet and low singlet density before the pulse (due to the long triplet lifetimes), there is a quenched triplet and an enhanced singlet density after the pulse. This causes a quickly decaying current *enhancement* due to decay of the enhanced singlet pair density, and a slowly decaying current *quenching* due to the recovery of the quenched triplet population. A pulse-induced inversion of the spin ensembles is the simplest form of a coherent pEDMR experiment. As shown in the following section, using this and other pulsed excitation sequences, a variety of experimental parameters become accessible that would remain elusive with conventional cw spectroscopies.

6.3.3.2   Pulsed EDMR and ODMR

A first theoretical description of pEDMR was given by Boehme and Lips in 2003,[57] following a number of studies exploring pODMR since the late 1970s.[23,56,132] Significant experimental work demonstrating and developing pEDMR has since been conducted, mostly on inorganic semiconductors, particularly various morphologies of silicon (crystalline,[30] amorphous,[52] microcrystalline,[121] etc.). While all these experiments differ profoundly in excitation schemes, and thus the information that they reveal, they all share the coherent spin pair detection scheme illustrated in Figure 6.12: Initially the ensemble of spin pairs is in a steady state of pairs in spin eigenstates with a high density of long-lived triplet pairs. After coherent excitation of the ensemble, the spin pairs assume superpositions of eigenstates. The ensemble of coherent spin pairs that exists at the end of the excitation

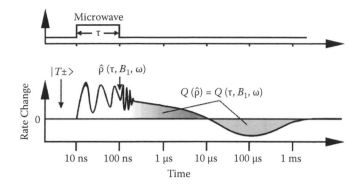

FIGURE 6.12 Sketch of the measurement principle for coherent spin pair states. Note that the excitation sequence illustrated in this sketch is a single pulse. In real experiments this can be a single pulse or a sequence of pulses. See text for details. (Adapted from Boehme and Lips, in *Charge Transport in Disordered Solids with Applications in Electronics*, ed. S. Baranovski [Wiley, New York, 2006]. With permission.)

sequence will gradually (on a much longer timescale than the excitation) relax back to the steady state, as will the measurable spin-dependent rate it controls. It can be shown[57] that integration of the spin-dependent rate change (reflecting the number of transitions that take place due to the pulsed excitation) is directly proportional to the singlet projection of the pair state at the end of the excitation sequence. Thus, measuring the transient after the excitation sequence is a direct way to probe the permutation symmetry (the singlet content) of a spin pair ensemble. Note that this measurement approach is very similar to optical pump/probe experiments with the difference that the pump pulse is a coherent magnetic resonant excitation and the spin state is probed by a current or luminescence intensity measurement. From a conventional pulsed magnetic resonance spectroscopy point of view, the experiment could be referred to as an electrically or optically detected free induction decay experiment.

The ability to measure a coherent spin state of electron spin pairs that control spin-dependent rates using the pEDMR pump/probe scheme sketched in Figure 6.12 opens many experimental possibilities. However, in contrast to conventional pEPR, it does not allow the direct real-time electrical measurement of coherent spin motion. This limitation is often the result of the long dielectric relaxation times of the materials and the conditions under which spin-dependent rates become relevant for conductivity. This is one of the reasons why pEDMR experiments are conducted with modified pEPR pulse sequences. Some of the pEDMR pulse sequences that have been demonstrated experimentally are discussed in the following sections, along with the experimental parameters that can be accessed using them.

6.3.3.3 Spin Rabi Nutation and Intrapair Coupling

The most simple and straightforward experiment for the electrical or optical detection of coherent spin motion is a single pulse transient nutation experiment. During the application of a coherent radiation pulse, a spin will nutate around the excitation field. In spin pairs whose Landé factors differ significantly (the difference of Larmor frequencies must be larger than γB_1), only one of the two spin pair partners will nutate, and thus the permutation symmetry will oscillate, resulting in periodic modulation of the spin-dependent rate, which can be observed. Due to the simplicity of this experiment, the first pEDMR experiments made on organic semiconductors were simple one-pulse transient Rabi nutation measurements.[67,68] Figure 6.13 displays the electrically detected spin nutation of optically induced excess polaron pairs in thin fullerene C_{60} films at room temperature.[67] The spin-dependent process used for the detection was polaron pair recombination.[67] Figure 6.13 shows an oscillation of the detected charges, as a

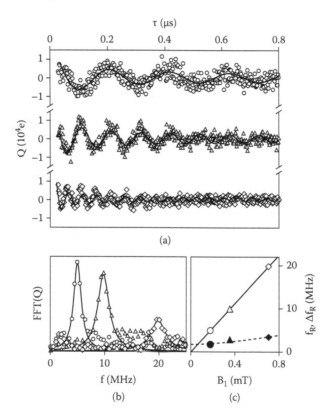

(a)

(b)   (c)

FIGURE 6.13 (a) The plot of the integrated current response of carbon C_{60} films after a brief resonant excitation at $g \approx 2.0018$ for three different B_1 fields shows an oscillatory behavior that can be fit excellently with first-kind Bessel functions (solid lines). (b) The fast Fourier transforms of the data shown in (a) and fits with Lorentzian functions (solid lines). (c) The open symbols represent a plot of the nutation frequencies obtained from the data in (b) as a function of the B_1 field. An excellent agreement with a linear function through the origin (solid line) is given. The closed symbols and the dotted lines represent the line width of the data in (b) as well as a fit with a dephasing function, respectively. (From Harneit et al., *Phys. Rev. Lett.* **98**, 216601 [2007]. Copyright © 2007 by the American Physical Society. With permission.)

function of the pulse duration, whose frequency is directly proportional to the applied B_1 field. The displayed data were the first electrical detection of coherent spin motion in a carbon-based semiconductor and show the applicability of pEDMR to this materials class.

6.3.3.4 Spectroscopy of Rabi Nutation Frequencies

The electrical observation of the Rabi nutation of a spin ($s = \frac{1}{2}$), as displayed in Figure 6.13, is a rather trivial example where one spin is resonantly excited while the other spin in the pair remains unchanged by the applied radiation. Oftentimes, spin pairs involved in spin-dependent transitions have very similar g-factors and, as explained in Section 6.2, due to their individual physical nature, they may also have various kinds of intrapair spin coupling. Thus, the coherent excitation of one pair partner alone may not be possible for some pairs, and during the application of a resonant pulse, both pair partners may propagate. Since the measurement of spin-dependent rates is a permutation symmetry measurement and not a polarization measurement, the nutation of both spins within a pair leads to nutation frequency components in the observed spin-dependent rates that correspond to either one of the two individual spins and also to the beat oscillations between the spins.[57,103] As the individual nutation components are determined by the excitation frequency and field strength, as well as the coupling parameters of the pairs, the spectroscopy of Rabi frequencies can give insights into the physics of a particular spin pair system. Rabi nutation spectroscopy is based on the measurement of spin-resonance-induced Rabi nutation followed by an analysis of its frequency components via Fourier transformation. Based on a variety of numerical simulations that have been conducted for such experiments,[57,103–105] the characteristic behavior of a range of effects such as exchange coupling, disorder, and Larmor separation regimes are well understood, and this knowledge can be applied to the interpretation of experimentally recorded nutation data.

Figure 6.14 displays a set of numerical simulations[104] that were conducted for spin pairs with different exchange coupling, J, and Larmor separation, $\Delta\omega$, for combinations of parameters that are smaller, larger, and about equal to the spin ($s = \frac{1}{2}$) nutation frequency, γB_1. The data sets are color plots showing the pEDMR signal intensity in arbitrary (but for all data sets equal) units, as a function of both the applied radiation frequency and the resulting nutation frequency. Notably, all nine plots show hyperbola-like features that appear due to Rabi's frequency formula, $\Omega = \sqrt{(\gamma B_1)^2 + (\omega - \omega_i)^2}$, which for spin $s = \frac{1}{2}$ leads to a nutation frequency $\Omega \propto (\omega - \omega_i)$ when $(\omega - \omega_i) \gg \gamma B_1$ and a nutation frequency minimum of $\Omega = \gamma B_1$ when $(\omega - \omega_i) \ll \gamma B_1$.

The plots in Figure 6.14 show that there are a variety of different qualitative features that can be attributed to the different Rabi nutation components, such as different exchange and Larmor separation regimes.[57,103–105] Thus, simple Rabi nutation spectroscopy experiments are well suited to the exploration of the physical nature of intermediate spin pairs.

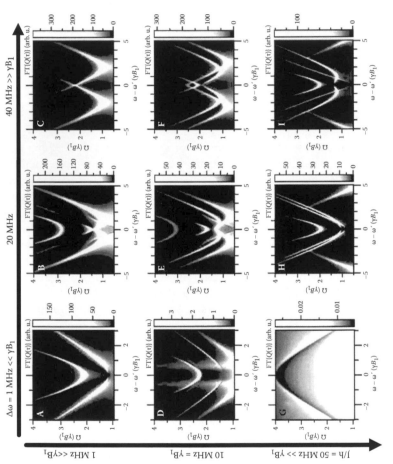

FIGURE 6.14   The data sets display the simulation of the frequency components, Ω, in units of γB_1 of pEDMR-detected transient nutation as a function of the excitation frequency, Ω. The nine plots correspond to different cases of the separation of Larmor frequencies, $\Delta\omega$, and exchange coupling, J. The plots show the presence of various nutation frequencies under the different conditions. (From Gliesche et al., *Phys. Rev. B* 77, 245206 [2008]. Copyright © 2008 by the American Physical Society. With permission.)

6.3.3.5   Spin Rabi Nutation and Geminate Recombination

Utilizing spin nutation spectroscopy allows the study of geminate recombination in disordered materials. Geminate recombination is recombination of correlated charge carriers. Recombining pair partners are correlated when they originate from the same generation process. Geminate recombination does not contribute to conductivity, and because of this, understanding whether a known recombination process is geminate or nongeminate is of great importance for understanding conductivity in many materials.

Figure 6.15 displays pEDMR- and pODMR-detected measurements[52] of the spin nutation spectra of hydrogenated amorphous silicon at low temperatures ($T = 10$ K) as a function of the applied magnetic fields, B_0. The data were collected via detection of both the integrated photoluminescence (pODMR) and the short-circuit photocurrent (pEDMR) on identical intrinsic materials under identical environmental

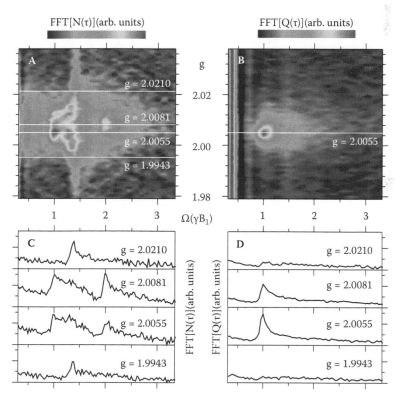

FIGURE 6.15   Plots of Fourier-transformed transient nutation data as a function of the applied magnetic field, B_0, measured with (a) pODMR and (b) pEDMR on hydrogenated amorphous silicon. Note that the frequency scale of the nutation frequency, Ω, has units of the spin ($s = \frac{1}{2}$) nutation frequency, γB_1. The pEDMR measurement reveals only a single significant process due to spin with $g \approx 2.0055$, which nutates with γB_1. The pODMR measurement shows a variety of different signals, including the $g = 2.0055$ process. The plots in (c) and (d) display data from (a) and (b), respectively: four selected g-factors. (From Herring et al., *Phys. Rev. B* 79, 195205 [2009]. Copyright © 2009 by the American Physical Society. With permission.)

conditions.[52] Figure 6.15 shows the spin-dependent signals observed using both pODMR and pEDMR. Only one process appears in both data, observed at a spin nutation frequency of γB_1 (weakly coupled spins $s = \frac{1}{2}$) and a Landé factor of $g \approx$ 2.0055 corresponding to silicon dangling bonds. There are a variety of features visible in the pODMR plot that do not appear in the pEDMR plot. As these influence the optical properties but not the electrical properties, they are able to be assigned to geminate recombination processes.

The Rabi nutation experiment performed here on hydrogenated amorphous silicon is equally applicable to organic semiconductors, although so far not yet reported. One of the reasons this experiment has not yet been performed is the technical challenge associated with conducting pEDMR and pODMR on the same materials at the same time by introducing an optical detection setup as well as appropriate electrical contacts into a microwave resonator.

6.3.3.6 Measurement of Intrapair Coupling

The data displayed in Figure 6.15 are also an example of the detection of different intrapair spin-coupling regimes using pEDMR- or pODMR-detected transient nutation measurements. The geminate recombination processes identified in the pODMR spectra appear at nutation frequencies of (1) $\Omega = 2\gamma B_1$ for a Landé factor of $g \approx 2.008$ and (2) $\Omega \approx \sqrt{2}\gamma B_1$ for a range of Landé factors in a broad symmetrical structure also centered around $g \approx 2.008$.[52] These pODMR features have been observed previously by Lips et al.[100] and can be attributed to (1) spin beating due to exchange-coupled spin pairs of conduction and valence band tail states and (2) dipolar coupling between weakly exchange-coupled pairs of conduction and valence band tail states. The exchange interaction is understood to lead to a beat frequency in two spin $s = \frac{1}{2}$ pair systems that, according to spin nutation simulations discussed above, are independent of the applied B_1 field (not shown here). The Landé factor is precisely the average of the known Landé factors of electrons and holes in band tail states that are $g_e \approx 2.004$ and $g_h \approx 2.012$, respectively. This is also predicted by the nature of exchange coupling.[100] The presence of dipolar coupling is the current interpretation of the nutation component with $\Omega \approx \sqrt{2}\gamma B_1$. In the absence of simulations of strongly dipolar coupled but weakly exchange-coupled spin pairs, this interpretation is based on predictions of pEPR transient nutation signals,[136] which would suggest that the symmetric spectral features centered around an intensity minimum at $g \approx 2.008$ are due to the convolution of Pake-like power spectra[101] of the anisotropic dipolar interaction within the spin pairs.

6.3.3.7 Coherence Decay of Spin Pairs vs. Coherent Dephasing

Sections 6.3.3.4 to 6.3.3.6 discussed various applications of electrically and optically detected spin Rabi nutation measurements for the exploration of electronic processes. Additionally, electrically or optically detected spin nutation can provide insight into the transition times of spin-dependent processes. During the pulse that induces the spin nutation, the ensemble will continuously undergo spontaneous transitions due to the various processes sketched in the rates picture of Figure 6.4. These random transitions contribute to a decay of the coherent spin pair ensemble, and thus an intensity decrease of the observed coherent spin motion. Note that as long as these processes are slow, they will only marginally influence observed signals. For very fast pair decay (on the timescale of the nutation experiment), rapid decay of the Rabi oscillation will be observed. Recently, this effect was used for the study of spin mixing rates of polaron pairs in MEH-PPV,[68] where spin Rabi nutation in excess of 500 ns was observed at low temperatures. It is important to note, however, that the investigation of spin mixing by observation of spin Rabi oscillation decay can provide only an upper boundary for the spin mixing rate: As various rates influence the pair ensemble decay, it is always the fastest rate that determines the decay rate coefficient. Moreover, the attenuation of the observed Rabi nutation may also be influenced by coherent dephasing where the pairs actually do not decay at all, but instead nutate at different frequencies due to the inhomogeneity of the driving field, B_1. This coherent dephasing is observed as a decay of the oscillation similar to coherence decay, even though it has a profoundly different physical cause. The different spin echo sequences discussed in the following sections can be used to distinguish coherent dephasing processes from coherence loss, and also to identify and discriminate between the different transition processes that act on a spin pair ensemble.

6.3.3.8   Rotary Echoes and Pair Generation

The rate pictures illustrated in Figures 6.4 and 6.5 include various pair decay and transition processes (recombination, dissociation, spin relaxation [mixing]), and also a generation term for spin pairs, which makes a rate equation describing this system inhomogeneous.[57] The generation rate G is a real rate, and not a rate coefficient like the generation, spin mixing, or dissociation coefficients. A consequence of this difference is that spin pair generation does not influence the coherence decay constants of coherent spin ensembles. Thus, measuring the dynamics of spin pair generation using pEDMR or pODMR is not straightforward.

Figure 6.16a to c illustrates a pEDMR/pODMR pulse sequence that allows the observation of the generation dynamics of spin pair ensembles.[118] The experiment consists of two resonant pulse sequences. The length of the first pulse is chosen such that a coherently dephasing spin Rabi nutation with nutation angle $(2n + 1)\pi$ will be produced (with n an integer). The length of the second pulse will be gradually increased to achieve nutation angles of up to $2(2n + 1)\pi$ (i.e., twice the length of the first pulse). If the second pulse has a 180° phase shift with respect to the first pulse, it is able to rephase the dephased spin ensemble when the second pulse has the same length as the first.[57,118] This effect is called a rotary echo. Note that rotary echoes were used for the first electrical detection of coherent spin motion in 2003.[59] With the choice of the $(2n + 1)\pi$-pulse length, the rephasing Rabi nutation caused by the spin ensemble that is coherently prepared by the first pulse will have a 180° phase compared with the coherent dephasing that appears during the second pulse due to the spin pairs that are generated during the time between the first and second pulses.[118] Hence, measuring the two opposite phase components of the Rabi nutation during the second pulse as a function of the time τ_{off} between the first and second pulses (when no radiation is applied to the system) will reveal the dynamics of the spin pair generation.

Figure 6.16d displays the measurement of the generation transient of spin pairs consisting of localized charge carriers at the semiconductor heterointerface between

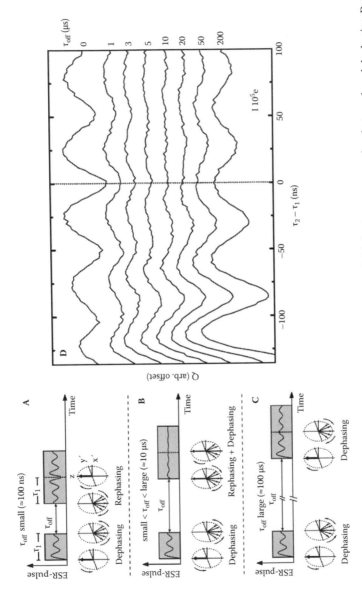

FIGURE 6.16 (a–c) Sketches of a rotary-spin-echo experiment consisting of an initial spin resonant pulse that produced dephasing Rabi nutation and a second rephasing pulse. During the pulse separation time, τ_{off} between the two pulses, the dephased spin pair ensemble remains in its coherently prepared state except for those pairs that decay. At the same time, new spin pairs in eigenstates are randomly generated. These pairs cause a dephasing Rabi nutation that is 180° out of phase with the rephasing Rabi nutation produced by the pairs that were present before the first pulse. Hence, the observation of the phase 180° inversion of the transient nutation data as a function of τ_{off} reveals the generation dynamics of spin pairs. (d) Experimental observation of spin pair generation for spin-dependent transport and recombination transitions at the crystalline silicon to amorphous silicon heterointerface. (From Behrends et al., *Phys. Rev. B* 80, 045207 [2009]. Copyright © 2009 by the American Physical Society. With permission.)

crystalline silicon and hydrogenated amorphous silicon.[118] The plots display a train of several rotary echo measurements made for different pulse separation times, τ_{off}, under otherwise identical sample conditions. One can clearly see the gradual phase inversion that is due to the gradually growing ensemble of newly generated charge carrier pairs. The quantitative interpretation of the displayed data can be found in Behrends et al.[118]

6.3.3.9   Hahn Echoes and T_2 Relaxation

The nature of spin relaxation rates is described by various random processes that can be quantified statistically via rate coefficients in the Redfield matrix (see Equation 6.10). As pointed out above, the attenuation of Rabi nutation can be due to a certain spin relaxation process, but it can also be due to other electronic relaxation processes, or dephasing. This problem also exists for pEPR measurements, and has been solved there by the use of spin excitation (pulse) sequences that allow discrimination between dephasing and coherence decay, and also between different coherence decay processes. Probably the most widely used experiment in this category is the two-pulse Hahn echo sequence for the measurement of T_2 transverse spin-spin relaxation rates, first demonstrated in 1950 for pNMR excitation.[137] The Hahn echo consists of an initial $\frac{\pi}{2}$-pulse that turns spins from eigenstates parallel to the $\hat{z}$-direction (also parallel to the direction of B_0) into states in the $\hat{x}$-$\hat{y}$-plane. After the pulse, the spin ensemble undergoes Larmor procession about B_0, and inhomogeneities will cause a fast coherent dephasing of the spins. The second pulse is applied a time, τ, after the first pulse, and it is twice as long as the first pulse. This π-excitation turns the spin ensemble that is dephased from the $\hat{x}$-$\hat{y}$-plane into the $\hat{x}$-$\hat{y}$-plane; however, after the excitation, the phase of faster precessing spins will be behind slower spins, and the faster spins will pass the slower spins at a time 2τ, when the entire spin ensemble rephases. It is at this moment that a brief Hahn echo occurs.[102,137] The Hahn echo sequence probes only spin propagation in the $\hat{x}$-$\hat{y}$-plane. Coherence decay is observed by monitoring the decay of the echo intensity with increasing pulse separation time τ, which reflects solely T_2 but not longitudinal spin-lattice (T_1) relaxation.[21]

Hahn echoes are detected with pEPR and pNMR through transient real-time measurements of the echo-induced radiation pulse, which comes from the rephasing of the sampled spin systems. As real-time measurement of the spin dynamics is not possible with pEDMR and pODMR, electrically or optically detected Hahn echo sequences must be conducted by scanning an additional $\frac{\pi}{2}$ detection pulse through the echo transient. This pulse projects the measured spin polarization in the $\hat{x}$-$\hat{y}$-plane onto the $\hat{z}$-axis, which then determines the pair permutation symmetry by interacting with the spin pair partner.[62,63] Figure 6.17a displays a sketch of the pulse sequence along with a high field ($B_0 \approx 8.5\text{T}$) pEDMR-detected Hahn echo that was recorded using a spin trap process of free electrons in crystalline silicon at phosphorous donor impurities. Due to the modified Hahn echo pulse sequence necessary for pEDMR detection, one can question whether the echo decay still represents the transverse T_2 spin relaxation process. Figure 6.17b displays the comparison of the electrically detected Hahn echo decay with a conventional radiation detected two-pulse Hahn echo decay. Both data sets show a remarkably (almost indistinguishable) agreement, and the T_2 times obtained from both data sets as plotted in the inset table

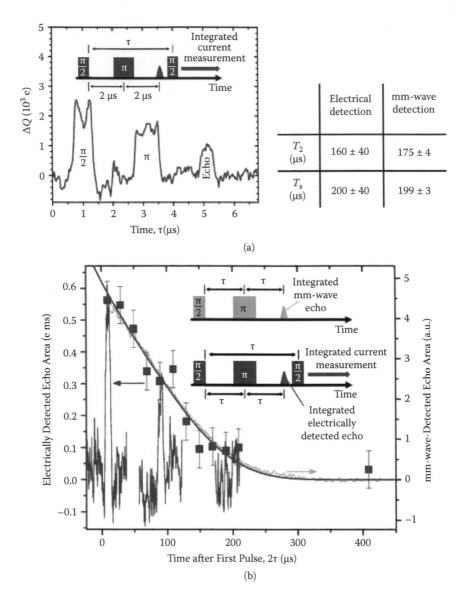

FIGURE 6.17 Electrical detection of the spin phase coherence time, T_2. (a) A pEDMR-detected Hahn echo sequence showing the readout pulse being swept through a traditional Hahn echo pulse sequence. The echo is observed by integrating the current change following the readout pulse. (b) The echo area as a function of time after the first pulse, for both pEDMR and pEPR. Note the high level of agreement between the coherence times (see table) obtained with the two methods. (From Morley et al., *Phys. Rev. Lett.* 101, 207602 [2008]. Copyright © 2008 by the American Physical Society. With permission.)

of Figure 6.17 confirm this. Thus, the three-pulse Hahn echo sequence allows the measurements of T_2 times with pEDMR and pODMR experiments, with the limitation that additional electronic processes (pair dissociation, relaxation) must be slower than the spin relaxation processes (see discussion of this process below).

6.3.3.10 Spin-Inversion Recovery Experiments and T_1 Relaxation

The Hahn spin echo sequence reveals T_2 times by measurement of the spin ensemble's polarization in the $\hat{x}$ - $\hat{y}$-plane, after the spin ensemble was turned from its equilibrium position (polarization parallel to $\hat{z}$-axis) by the initial $\frac{\pi}{2}$-pulse. Thus, one can also view the Hahn echo sequence as an equilibrium polarization measurement, as long as the experiment is started from the equilibrium. This realization opens the door to an extension of the Hahn echo sequence to an experiment, which allows the measurement of transverse spin relaxation (T_1 relaxation), which is, for instance, the determining process behind the spin mixing of excitonic precursor pairs in organic semiconductors. The experiment is called an inversion recovery experiment. The pulse sequence as sketched in Figure 6.18 consists of a Hahn echo sequence that is preceded by a π-pulse that inverts the equilibrium polarization before the Hahn echo sequence begins. The consequence of this inversion is the inversion of the Hahn echo. However, as the time τ'' between the inversion pulse and the Hahn echo sequence is increased while the pulse separation time, τ, within the Hahn echo sequence remains constant, the spin ensembles polarization before the $\frac{\pi}{2}$-pulse will gradually return to its equilibrium polarization due to the continuously occurring T_1 events. Thus, the measurement of the Hahn echo amplitude as a function of τ'' reveals the dynamics of the T_1 relaxation.

Similar to the Hahn echo sequence discussed above, the pEDMR or pODMR detection of an inversion recovery requires scanning through the echo transient with additional projection pulses. Figure 6.18 displays two Hahn echoes detected with an inversion recovery pulse sequence for identical conditions except for different values of τ''. One can clearly see how an inverted echo at a small value of τ'' recovers into a positive echo for large values of τ''. The data shown in Figure 6.18 were measured on a spin-dependent interface transition at a phosphorous-doped crystalline silicon–silicon dioxide interface. The spin pairs consisted of phosphorous donor electron states and silicon dangling bonds.[64]

6.3.3.11 Determination of Electronic Pair Decay

As in the determination of T_2 spin relaxation discussed above, the determination of T_1 spin relaxation with inversion recovery experiments can only work for electrical or optical detection if the spin relaxation rate is faster than the spin-dependent electronic transitions used to measure them. If the spin pair decay (due to either pair dissociation or pair recombination) is faster than the spin relaxation rates, the decay rates from Hahn echo decay measurements will not be determined by T_2, but instead the fast electronic rates, and equally, the decay rates from inversion recovery measurements will not be determined by T_1, but by the fast electronic rates also. This realization is crucial when determining whether the measured rates from pEDMR/pODMR-detected Hahn echo decay or inversion recovery experiments are electronic transition rates or spin relaxation rates. If the two measured times' constants

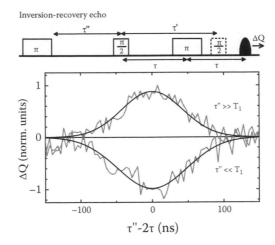

FIGURE 6.18   Electrical detection of the spin flip time, T_1, of phosphorus donors in silicon. By adding a preceding π-pulse to the pEDMR Hahn echo sequence, the pEDMR inversion recovery sequence is obtained. The panel shows two echoes, obtained by sweeping the read-out $\pi/2$-pulse through the echo, with different wait times, τ'', much shorter and much longer than T_1. (Adapted from Paik et al., arXiv, 0905.0416v1 [2009].)

(T_1 and T_2) are equal, then they are likely due to electronic transitions; if they are significantly different, at least the faster of the two rates (which usually means the T_2 process) is due to spin relaxation.

6.4   CONCLUSIONS

Much progress has been made in the past twenty years toward understanding spin-dependent processes in organic semiconductors. In spite of this progress, there are still many open questions remaining even on materials systems that have been used for decades in technical applications. One of the most important questions that has remained unanswered is how spin mixing processes of polaron pairs influence the efficiency limitations of OLEDs, and whether and how spin mixing within these pairs is a determining factor for organic magnetoresistance effects. These and other open questions could potentially be resolved by the recently developed coherent electrically and optically detected spin spectroscopy techniques (pEDMR and pODMR), especially given their very recent (in the past two years) application to organic semiconductors. While these methods are still the subject of exploration and investigation, by using those pulse sequences that have previously been applied to pEPR and pNMR for pEDMR and pODMR, it is expected that a complete picture of the relevant transitions and relaxation processes in organic materials could evolve in the coming years. Such progress could pave the way for a systematic development of organic materials that overcome the limitations of spin statistics. As the methods discussed here provide a reliable tool for the observation of spin-dependent processes, systematic strategies for the manipulation of these effects could be developed and tested.

6.5 SUMMARY

In this chapter a review of our current knowledge regarding the nature of spin-dependent processes in semiconductor materials and in particular in organic semiconductors was given, and the current state of experimental techniques for the investigation of spin-dependent processes, namely, the electrically and optically detected magnetic resonance spectroscopies (pEDMR and pODMR) were discussed. The spin-dependent processes have been discussed with regard to their relevance for organic magnetoresistance effects and internal quantum efficiency in OLED materials. Using recent pEDMR and pODMR examples from experiments investigating the coherent spin motion in inorganic materials, the range of experimental capabilities was outlined and the potential for the exploration of spin-dependent processes in organic materials was discussed.

ACKNOWLEDGMENTS

We acknowledge fruitful discussions with Z. V. Vardeny, E. Ehrenfreund, J. M. Lupton, W. J. Baker, S.-Y. Paik, S.-Y. Lee, T. W. Herring, K. J. van Schooten, M. J. Walter, N. J. Borys, and H. A. Seipel. This material is based upon work supported by the Department of Energy under Award Number DE-SC0000909. Acknowledgment is also made to the donors of the American Chemical Society Petroleum Research Fund under Grant PRF 48916-DNI10 for support of this research.

REFERENCES

1. J. H. Burroughes, D. D. C. Bradley, A. R. Brown, R. N. Marks, K. Mackay, R. H. Friend, P. L. Burns, and A. B. Holmes, *Nature* 347, 539 (1990).
2. T. Hamilton, *Clean Break* (November 26, 2008).
3. O. Prache, *Displays* 22, 49 (2001).
4. D. Fyfe, *Nature Photonics* 3, 453 (2009).
5. Sony XEL-1 OLED digital HDTV specifications, 2009.
6. Samsung A877 impression specifications, 2009.
7. C. W. Tang, *Appl. Phys. Lett.* 48, 183 (1986).
8. A. C. Mayer, S. R. Scully, B. E. Hardin, M. W. Rowell, and M. D. McGehee, *Mater. Today* 10, 28 (2007).
9. J. Y. Kim, K. Lee, N. E. Coates, D. Moses, T.-Q. Nguyen, M. Dante, and A. J. Heeger, *Science* 317, 222 (2007).
10. Konarka Technologies (http://www.konarka.com).
11. V. Shrotriya, *Nature Photonics* 3, 447 (2009).
12. M. A. Green, K. Emery, Y. Hishikawa, and W. Warta, *Progress Photovoltaics Res. Appl.* 17, 320 (2009).
13. M. Pope and C. E. Swenberg, *Electronic Processes in Organic Crystals and Polymers* (Oxford University Press, New York, 1999).
14. M. D. McGehee, *Nature Photonics* 3, 250 (2009).
15. G. B. Silva, F. Nüesch, L. Zuppiroli, and C. F. O. Graeff, *Phys. Stat. Sol. C* 2, 3661 (2005).
16. V. Dediu, M. Murgia, F. C. Matacotta, C. Taliani, and S. Barbanera, *Solid State Commun.* 122, 181 (2002).
17. Z. Xiong, D. Wu, Z. Vardeny, and J. Shi, *Nature* 427, 821 (2004).
18. S. Majumdar, H. S. Majumdar, D. Tobjörk, and R. Österbacka, *Phys. Stat. Sol. A*, published online (2009).

19. N. M. Atherton, *Principles of Electron Spin Resonance*, Physical Chemistry Series (Ellis Horwood and PTR Prentice Hall, Ellis Horwood, Chichester, 1993).
20. C. P. Slichter, *Principles of Magnetic Resonance*, Vol. 1 of Springer Series in Solid-State Sciences (Springer, Berlin, 1996).
21. A. Schweiger and G. Jeschke, *Principles of Pulse Electron Paramagnetic Resonance* (Oxford University Press, Oxford, 2001).
22. C. P. Poole, *Electron Spin Resonance* (Interscience, New York, 1967).
23. B. C. Cavenett, *Adv. Phys.* 30, 475 (1981).
24. S. Geschwind, R. J. Collins, and A. L. Schawlow, *Phys. Rev. Lett.* 3, 545 (1959).
25. A. Honig, *Phys. Rev. Lett.* 17, 186 (1966).
26. R. Maxwell and A. Honig, *Phys. Rev. Lett.* 17, 188 (1966).
27. D. J. Lepine, *Phys. Rev. B* 6, 436 (1972).
28. D. Kaplan, I. Solomon, and N. F. Mott, *J. Phys. Lett.* 39, L51 (1978).
29. C. Boehme and K. Lips, *Phys. Stat. Sol.B* 233, 427 (2002).
30. A. R. Stegner, C. Boehme, H. Huebl, M. Stutzmann, K. Lips, and M. S. Brandt, *Nature Phys.* 2, 835 (2006).
31. J. Brossel, S. Geschwind, and A. L. Schawlow, *Phys. Rev. Lett.* 3, 548 (1959).
32. J. Schmidt and I. Solomon, *C. R. Acad. Sci.* B 263, 169 (1966).
33. P. A. Lane, L. S. Swanson, Q.-X. Ni, J. Shinar, J. P. Engel, T. J. Barton, and L. Jones, *Phys. Rev. Lett.* 68, 887 (1992).
34. G. B. Silva, L. F. Santos, R. M. Faria, and C. F. O. Graeff, *Mater. Res. Soc. Symp. Proc.* 725, 4.18 (2002).
35. F. A. Castro, G. B. Silva, L. F. Santos, R. M. F. F. Nüesch, L. Zuppiroli, and C. F. O. Graeff, *J. Non-Cryst. Solids* 338–340, 622 (2004).
36. G. Li, C. H. Kim, P. A. Lane, and J. Shinar, *Phys. Rev. B* 69, 165311 (2004).
37. M.-K. Lee, M. Segal, Z. G. Soos, J. Shinar, and M. A. Baldo, *Phys. Rev. Lett.* 94, 137403 (2005).
38. C. F. O. Graeff, G. B. Silva, F. Nesch, and L. Zuppiroli, *Eur. Phys. J. E* 18, 21 (2005).
39. C. G. Yang, E. Ehrenfreund, and Z. V. Vardeny, *Phys. Rev. Lett.* 99, 157401 (2007).
40. C. G. Yang, E. Ehrenfreund, F. Wang, T. Drori, and Z. V. Vardeny, *Phys. Rev. B* 78, 205312 (2008).
41. A. Honig and M. Moroz, *Rev. Sci. Instruments* 49, 183 (1978).
42. K. Murakami, S. Namba, N. Kishimoto, K. Masuda, and K. Gamo, *J. Appl. Phys.* 49, 2401 (1978).
43. Dersch, H., L. Schweitzer, and J. Stuke, *Phys. Rev. B* 28, 4678 (1983).
44. P. M. Lenahan and W. K. Schubert, *Phys. Rev. B* 30, 1544 (1984).
45. M. S. Brandt and M. Stutzmann, *Phys. Rev. B* 43, 5184 (1991).
46. R. N. Ghosh and R. H. Silsbee, *Phys. Rev. B* 46, 12508 (1992).
47. M. S. Brandt, S. T. B. Goennenwein, T. Graf, H. Huebl, S. Lauterbach, and M. Stutzmann, *Phys. Stat. Sol.C* 1, 2056 (2004).
48. M. Stutzmann, M. S. Brandt, and M. W. Bayerl, *J. Non-Cryst. Solids* 266–69, 1 (2000).
49. D. R. McCamey, H. Huebl, M. S. Brandt, W. D. Hutchison, J. C. McCallum, R. G. Clark, and A. R. Hamilton, *Appl. Phys. Lett.* 89, 182115 (2006).
50. M. Xiao, I. Martin, E. Yablonovitch, and H. W. Jiang, *Nature* 430, 435 (2004).
51. F. Jelezko, T. Gaebel, I. Popa, A. Gruber, and J. Wrachtrup, *Phys. Rev. Lett.* 92, 076401 (2004).
52. T. W. Herring, S.-Y. Lee, D. R. McCamey, P. C. Taylor, K. Lips, J. Hu, F. Zhu, A. Madan, and C. Boehme, *Phys. Rev. B* 79, 195205 (2009).
53. H. Morishita, L. S. Vlasenko, H. Tanaka, K. Semba, K. Sawano, Y. Shiraki, M. Eto, and K. M. Itoh, *Phys. Rev. B* 80, 205206 (2009).
54. L. H. Willems van Beveren, H. Huebl, D. R. McCamey, T. Duty, A. J. Ferguson, R. G. Clark, and M. S. Brandt, *Appl. Phys. Lett.* 93, 072102 (2008).

55. J. M. Spaeth and H. Overhof, *Point Defects in Semiconductors and Insulators* (Springer, Berlin, 2003).
56. D. J. Gravesteijn and M. Glasbeek, *Phys. Rev. B* 19, 5549 (1979).
57. C. Boehme and K. Lips, *Phys. Rev. B* 68, 245105 (2003).
58. C. Boehme, P. Kanschat, and K. Lips, *Nucl. Instr. Meth. B* 186, 30 (2002).
59. C. Boehme and K. Lips, *Phys. Rev. Lett.* 91, 246603 (2003).
60. C. Boehme and K. Lips, *Physica B* 376–377, 930 (2006).
61. S. Schaefer, S. Saremi, K. Fostiropoulos, J. Behrends, K. Lips, and W. Harneit, *Phys. Stat. Sol. B* 245, 2120 (2008).
62. H. Huebl, F. Hoehne, B. Grolik, A. R. Stegner, M. Stutzmann, and M. S. Brandt, *Phys. Rev. Lett.* 100, 177602 (2008).
63. G. W. Morley, D. R. McCamey, H. Seipel, L. C. Brunel, J. van Tol, and C. Boehme, *Phys. Rev. Lett.* 101, 207602 (2008).
64. S.-Y. Paik, S.-Y. Lee, W. J. Baker, D. R. McCamey, and C. Boehme, arXiv, 0905.0416v1 (2009). To be published in *Phys. Rev B*.
65. K. Murata, Y. Shimoi, S. Abe, S. Kuroda, T. Noguchi, and T. Ohnishi, *Chem. Phys.* 227, 191 (1998).
66. M. Segal, M. Singh, K. Rivoire, S. Difley, T. V. Voorhis, and M. A. Baldo, *Nature Mater.* 6, 374 (2007).
67. W. Harneit, C. Boehme, S. Schaefer, K. Huebener, K. Fostiropoulos, and K. Lips, *Phys. Rev. Lett.* 98, 216601 (2007).
68. D. R. McCamey, H. A. Seipel, S. Y. Paik, M. J. Walter, N. J. Borys, J. M. Lupton, and C. Boehme, *Nature Mater.* 7, 723 (2008).
69. E. A. Thomsen, PhD thesis, University of St. Andrews, 2008.
70. Y. Wu, Z. Xu, B. Hu, and J. Howe, *Phys. Rev. B* 75, 035214 (2007).
71. Z. V. Vardeny and X. Wei, *Mol. Cryst. Liquid Cryst.* 256, 465 (1994).
72. J. Orenstein, Z. Vardeny, G. L. Baker, G. Eagle, and S. Etemad, *Phys. Rev. B* 30, 786 (1984).
73. B. Hu and Y. Wu, *Nature Mater.* 6, 85 (2007).
74. P. A. Bobbert, T. D. Nguyen, F. W. A. van Oost, B. Koopmans, and M. Wohlgenannt, *Phys. Rev. Lett.* 99, 216801 (2007).
75. M. Wohlgennant, K. Tandon, S. Mazumdar, S. Ramasesha, and Z. Vardeny, *Nature* 409, 494 (2001).
76. M. Reufer, M. J. Walter, P. G. Lagoudakis, A. B. Hummel, J. S. Kolb, H. G. Roskos, U. Scherf, and J. M. Lupton, *Nature Mater.* 4, 340 (2005).
77. R. Haberkorn and W. Dietz, *Solid State Commun.* 35, 505 (1980).
78. L. S. Vlasenko, Y. V. Martynov, T. Gregorkiewicz, and C. A. J. Ammerlaan, *Phys. Rev. B* 52, 1144 (1995).
79. Barabanov, A. V., O. V. Tretiak, and V. A. L'vov, *Phys. Rev. B* 54, 2571 (1996).
80. M. Knupfer, *Appl. Phys. A* 77, 623 (2003).
81. M. A. Baldo, D. F. O'Brien, M. E. Thompson, and S. R. Forrest, *Phys. Rev. B* 60, 14422 (1999).
82. V. Dyakonov, G. Rösler, M. Schwoerer, and E. L. Frankevich, *Phys. Rev. B* 56, 3852 (1997).
83. J. Partee, E. L. Frankevich, B. Uhlhorn, J. Shinar, Y. Ding, and T. J. Barton, *Phys. Rev. Lett.* 82, 3673 (1999).
84. M. Segal, M. A. Baldo, R. J. Holmes, S. R. Forrest, and Z. G. Soos, *Phys. Rev. B* 68, 075211 (2003).
85. I. Samuel, F. Raksi, D. Bradley, R. Friend, P. Burn, A. Holmes, H. Murata, T. Tsutsui, and S. Saito, *Synth. Metals* 55, 15 (1993).
86. B. Kraabel, D. Moses, and A. J. Heeger, *J. Chem. Phys.* 103, 5102 (1995).
87. H. D. Burrows, J. S. de Melo, C. Serpa, L. G. Arnaut, A. P. Monkman, I. Hamblett, and S. Navaratnam, *J. Chem. Phys.* 115, 9601 (2001).

88. M. Orchin, R. S. Macomber, A. R. Pinhas, and R. M. Wilson, *The Vocabulary and Concepts of Organic Chemistry* (John Wiley & Sons, Hoboken, NJ, 2005).

89. D. R. McCamey, G. W. Morley, H. A. Seipel, L. C. Brunel, J. van Tol, and C. Boehme, *Phys. Rev. B* 78, 045303 (2008).

90. C. Boehme and K. Lips, in *Charge Transport in Disordered Solids with Applications in Electronics*, ed. S. Baranovski (Wiley, New York, 2006).

91. I. Dzyaloshinsky, *J. Phys. Chem. Solids* 4, 241 (1958).

92. T. Moriya, *Phys. Rev.* 120, 91 (1960).

93. M. Born, R. J. Blin-Stoyle, and J. M. Radcliffe, *Atomic Physics*, 8th ed. (Dover Books on Physics and Chemistry, Mineola, New York, 1989).

94. R. A. Street, *Hydrogenated Amorphous Silicon* (Cambridge University Press, Cambridge, England, 1991).

95. D. L. Staebler and C. R. Wronski, *J. Appl. Phys.* 51, 3262 (1980).

96. U. E. Steiner and T. Ulrich, *Chem. Rev.* 89, 51 (1989).

97. E. Frankevich, *Chem. Phys.* 297, 315 (2004).

98. K. Schulten and P. G. Wolynes, *J. Phys. Chem.* 68, 3292 (1978).

99. J. M. Lupton and C. Boehme, *Nature Mater.* 7, 598 (2008).

100. K. Lips, C. Boehme, and T. Ehara, *J. Optoelectronics Adv. Mater.* 7, 13 (2005).

101. A. Weber, O. Schiemann, B. Bode, and T. Prisner, *J. Magn. Reson.* 157, 277 (2002).

102. C. Boehme, PhD thesis, Fachbereich Physik der Philipps-Universitat Marburg, 2002.

103. V. Rajevac, C. Boehme, C. Michel, A. Gliesche, K. Lips, S. D. Baranovskii, and P. Thomas, *Phys. Rev. B* 74, 245206 (2006).

104. A. Gliesche, C. Michel, V. Rajevac, K. Lips, S. D. Baranovskii, F. Gebhard, and C. Boehme, *Phys. Rev. B* 77, 245206 (2008).

105. C. Michel, A. Gliesche, S. D. Baranovskii, K. Lips, F. Gebhard, and C. Boehme, *Phys. Rev. B* 79, 052201 (2009).

106. J. J. Sakurai, *Modern Quantum Mechanics*, rev. ed. (Addison Wesley, New York, 1993).

107. A. G. Redfield, *IBM J. Res. Dev.* 1, 19 (1957).

108. H. Huebl, PhD thesis, Technischen Universitat Munchen, 2007.

109. T. Eickelkamp, S. Roth, and M. Mehring, *Mol. Phys.* 95, 967 (1998).

110. P. A. Bobbert, W. Wagemans, F. W. A. van Oost, B. Koopmans, and M. Wohlgenannt, *Phys. Rev. Lett.* 102, 156604 (2009).

111. T. D. Nguyen, Y. Sheng, J. Rybicki, G. Veeraraghavan, and M. Wohlgenannt, *J. Mater. Chem.* 17, 1995 (2007).

112. T. L. Francis, O. Mermer, G. Veeraraghavan, and M. Wohlgenannt, *New J. Phys.* 6, 185 (2004).

113. R. E. Merrifield, *J. Chem. Physics* 48, 4318 (1968).

114. R. Belaid, T. Barhoumi, L. Hachani, L. Hassine, and H. Bouchriha, *Synth. Metals* 131, 23 (2002).

115. Y. Sheng, T. D. Nguyen, G. Veeraraghavan, O. Mermer, M. Wohlgenannt, S. Qiu, and U. Scherf, *Phys. Rev. B* 74, 045213 (2006).

116. J. Kalinowski, M. Cocch, D. Virgili, P. D. Marco, and V. Fattori, *Chem. Phys. Lett.* 380, 710 (2003).

117. J. D. Bergeson, V. N. Prigodin, D. M. Lincoln, and A. J. Epstein, *Phys. Rev. Lett.* 100, 067201 (2008).

118. J. Behrends, K. Lips, and C. Boehme, *Phys. Rev. B* 80, 045207 (2009).

119. H. Huebl, R. P. Starrett, D. R. McCamey, A. J. Ferguson, and L. H. Willems van Beveren, *Rev. Sci. Instrum.* 80, 114705 (2009).

120. C. Boehme, D. R. McCamey, K. J. van Schooten, W. J. Baker, S.-Y. Lee, S.-Y. Paik, and J. M. Lupton, *Phys. Stat. Sol., B* 246, 2750 (2009).

121. C. Boehme and K. Lips, *Phys. Stat. Sol.C* 1, 1255 (2004).

122. Bruker Biospin EPR (www.bruker-biospin.com/epr_systems.html).

123. S. Kuroda, T. Noguchi, and T. Ohnishi, *Phys. Rev. Lett.* 72, 286 (1994).
124. V. I. Krinichnyi, *Synth. Metals* 108, 173 (2000).
125. G. Li, J. Shinar, and G. E. Jabbour, *Phys. Rev. B* 71, 235211 (2005).
126. J. Köhler, J. A. J. M. Disselhorst, M. C. J. M. Donckers, E. J. J. Groenen, J. Schmidt, and W. E. Moerner, *Nature* (London) 363, 242 (1993).
127. C. G. Yang, E. Ehrenfreund, M. Wohlgenannt, and Z. V. Vardeny, *Phys. Rev. B* 75, 246201 (2007).
128. B. Movaghar, B. Ries, and L. Schweitzer, *Philos. Mag. B* 41, 159 (1980).
129. S.-Y. Lee, D. R. McCamey, S.-Y. Paik, and C. Boehme, to be published (2010).
130. A. Matsuyama, K. Maeda, and H. Murai, *J. Phys. Chem. A* 103, 4137 (1999).
131. I. Hiromitsu, Y. Kaimori, M. Kitano, and T. Ito, *Phys. Rev. B* 59, 2151 (1999),
132. R. Furrer and F. Sciff, *Z. Fr. Phys. B* 29, 189 (1978).
133. M. Yoshida and K. Morigaki, *J. Phys. Soc. Jpn.* 59, 224 (1990).
134. L. Langof, E. Ehrenfreund, E. Lifshitz, O. I. Micic, and A. J. Nozik, *J. Phys. Chem. B* 106, 1606 (2002).
135. D. R. McCamey, S.-Y. Lee, and C. Boehme, to be published (2010).
136. A. V. Astashkin and A. Schweiger, *Chem. Phys. Lett.* 174, 595 (1990).
137. E. L. Hahn, *Phys. Rev.* 80, 580 (1950).
138. T. W. Herring, PhD thesis, University of Utah, 2008.

7 Organic Spintronics
Toward Sensor and
Memory Applications

Alberto Riminucci, Mirko Prezioso,
and Valentin Dediu

CONTENTS

Since the first demonstration of an organic spintronic device in 2002, a great deal of work has been done on the subject of organic spintronics. Research groups experimented with a wide variety of materials to improve the performance of organic spintronic devices and to understand the fundamental physical mechanisms that underpin spin injection and transport in organic semiconductors. In this chapter we will take a different perspective and look at how organic spintronics can be used for applications. We will focus on memory and sensor devices, and we will compare their performance to that of their inorganic counterparts. We will show that, although organic spintronics hasn't reached the level of performance of inorganic devices, it is approaching a stage at which it could enter into mainstream applications.

7.1   INTRODUCTION

Organic spintronics stands at the crossroads between plastic electronics and conventional inorganic spintronics. In common with plastic electronics it has the transport medium. With the inorganic spintronics it shares the manipulation of the spin polarization of the flowing charges. Historically, early efforts to study spin injection and transport in organic semiconductors (OSCs) were motivated by the possibility of increasing organic light-emitting diodes' (OLEDs) efficiency by manipulating the singlet and triplet exciton formation rates in the organic emitter.[1,2] However, the spin manipulation in OLEDs still remains a controversial issue, while the field achieved its strongest results with magnetoresistive devices. Although many fundamental issues have yet to be understood, we would like to propose a detailed analysis of

301

the prospects for the application of recent and future organic spintronic devices for memory and sensor applications.

The first organic spintronic device was reported in 2002.[3] There, injection and transport of a spin-polarized current was demonstrated in planar devices with sexithiophene (T_6) deposited between two $La_{0.7}Sr_{0.3}MnO_3$ (LSMO) electrodes separated by gaps as wide as 70 to 500 nm[3]. The authors reached a 30% room temperature magnetoresistance and reported spin diffusion times of about 1 μs. Since then, a great deal of work has been done, and several review articles on the subject have already appeared.[1,4]

Further steps to investigate the spin injection and transport in OSCs were based on vertical spin transport devices. A sketch of one such device is shown in Figure 7.1a. It consists of a multilayered structure in which two ferromagnetic (FM) electrodes (in this case, a bottom $La_{0.7}Sr_{0.3}MnO_3$ one and a top Co one) are separated by a nonmagnetic medium, which in this instance is made of an OSC plus, in some cases, a tunnel barrier. The spin-polarized current is injected from one electrode into the OSC, in which transport occurs, until carriers reach the other electrode. The magnetization of the electrodes is manipulated by the application of an in-plane external magnetic field, which normally is parallel to either electrode. The resistance between the electrodes changes depending upon the relative orientation of their magnetizations. Figure 7.1b shows the magnetoresistance (MR) starting from a high positive field, which is swept past 0 to reach a high negative field. The arrows at the bottom of the figure show the orientation of the magnetization of the electrodes. Figure 7.1c shows the MR when the field is swept starting from a high negative magnetic field, and Figure 7.1d shows the MR curve of a complete MR measurement. Such a device is usually called organic spin valve (OSV). While usually the spin valve effect is described by a standard magnetoresistance definition (see the Section 7.3), we consider it useful to introduce the spin valve strength notion (SV-S), namely, the percent difference in resistance between the parallel and antiparallel states. In more mathematical terms:

$$SV-S = \frac{R(\tilde{H})_{AP} - R(\tilde{H})_P}{R(\tilde{H})_P} \text{ with } \tilde{H} \text{ such that } \left| R(\tilde{H})_{AP} - R(\tilde{H})_P \right| \text{ is maximum}$$

where $R(\tilde{H})_P$ and $R(\tilde{H})_{AP}$ are the resistances in the parallel and antiparallel states, respectively, and $\tilde{H}$ is the applied magnetic field. Referring to Figure 7.1d, these two quantities are defined simultaneously only where the magnetic field allows for the existence of both the parallel and antiparallel states; SV-S is negative when $R(\tilde{H})_{AP} < R(\tilde{H})_P$. We will use this definition in the remainder of the chapter. We believe its usefulness lies in the fact that it focuses on magnetic hysteresis-based effects, and it cancels out some of the other magnetoresistive effects.

In 2004 Xiong et al.[5] demonstrated the first OSV with a vertical structure. There they sandwiched a Tris(8-hydroxyquinolinato)aluminum (Alq_3) layer between a bottom LSMO electrode and a top Co electrode, reporting a giant MR (GMR) of 40% at 11 K and of about 4% at 80 K. Unexpectedly, the spin valve effect was inverse; that is, the lowest resistance corresponded to antiparallel orientation of the electrodes.

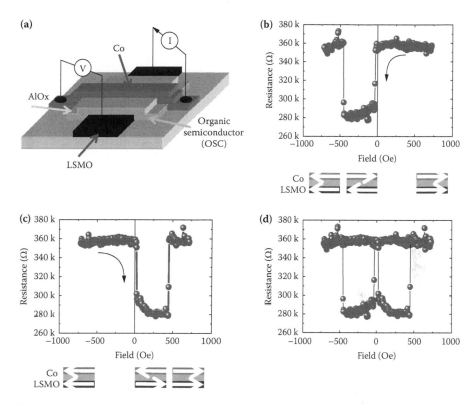

FIGURE 7.1 Schematic drawing and working principle of an OSV. (a) Schematic drawing of the device. Starting from the bottom the layers are: LSMO 20 nm/OSC/AlO$_X$ 2 nm/Co 40 nm. If not stated otherwise, LSMO, AlO$_X$, and Co thicknesses are the same throughout the chapter. (b) MR sweep starting from a high positive magnetic field. The arrows at the bottom show schematically the orientation of the LSMO and Co magnetizations. (c) MR sweep starting from a high negative magnetic field. (d) The complete MR curve. The MR traces are taken from a LSMO/Alq$_3$ 250 nm/AlO$_X$/Co 30 OSV at 100 K, R(0 Oe) = 360 kΩ.

The effect was constrained in a ±1 V interval and vanished above about 240 K to 250 K. Starting from this work, the vertical geometry has acquired a preeminent role in organic spintronics, mainly because of its technological simplicity compared to planar devices, which require a substantial involvement of lithography. Several groups have confirmed the so-called inverse spin valve effect in LSMO/Alq$_3$/Co devices.[6,7] Nevertheless, its nature has not been clearly understood yet, although at least two different explanations have been proposed.[5,7]

In order to shed light on the application potential of organic spintronics devices, we will try to define the most important quantitative parameters of OSVs and will catalog the main achievements in the field as far as these aspects are concerned. The most important parameter is probably the SV-S. The first report on GMR in an OSV is that of 40% by Xiong et al.[5] Majumdar et al. in 2006 reported a GMR of up to 80% at 5 K in their LSMO/regioregular poly 3-hexylthiophene (RRP3HT) 100 nm/Co OSV.[8] By our definition, the SV-S is −30% in the former case and 60% in the latter.

These values are smaller (in absolute terms) than the ones reported above because our SV-S definition eliminates MR effects other than those associated with the hysteresis of the electrodes' magnetization.

The best SV-S results were obtained at low temperature. This is understandable, since LSMO loses its surface spin polarization quickly as it approaches its Curie temperature, T_C, which is 325 K to 330 K, just above room temperature. Attempts were made and are still being made to increase the working temperature of OSVs by using high T_C materials.[9]

Of course, for practical purposes, it is important to achieve at least room temperature operation. In this respect OSVs made considerable progress over time, achieving room temperature operation in various systems. In 2008 Santos et al. demonstrated a tunneling magnetoresistance of 4% at room temperature across an Alq_3 layer.[10] The device structure was Co 8 nm/Al_2O_3 ~0.6 nm/Alq_3/$Ni_{80}Fe_{20}$ 10 nm. The thickness of the Alq_3 layer ranged from 1 to 4 nm. It must be stressed that this was a tunneling experiment, and therefore no injection or transport of a spin-polarized current inside the organic layer occurred.

Room temperature results were reported for spin injection and transport in regioregular and regiorandom poly (3-hexylthiophene) (RR-P3HT and RRa-P3HT), with an MR of 0.02% in RRa-P3HT[11] and 0.1% in RR-P3HT.[12]

Recently room temperature spin injection and transport in an Alq_3-based spin valve was demonstrated for LSMO/Co[7] and Co/Fe[13] electrodes. In the first case a 2-nm thick AlO_X film was introduced on the Alq_3 layer prior to the deposition of Co in order to provide a mechanical and chemical barrier.

The two experiments described above had in common considerable efforts dedicated to the interface improvement. Indeed, for an efficient spin injection the quality of the interfaces between the electrodes and the OSC is of course of the utmost importance. The good quality of LSMO as a spin injector at low temperature has been proven in both devices, as the ones reported above, and in studies that focused solely on its surface properties.[14] LSMO is grown at about 880°C and therefore has to be the bottom electrode. The OSC is grown on top of it, and although it can interact with LSMO,[15] it does not damage it. The top interface is formed on top of the organic layer by an inorganic material; since OSCs are generally soft materials, it is widely recognized that the direct growth of a metallic electrode on top of it could cause considerable damage.[5,16] To circumvent this problem, a tunnel barrier (AlO_X or LiF) was proposed as a chemical and mechanical barrier between the soft OSC and the energetic Co being deposited in LSMO/Alq_3/AlO_X/Co OSVs was investigated in detail by Dediu et al.[7,17] The importance of the microstructure was demonstrated also in Fe/OSC/Co devices: by improving the quality of the interfaces Liu et al.[13] were able to achieve room temperature operation in such spin valves. Particularly interesting is the Co/CuPc interface, in which spin injection efficiency of 85 to 90% was demonstrated with two-photon photoemission,[18] which makes this system particularly interesting for organic spintronics. Also, the Pentacene/Co interface is interesting for its quality.[19]

Once the spin-polarized current is injected, it is of course important to know what happens to it while it is in the OSC. There are very few techniques that can give this information. Muon spin rotation (μSR) is one such technique and was used by Drew et al.[20] to study the spin diffusion length in a working NiFe/LiF/Alq$_3$/TPD/ FeCo OSV. They were able to show experimentally a strict correlation between the spin diffusion length and the MR as a function of temperature.

We would like to mention, as an alternative approach, the works where FM nanoparticles were included in an OSC matrix. Significant work in this direction was reported by Luo et al.[21] in 2004. They embedded LCMO nanoparticles in polyparaphenyl derivatives with different conductivities. They attributed the observed large positive MR to the weakening of spin-polarized electron tunneling and to the magnetic scattering enhancement on polarized π electrons across the LCMO-polymer interfaces. Kusai et al.[22] embedded Co particles in rubrene, achieving an MR of 78% at 4.2 K. This is ascribed to spin-dependent transport via rubrene molecules. The same group also fabricated a Co/C$_{60}$ composite, where the spin-polarized current from the Co nanoparticles was believed to tunnel through the C$_{60}$ molecules.[23]

Interestingly for sensor applications, in the OSC layer, the magnetic moment of the carriers can also be affected by an external magnetic field that can induce precession. Yu et al.[24] predicted that in FM/OSC/FM structures the interplay between spin drift and spin precession can give rise to damped oscillating magnetoresistances when a transverse magnetic field is applied.

Finally, we would like to report some of the early efforts to demonstrate spin injection and transport in an organic semiconductor, which were directed at manipulating the singlet and triplet exciton formation rates. For example, in order to generate magnetic-field-dependent luminescence in OLEDs, Davis and Bussmann[25] engineered a number of OLEDs with different FM electrodes and OSC layers. They found that efficiencies were much lower and turn on voltages much greater than that of indium tin oxide (ITO)-based devices. In 2004 Salis et al.[26] used the structure permalloy (Py)/Alq$_3$/2,28,7,78-tetrakis(diphenylamino)-9,98-spirobifluorene(STAD)/Ni, where Alq$_3$ acted as the electron transporting layer and STAD as the hole transporting layer. They detected a 0.15% increase in the emission intensity when the relative magnetization direction of the two FM electrodes went from parallel to antiparallel. By taking into account the effect of stray magnetic fields from the electrodes, they gave an upper limit for the polarization of the carriers in their samples of $p_e \times p_h < 5 \times 10^{-5}$ for a voltage bias of 2.7 V. A model by Yunus et al.[27] predicts that spin-polarized injection increases the formation of singlet excitons.

The transfer of the organic spintronics technology into real applications is a challenging issue and, given the young age of the field, represents a difficult puzzle. Nevertheless, unveiling the potential of OS devices is of paramount importance not only for future applications, but also for the powerful thrust this could impart to fundamental research. We would like to propose in this chapter a brief analysis of two of the most direct applications of spintronic effects in organic-based devices: memory and sensor devices. We will present some general issues of these two options, give a brief overview of the most recent achievements in these fields by our and other groups, and compare the performance of our OSVs with that of inorganic devices. We will mostly leave out the very interesting problem of organic semiconductor magnetoresistance (OMAR)[28] since it involves no FM electrodes and is about the intrinsic properties of the OSC.

7.2 MAGNETORESISTIVE BISTABILITY

Hybrid organic-inorganic devices offer the unique opportunity of a simultaneous control of magnetic and electric bistability. On the same device it is possible to have both the spin valve effect and significant nonvolatile electrical bistability.[29] The coexistence of magnetic and electrical stable states constitutes a very interesting physical problem as well as a desirable property of materials and devices for information and communication technology (ICT). Such a combination is not totally new and is, of course, reminiscent of the field of multiferroics, which are materials that show simultaneous ferroelectric and magnetic ordering. These properties are very appealing for the information storage industry, and therefore there is great interest in their development. Accordingly, the scientific community is devoting to multiferroics a considerable amount of energy.[29,30]

Within the family of magnetically and electrically bistable materials and devices, organic spintronics is an interesting paradigm and it still poses many unanswered questions, such as the spin injection, transport, and scattering mechanisms.[31] From a technological point of view, the use of organic materials in ICT is advantageous because they are a cheaper, greener alternative to their inorganic counterparts.

Memory elements often exploit bistable properties. The SV-S is a typical example of a magnetically driven resistance bistability where a transition from a relatively stable state in high magnetic fields (parallel electrode configuration) into a metastable antiparallel configuration takes place. Indeed, this effect is widely used nowadays in magnetoresistive hard disk drives (HDD) read heads for information storage applications.

The existence of memory effects in organic devices has a long history.[32] There are three main working structures to which most devices can be linked: transistor-type organic memories, capacitor-type organic memories, and resistive-type organic memories.[33]

The transistor structure mainly used in organic electronics is that of the organic field effect transistor (OFET). For memory applications, the working principle is based on the storage of charge between the gate and the semiconductor channel.[34] This creates a voltage that shows as a shift in the drain current-gate voltage (I_D-V_G) transfer curve.

Capacitor-type organic memories are based on ferroelectric materials that retain their displacement field and therefore are nonvolatile.[35] Their electrical polarization can be switched between two stable states by the application of an external electric field. This kind of organic memory has the main disadvantage of destructive readout,[36] which limits the number of programming cycles in its lifetime.

Resistive-type memories are based on materials that can switch between two resistive states (bistability) upon the application of two different electric fields; their state is then read with the application of a read voltage. In ICT they are known as resistive random access memories (RRAMs). There is a wide variety of organic materials that exhibit resistive switching, including well-known organic semiconductors,[37] polymers,[38] charge transfer complexes,[39] mobile ions,[40] and nanoparticle blends.[41] There can be many different mechanisms underlying this behavior. One example is filamentary conduction in which highly conductive filaments are created or destroyed by the application of a voltage. Conformational changes in the organic with corresponding change in conductivity can occur under the application of an electric field. Changes in conductivity can also be caused by ionic migration within the organic matrix. An extensive review of these and other mechanisms can be found in Ling et al.[33]

There are several parameters to be considered when we want to evaluate the performance of a resistive memory device. The ON/OFF ratio is the ratio between the current in the low resistance state and the high resistance state at the read voltage. A greater ON/OFF ratio means that the two states are more different from each other and therefore are easier to read. The lifetime of the device in terms of programming cycles (write, read, erase, read) is also of paramount importance, and the standard in the industry is 10^{12}. Scalability is a must if a device is to be part of an integrated circuit; from this point of view, capacitor-type and resistor-type memories have a definite advantage over the transistor type. In fact, they are two-terminal devices and therefore can be scaled down to an area of $4f^2$, where f is the minimum feature size. On the other hand, transistors, being three-terminal devices, require more wiring and correspondingly occupy a greater area.

Of great importance is also the operating voltage. In this sense, the fact that the gate voltage is often high creates a disadvantage for OFET memories, while resistive memories have operating voltages of the order of 1 to 10 V. The read and write times are important too, and compared, for example, to that of static RAM (SRAM), the performance of organic memories is still lagging.

For the purpose of this section, we will concentrate attention on the small molecule Alq_3. This material started to be the focus of a vigorous research effort after the seminal paper by Tang and Van Slyke.[42] There, the authors reported on an efficient OLED based on Alq_3 as the emissive layer. Following this work, Alq_3 became a mainstay of organic electronics.[5,7,20,43] Resistive switching has already been proved in several structures that include Alq_3 as an active element. It was included in an Al/Alq_3/n-type Si structure, where the bistability was attributed to defects at the Al/Alq_3 interface[44] with an ON/OFF ratio of 10^6, or deposited on top of Si nanoparticles,[45] reaching an ON/OFF ratio of 10^5.

After this quick but important foray into resistive memories, we shall go back to the main topic of this section. The coexistence of bistable magnetic and electrical states was demonstrated in 2007 by Hueso et al.[29] They used the LSMO/Alq_3/AlO_X/Co structure, with an Alq_3 thickness ranging from 100 to 400 nm. The authors showed the coexistence of two distinct phenomena. First they demonstrated the spin valve MR (Figure 7.2b) at 200 K. Following this, they demonstrated that the device could also work as resistive memory (a behavior similar to that shown in Figure 7.3). In fact, its resistance could be controlled by the application of a voltage pulse. Crucially, the resistive state could be read nondestructively by the application of a read voltage. Also, the resistive memory was nonvolatile, meaning that the voltage across the electrodes could be put to 0 without altering the state of the device.

The structure of the devices reported on in this section is, starting from the bottom, LSMO/Alq_3/AlO_X/Co. The bottom electrode of the device was a 1×5 mm^2 strip LSMO, 20 nm thick. Its magnetic properties were reported in Dediu et al.[7] This film was grown on a matching $SrTiO_3$ 5×10 mm^2 substrate in a pulsed plasma deposition (PPD) machine in a 4×10^{-2} mbar oxygen atmosphere, with the substrate kept at

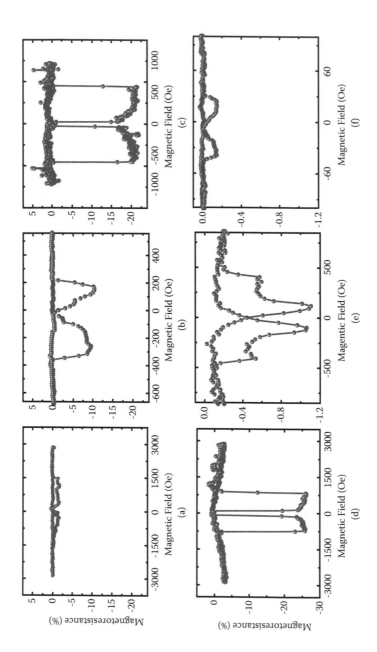

FIGURE 7.2 Examples of MR traces from different devices. The negative voltage bias means electrons are injected by the LSMO electrode. (a) LSMO/Alq$_3$ 70 nm/AlO$_x$/Co 40 nm, voltage bias –200 mV, 100 K, R(0 Oe) = 142 kΩ. (b) LSMO/Alq$_3$ 100 nm/AlO$_x$/Co, voltage bias –100 mV, 20 K, R(0 Oe) = 1.61 kΩ. (c) LSMO/Alq$_3$ 250 nm/AlO$_x$/Co 30 nm, voltage bias –100 mV, 100 K R(0 Oe) = 360 kΩ, sensitivity 4.4%/Oe. (d) Same as previous but at 50 K. (e) LSMO/Alq$_3$ 2 nm/Pentacene 100 nm/Alq$_3$ 2 nm/AlO$_x$ 2 nm/Co 40 nm, voltage bias –10 mV at room temperature, R(0 Oe) = 1.83 kΩ, sensitivity 1.2 × 10^{-2}%/Oe. (f) LSMO/Alq$_3$ 100 nm/AlO$_x$/Co, voltage bias –100 mV at 297 K.

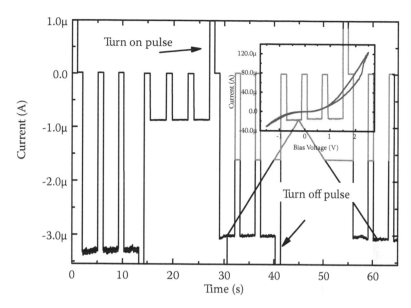

FIGURE 7.3 Programming (write-read-erase-read) cycles, taken at 100 K, of the device in Figure 7.2c. The write (ON pulse) voltage was 1.5 V, while the erase (OFF pulse) voltage was –2 V. The read voltage (–100 mV) was applied three times at each resistance state to show the nonvolatility of the effect. There are two clear resistive states, with an ON/OFF ratio of about 4.

880 ÷ 900°C. The sample was then exposed to air and moved to the OSC deposition chamber (base pressure 2×10^{-8} mbar). Prior to OSC deposition, the LSMO sample was heated to 250°C for 20 min to recover its surface magnetization properties. A 250-nm thick Alq_3 layer was evaporated on top of it at a rate of 0.06 Å/s, with the sample kept at room temperature. Subsequently, the sample was exposed to air and brought back to the PPD machine where a 2-nm thick AlO_x tunnel barrier was deposited at a 2.5×10^{-2} oxygen pressure.[46] Finally, the sample was exposed to air and moved to the metals deposition chamber, where a 40 nm thick Co film was evaporated with an electron gun at a 5×10^{-8} mbar base pressure. Its properties were reported in Bergenti et al.[47]

The measurements were carried out in a four-probe cross-bar configuration (see Figure 7.1a). The resistance of the LSMO electrode was 8 kΩ, and that of the Co one was 1.4 kΩ, both of them at least two orders of magnitude smaller than the resistance of the device, which was 360 kΩ at 100 K. Therefore, we exclude any geometrical effects due the finite size of the device, as well as any contribution of the MR of the electrodes to the device MR. A voltage bias is applied to the LSMO electrode, while the Co is kept at ground. In what follows, if not stated otherwise, the device MR measurements are carried out at a bias voltage of –100 mV (i.e., electrons were injected from the LSMO electrode into the device) at 100 K.

The sample was placed in a gas exchange cryostat, which in turn was mounted between the poles of a magnet that could sweep the field from –2.88 kOe to

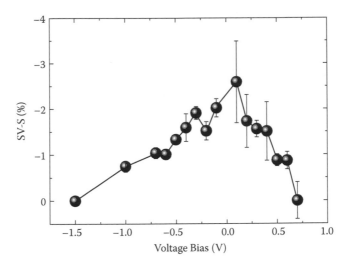

FIGURE 7.4 SV-S voltage bias dependence of an LSMO/Alq$_3$ 200 nm/AlO$_x$/Co device at 100 K.

+2.88 kOe. When measured with a hysteresis loop, the device showed a SV-S of −22% at 100 K (Figure 7.2c) (and up to −25.4% at 50 K, Figure 7.2d) at −100 mV. In general, the SV-S depends on the measuring voltage bias. Figure 7.4 shows this dependence on an LSMO/Alq$_3$ 200 nm/AlO$_x$/Co OSV. An interesting issue that arises from these data is how far we can carry the interplay between the electrical and magnetic bistability.

In order to investigate the resistive memory capabilities write/read/erase/read programming cycles were performed fifteen times, applying voltage pulses for 1 s. The write voltage bias (ON pulse) was 1.5 V, the erase voltage bias (OFF pulse) was −2 V, and the read voltage bias −100 mV. As shown in Figure 7.3, there are two reproducible resistive states with an ON/OFF ratio of about 4. The inset in Figure 7.3 shows the IV curve of the device at 100 K. Its hysteretic behavior is of course directly linked to the memory effect. The programming cycle can be repeated with only marginal degradation of the device.

As we mentioned earlier when dealing with organic resistive memories, the organic layer might well be the site of the resistive bistability, but it might not be the only part of the device responsible for it. Switching has been already observed in LSMO and in ferromagnetic thin films. Electric field control of magnetization is also theoretically possible in ferromagnetic Co, Ni, and Fe thin films near their T_C,[48] and a simple model that would explain the existence of high electric fields near the electrodes is given by Simmons and Verderber.[49]

To conclude this section, we have illustrated the coexistence of the spin valve effect and electrical bistability. The interest to the ICT industry is due to the fact that these OSVs could act as both resistive memories and magnetic ones, leading to multifunctional devices. From a physical point of view, a very interesting prospect arises from the possible interplay of the two bistabilities. This could elucidate further the workings of spin transport in organics, and also pave the way for the development of four-fold-stable resistive memories.

7.3 SENSOR APPLICATIONS

In this section we will analyze the OSV stack from the point of view of magnetic field sensing. There is a huge amount of work on comparable inorganic sensor devices, such as inorganic spin valves (SVs)[50] and magnetic tunnel junctions (MTJs).[51] On the contrary, there is very little published literature in the field of organic electronics for magnetic field sensing. Veeraraghavan et al.[52] demonstrated an 8×8 array device-based organic semiconductor magnetoresistance effect (OMAR), rather than in SV stacks. Yu et al.[53] suggested a magnetic OFET as a very sensitive magnetic sensor. Meruvia et al.[54] fabricated a magnetic multilayer metal-base transistor with an organic emitter whose output characteristics were sensitive to those of an applied magnetic field. With the definition of the magnetocurrent as $M_C = I_C(H = 2 \text{ kOe}) - I_C(H = 0 \text{ Oe})/I_C(H = 0 \text{ Oe})$, where I_C is the collector current, they achieved a room temperature, M_C, of 99% with an emitter current $I_E = 10$ mA.

In general, SV and MTJ spintronic sensors are based on the fact that their electrical resistance depends on the relative orientation of the magnetization of the two magnetic electrodes. The change in resistance, $\Delta R/R$, is proportional to $\cos\theta$, where θ is the angle between the magnetizations of the two electrodes. In the most common sensor geometry, one of the electrodes has its magnetization pinned by an antiferromagnetic layer and is not expected to be affected by the magnetic field in the range of strengths of intended use. The other layer is made from a soft magnetic material and is free to rotate. The highest rate of resistance change (maximum of the cosine derivative) is for $\theta = \frac{\pi}{2}$; therefore, it is advantageous to have the easy magnetization axis for the free layer oriented perpendicular to the magnetization of the pinned electrode.[55]

Before starting to compare the performance of organic and inorganic spintronic sensors, let's define which is the parameter of interest. The sensor readout often consists often in a simple resistance measurement: the more the resistance changes with the magnetic field (i.e., the greater the absolute MR), the more sensitive the sensor is. In general, the magnetoresistance at a given field $MR(H)$ is given by

$$\text{MR}(H) = \frac{R(H) - R(0)}{R(0)}$$

where $R(H)$ is the resistance at field H and $R(0)$ is the resistance with no applied field. In our devices, in general $R(H) < R(0)$ and the maximum possible absolute value is reached when $\text{MR}(H) = -100\%$, which happens when $R(H) \ll R(0)$; these are negative MR devices. In positive MR devices, $R(H) > R(0)$, and when $R(H) \gg R(0)$, $\text{MR}(H) = +\infty$. In other words, the best-performing negative MR spin valve has $\text{MR}(H) = -100\%$, whereas the best-performing positive MR spin valve has $\text{MR}(H) = +\infty$. Quite clearly, the positive and negative MR values cannot be compared straightaway, with -100% MR being as good as $+\infty$ MR. From an application point of view, a better figure can be obtained by remembering that on these MR measurements a voltage bias is applied and a current is read. The maximum and minimum $R(H)$ value in a magnetic field sweep corresponds to a minimum and maximum measured current. Therefore, the difference between two $R(H)$ states in an MR sweep can be better expressed as the ratio between the two different resistances, i.e., as an ON/OFF ratio. Therefore, when comparing negative and positive MR spin valves, we should consider:

$$\frac{R(0)}{R(H)} \text{ for negative MR; } \frac{R(H)}{R(0)} \text{ for positive MR}$$

Both of these quantities can go from 0 to $+\infty$.

Figure 7.5 shows the working principle of an OSV sensor. Figure 7.5a shows two stylized magnetization loops of the Co (red) and the LSMO (blue) electrode. Figure 7.5b shows the corresponding stylized MR curve, where the resistance is higher when the magnetizations of the electrodes are parallel to each other (at the bottom of Figure 7.5a, P marks the parts in which this happens, and AP is for the antiparallel case). The MR is very sensitive to the applied magnetic field on each of the four switching events, where the dependence of the resistance on the magnetic field is steepest. The switching events (which one depends on the details of the device) are therefore the best working points of the device (Figure 7.5c shows a zoom in of one such switching event). To keep the device on its working point, a static magnetic field must be applied. Since the switching events are not necessarily reversible, after each reading the OSV must be refreshed by applying to it a field of opposite sign and greater than the coercive field of both electrodes.

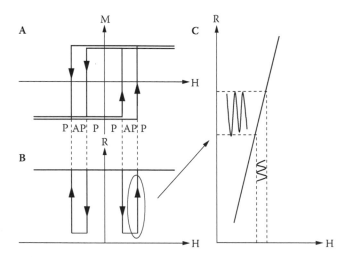

FIGURE 7.5 Schematic diagram of the working principle of an OSV magnetic field sensor. (a) Stylized drawing of the magnetization curves of the two magnetic electrodes. The lower coercive field curve corresponds to the LSMO electrode in our devices (b), resulting in magnetoresistance of the OSV. The sections of interests are the four switching events (marked by the arrows), where the dependence of the resistance on the magnetic field is at its steepest. (c) Zoom in of one of the four possible working points of the device as a magnetic sensor.

Based on these considerations, the sensitivity, S, of the sensor can be defined as the maximum percent change in resistance divided by the change of magnetic field that causes it:

$$S = \max\left\{\left|\frac{R(H) - R(\Delta H + H)}{\Delta H}\right| \times \frac{1}{R\text{min}} \times 100\right\}$$

where

$$R\text{min} = \min\{R(H)\}$$

Its units are %/Oe. The sensitivity is related to the SV-S of a device, but it doesn't coincide with it.

In order to calculate the magnetic field sensitivity, S_H, i.e., the minimum field that gives a nonzero readout, it is necessary to consider noise. Noise can be thought of as a stochastic signal added to the quantity that we are measuring. There are several kinds of noise, such as $1/f$, or flicker noise, Johnson noise, and shot noise. Before describing in greater detail these kinds of noise, we will give a brief explanation on how noise is measured. In order to measure noise, usually the power of the signal in the time domain is Fourier transformed into the power spectral density, $S(f)$.

If the signal is measured in V, $S(f)$ is measured in V^2/Hz and represents the power of the measured signal per unit bandwidth, at a frequency f. The bandwidth, measured in Hz, is the range of frequencies of interest. To obtain the power of the noise in a chosen frequency range, it is sufficient to integrate $S(f)$ over that range. For example, in order to measure signals slower than 1 ms, the noise power will amount to the integral of $S(f)$ between 0 and 1,000 Hz. The square root of this integral is the root mean square voltage noise signal, V_{rms}, in $V/\sqrt{Hz}$. The same arguments also hold if other quantities are measured, for example, the resistance, which is the signal measured in MR measurements, or the magnetic field.

$1/f$ noise originates from fluctuations of energy around equilibrium, and it is the prevalent noise at low frequencies. In the case of spin valves, the main contribution to noise is given by the magnetic noise, which shows as $1/f$ noise. The general formula for its power density spectrum was given by Hooge in 1969[56]:

$$S_V(f) = \frac{\gamma_H V^2}{N_c f}$$

where $S_V(f)$ is the power spectral density of the voltage signal, γ_H is the device-dependent Hooge constant, V is the volume of the magnetic material, N_c is the number of charge carriers, and f is the frequency at which the noise is measured. Due to its detrimental effect on sensitivity, effort has been made to reduce the amount of $1/f$ noise in magnetic sensors.[57,58]

The Hooge constant for metallic systems normally ranges between 10^{-2} and 10^{-3} but can be much greater when there are magnetic fluctuations. Magnetic noise is inversely proportional to the volume of the magnetic component on the device;

therefore, thin films can be very noisy. It is possible to reduce magnetic noise at low frequency in GMR/TMR sensors by a combination of cross anisotropies, window frame shapes, and a suitably designed magnetoresistive stack.[59]

As we said, other types of noise are Johnson noise and shot noise. Johnson noise sets a lower limit to voltage noise. It results from fluctuations due to thermal excitations and its power spectral density is given by[60,61]

$$S_V(f) = 4k_B T R$$

where k_B is the Boltzmann constant, T is the temperature of the device in Kelvin, R is its resistance, and B is the bandwidth. Shot noise occurs instead when a current flows through the device and is caused by the discrete nature of charge. It can be calculated with[61]

$$I_{rms} = \sqrt{2q I_{DC} B}$$

where q is the charge of the carrier, I_{DC} is the dc flowing in the system, and B is the bandwidth.

Once the *rms* amplitudes of the signal and noise (*Si* and *N*, respectively) are known, it is possible to calculate the signal-to-noise ratio (SNR):

$$SNR = 10 \log_{10} \frac{Si^2}{N^2}$$

In our devices, S and N are important to calculate the field sensitivity, S_H:

$$S_H = \left(\frac{Si}{N} \times S \right)^{-1}$$

measured in Oe.

For practical purposes, the bandwidth of the device is limited by a variety of phenomena, for example, the time for injection and transport of the spin-polarized current across the nonmagnetic layer, the switching dynamics of the magnetization,[62] or capacitive effects. These mechanisms combine to give the time-resolved response of the sensor. Currently, most OSV sensors are still being fabricated with sizes of the order of the mm². With these sizes, several magnetization switching mechanisms add up to the effect of slowing the magnetoresistive switching, and the capacitance of the device is relatively big (~0.3 nF).

Transport across the organic layer can introduce a time lag in the response of the sensor. The low mobility of Alq_3 (typically $10^{-4} \div 10^{-6} cm^2 V^{-1} s^{-1}$ [63]), which is the organic material that has given the best results in OSVs, might at first seem to make Alq_3 unfit as the best choice. But the picture in which drift dominates the transport of carriers in the organic layer is at best incomplete. In fact, the spin valve effect can be measured even with an applied voltage of 1 mV at 100 K. With the above mobility,

TABLE 7.1

Relevant Quantities for Sensor Applications for Selected Devices

Device Type	Resistance (kΩ) at 0 Oe	Temperature (K)	SV-S (%)	Sensitivity (%/Oe)
Alq_3 2 nm/Pentacene 100 nm/Alq_3 2 nm	1.83 (Figure 7.2e)	297	−1	1.2×10^{-2}
Alq_3 2 nm/Pentacene 100 nm/Alq_3 2 nm	72.4	100	−7.1	1.8×10^{-1}
T_6 (100 nm)	46.4	40	−2.4	9.9×10^{-2}
Alq_3 (250 nm)	360 (Figure 7.2d)	50	−25.4	3.1×10^{-1}
Alq_3 (100 nm)	56.0	200	−14	8.9×10^{-1}
Alq_3 (250 nm)	294 (Figure 7.2c)	100	−22	4.4

an electron would take a time $t = \frac{d^2}{\mu \times V} = 1$ to 100 ms to cross the organic layer. This is a large time for the current to stay spin polarized, although in literature there are greater relaxation times reported for the same material.[64]

Having given an overview of the basic issues that are relevant to OSVs as magnetic field sensors, we will now turn our attention to their viability for such application. The analysis will be performed on the LSMO/Alq_3/AlO$_X$/Co devices produced recently in our group and schematically shown in Figure 7.1a. The OSC layer was made from various materials and thicknesses, as shown in Table 7.1. The AlO$_X$ barrier was 2 nm thick for all devices but one: the 100-nm thick Alq_3 device reported in Table 7.1 has a 5-nm thick AlO$_X$ barrier. The magnetic field was applied in plane, and for the resistance measurements we applied a −100-mV bias voltage to the LSMO electrode compared to the Co one. This kind of application can be extended to other devices, considering a few main parameters, which will be presented and described.

The greatest SV-S was obtained at 50 K for a 250-nm thick Alq_3 layer (Figure 7.2d). Its value was −25.4%, which is somewhat lower than the highest values reported.[5,8] The most striking characteristic of these spin valves is the great sharpness of the magnetic switching (see Figure 7.1), considerably exceeding that of all the reported OSVs. We will discuss these aspects in the following.

The highest value of SV-S at room temperature was obtained for a multilayered Alq_3 2 nm/pentacene 100 nm/Alq_3 2 nm OSC layer. When measured with a voltage bias of −10 mV, the SV-S was −1% (Figure 7.2e), and it decreased to −0.7% when measured at −100 mV. The 100-nm thick sexithiophene device shows a maximum SV-S of −2.4% at 40 K, whereas the 100-nm Alq_3 one has a maximum SV-S of −14% at 200 K. All OSVs parameters are reported in Table 7.1.

SV-S falls monotonically as the temperature increases, as shown in Figure 7.7b. This is attributed to the loss of surface magnetization on the LSMO electrode, and the spin-lattice relaxation time in the OSC plays no role in determining the temperature dependence of SV-S.[7] Sensitivity has a different behavior as a function of temperature; Figure 7.7a shows that it reaches its maximum at 100 K.

High sensitivities are of course desirable, and some recent work on inorganic MTJs and SVs achieved impressive results. Cardoso et al.[65] developed a MTJ stack and an SV sensor to detect magnetic nanoparticles (MNPs). They biased the sensors

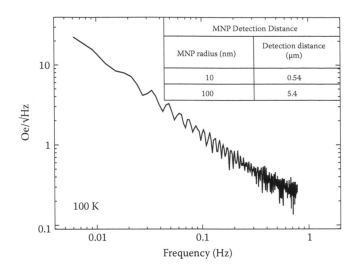

FIGURE 7.6 Field rms noise spectrum of the spin valve in Figure 7.2c. The sensitivity reaches 0.2 Oe/$\sqrt{\text{Hz}}$ at 100 K at 0.8 Hz. As an example of what use a sensor with such sensitivity can have, we show in the inset the distance at which such a sensor can detect a 10-nm diameter and a 100-nm diameter magnetite. These figures are of interest, for example, in the detection of magnetic tags used currently in medicine.

with a 25-Oe constant magnetic field and excited the MNPs with a 15-Oe rms ac field at a frequency of 375 Hz, achieving a sensitivity of 0.23%/Oe for the MTJ and 0.08%/Oe for the SV. Shen et al.[66] also used an MTJ stack as a magnetic field sensor, reaching a sensitivity of 1.0%/Oe over a field range of ±10 Oe. Also in this case, an ac magnetic field was applied (15 Oe rms at 100 Hz) to excite the MNPs. Guerrero et al.[67] achieved a sensitivity of 3%/Oe on similar devices. Regarding OMAR magnetic field sensors, the sensitivity reported by Veeraraghavan et al.[52] was of 0.1%/Oe at room temperature, with a detection threshold of 10 Oe.

The rightmost column of Table 7.1 shows the sensitivity of selected spin valves. The best performance is 4.4 %/Oe, which compares favorably with the data reported in the literature on their more conventional inorganic counterparts. This is very good news, but for most practical uses, a sensor should be able to perform at room or even higher temperatures. Table 7.1 also reports the room temperature sensitivity of an LSMO/Alq$_3$/pentacene/Alq$_3$/AlO$_X$/Co device. Its value, 1.2×10^{-2}%/Oe, is still quite small compared to the figures reported above.

In order to translate the %/Oe sensitivities in magnetic field sensitivities, S_H, it is necessary to include considerations about noise. Figure 7.6 shows the noise power spectrum of the device in Figure 7.2c. The field sensitivity reaches 0.2 Oe/$\sqrt{\text{Hz}}$ at 100 K at 0.8 Hz. This can be compared with that of more conventional tunneling magnetoresistance devices, such as the one presented by Pannetier et al.,[58] which is 4 μOe/$\sqrt{\text{Hz}}$ at room temperature.

When comparing the sensitivities of conventional inorganic magnetic sensors with organic ones, we should not forget that the operating temperature of the latter devices is basically limited by the Curie temperature of the LSMO electrode.[7] The organic

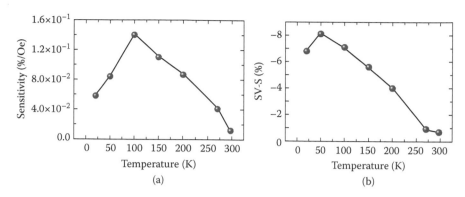

FIGURE 7.7 (a) SV-S as a function of temperature for the device in Figure 7.2e. (b) Sensitivity as a function of temperature of the device in Figure 7.2e.

basically plays no role in the degradation of the SV-S with increasing temperature. Therefore, the still lagging room temperature performance of organic magnetic sensors could be improved by using magnetic electrodes that preserve the surface spin polarization of their carriers at higher temperatures.

As an example of what use a sensor with such sensitivity can have, we show in the inset in Figure 7.6 the distance at which such a sensor can detect a 10-nm diameter and a 100-nm diameter magnetite. These figures are of interest, for example, in the detection of magnetic tags used currently in medicine.

The devices presented in this chapter were not made for the purpose of being used as magnetic sensors, and therefore there is ample room for improvement in the sensitivity. This could be done in several ways. The magnetic electrodes could be engineered in a way similar to that for MTJ sensors. In particular, thin LSMO has a low T_C, and higher T_C materials could be used to improve the room temperature performance of our devices. In addition, Alq_3 could not necessarily be the best organic semiconductor for this purpose. We already demonstrated the superior performance of a combination of pentacene and Alq_3 at room temperature, and we believe that more gains can be made by pursuing this line of investigation.

7.4 CONCLUSIONS

In this chapter we have shown that OVSs are approaching a stage at which they could enter the technological development phase. We have considered what we believe to be the two most promising fields of application: nonvolatile memories and magnetic field sensors. Concerning the former, in addition to the resistive switching shown by many organic materials, OSVs not only introduce magnetic memory effects but also offer a chance to generate multifunctional devices. In view of the current effort by ICT to develop the universal memory, which ideally would bring together the storage capacity of HDDs and the speed of RAM, the coexistence of resistive and magnetic switching looks very appealing.

Concerning sensors, we have shown that at low temperature (100 K) the sensitivity of our OSVs can reach 4.4%/Oe, which is comparable to that of equivalent inorganic devices at room temperature. Although this by no means constitutes a final result, it must be kept in mind that these devices were not fabricated with the purpose of making sensors. In particular, temperature limitations might be overcome by using electrodes with increased Curie temperature.

To conclude, in our opinion organic spintronics is about to extend its influence beyond fundamental research and reach the technological playground.

We acknowledge Carlo Taliani, Ilaria Bergenti, Luis Hueso, Patrizio Graziosi, and David Brunel for valuable discussions. We acknowledge also the financial support from EU-FP6-STRP 033370 OFSPIN and also the European project NMP3-LA-2008-214685-MAGISTER.

REFERENCES

1. V. A. Dediu, L. E. Hueso, I. Bergenti, et al., Spin routes in organic semiconductors, *Nature Materials* 8, 707 (2009).
2. Why going organic is good, *Nature Materials* 8, 691 (2009).
3. V. Dediu, M. Murgia, F. C. Matacotta, et al., Room temperature spin polarized injection in organic semiconductor, *Solid State Communications* 122, 181 (2002).
4. F. Wang and Z. V. Vardeny, Organic spin valves: The first organic spintronics devices, *Journal of Materials Chemistry* 19, 1685 (2009); W. J. M. Naber, S. Faez, and W. G. van der Wiel, Organic spintronics, *Journal of Physics D—Applied Physics* 40, R205 (2007).
5. Z. H. Xiong, D. Wu, Z. V. Vardeny, et al., Giant magnetoresistance in organic spin-valves, *Nature* 427, 821 (2004).
6. F. J. Wang, C. G. Yang, Z. V. Vardeny, et al., Spin response in organic spin valves based on La2/3Sr1/3MnO3 electrodes, *Physical Review B* 75, 245324 (2007); A. Riminucci, I. Bergenti, L. E. Hueso, et al., Negative spin valve effects in manganite/organic based devices, arXiv, 0701603 (2007); S. Majumdar, H. S. Majumdar, R. Laiho, et al., Comparing small molecules and polymer for future organic spin-valves, *Journal of Alloys and Compounds* 423, 169 (2006); W. Xu, G. J. Szulczewski, P. LeClair, et al., Tunneling magnetoresistance observed in La0.67Sr0.33MnO3/organic molecule/Co junctions, *Applied Physics Letters* 90, 072506 (2007).
7. V. Dediu, L. E. Hueso, I. Bergenti, et al., Room-temperature spintronic effects in Alq3-based hybrid devices, *Physical Review B* 78, 115203 (2008).
8. S. Majumdar, R. Laiho, P. Laukkanen, et al., Application of regioregular polythiophene in spintronic devices: Effect of interface, *Applied Physics Letters* 89, 122114 (2006).
9. F. J. Wang, Z. H. Xiong, D. Wu, et al., Organic spintronics: The case of Fe/Alq(3)/Co spin-valve devices, *Synthetic Metals* 155, 172, (2005).
10. T. S. Santos, J. S. Lee, P. Migdal, et al., Room-temperature tunnel magnetoresistance and spin-polarized tunneling through an organic semiconductor barrier, *Physical Review Letters* 98, 016601, (2007).
11. D. Dhandapani, N. A. Morley, A. Rao, et al., Comparison of room temperature polymeric spin-valves with different organic components, *IEEE Transactions on Magnetics* 44, 2670 (2008).
12. N. A. Morley, A. Rao, D. Dhandapani, et al., Room temperature organic spintronics, *Journal of Applied Physics* 103, 07F306 (2008).
13. Y. Liu, S. M. Watson, T. Lee, et al., Correlation between microstructure and magnetotransport in organic semiconductor spin-valve structures, *Physical Review B* 79, 075312 (2009); Y. H. Liu, T. Lee, H. E. Katz, et al., Effects of carrier mobility and

morphology in organic semiconductor spin valves, *Journal of Applied Physics* 105 07C708 (2009).

14. I. Bergenti, V. Dediu, E. Arisi, et al., Spin polarized La0.7Sr0.3MnO3 thin films on silicon, *Journal of Magnetism and Magnetic Materials*. 312, 453 (2007); M. P. de Jong, V. A. Dediu, C. Taliani, et al., Electronic structure of La0.7Sr0.3MnO3 thin films for hybrid organic/inorganic spintronics applications, *Journal of Applied Physics* 94, 7292 (2003).
15. Y. Q. Zhan, I. Bergenti, L. E. Hueso, et al., Alignment of energy levels at the Alq(3)/ La0.7Sr0.3MnO3 interface for organic spintronic devices, *Physical Review B* 76, 045406 (2007).
16. H. Vinzelberg, J. Schumann, D. Elefant, et al., Low temperature tunneling magnetoresistance on (La,Sr)MnO3/Co junctions with organic spacer layers, *Journal of Applied Physics* 103, 093720 (2008).
17. Y. Q. Zhan, X. J. Liu, E. Carlegrim, et al., The role of aluminum oxide buffer layer in organic spin-valves performance, *Applied Physics Letters* 94, 053301 (2009).
18. M. Cinchetti, K. Heimer, J. P. Wustenberg, et al., Determination of spin injection and transport in a ferromagnet/organic semiconductor heterojunction by two-photon photoemission, *Nature Materials* 8, 115 (2009).
19. M. Popinciuc, H. T. Jonkman, and B. J. van Wees, Energy level alignment symmetry at Co/pentacene/Co interfaces, *Journal of Applied Physics* 100, 093714 (2006).
20. A. J. Drew, J. Hoppler, L. Schulz, et al., Direct measurement of the electronic spin diffusion length in a fully functional organic spin valve by low-energy muon spin rotation, *Nature Materials* 8, 109 (2009).
21. F. Luo, W. Song, Z. M. Wang, et al., Tuning negative and positive magnetoresistances by variation of spin-polarized electron transfer into pi-conjugated polymers, *Applied Physics Letters* 84, 1719 (2004).
22. H. Kusai, S. Miwa, M. Mizuguchi, et al., Large magnetoresistance in rubrene-Co nanocomposites, *Chemical Physics Letters* 448, 106 (2007).
23. S. Miwa, M. Shiraishi, M. Mizuguchi, et al., Spin-dependent transport in C-60-Co nanocomposites, *Japanese Journal of Applied Physics Part 2* 45, L717 (2006).
24. Z. G. Yu, M. A. Berding, and S. Krishnamurthy, Spin drift, spin precession, and magnetoresistance of noncollinear magnet-polymer-magnet structures, *Physical Review B* 71, 060408(R) (2005).
25. A. H. Davis and K. Bussmann, Organic luminescent devices and magnetoelectronics, *Journal of Applied Physics* 93, 7358 (2003).
26. G. Salis, S. F. Alvarado, M. Tschudy, et al., Hysteretic electroluminescence in organic light-emitting diodes for spin injection, *Physical Review B* 70, 085203 (2004).
27. M. Yunus, P. P. Ruden, and D. L. Smith, Spin injection effects on exciton formation in organic semiconductors, *Applied Physics Letters* 93, 123312 (2008).
28. T. L. Francis, O. Mermer, G. Veeraraghavan, et al., Large magnetoresistance at room temperature in semiconducting polymer sandwich devices, *New Journal of Physics* 6, 185 (2004).
29. L. E. Hueso, I. Bergenti, A. Riminucci, et al., Multipurpose magnetic organic hybrid devices, *Advanced Materials* 19, 2639 (2007).
30. M. Gajek, M. Bibes, S. Fusil, et al., Tunnel junctions with multiferroic barriers, *Nature Materials* 6, 296 (2007); W. Eerenstein, N. D. Mathur, and J. F. Scott, Multiferroic and magnetoelectric materials, *Nature* 442, 759 (2006); R. Ramesh and N. A. Spaldin, Multiferroics: Progress and prospects in thin films, *Nature Materials* 6, 21 (2007).
31. P. A. Bobbert, W. Wagemans, F. W. A. van Oost, et al., Theory for spin diffusion in disordered organic semiconductors, *Physical Review Letters* 102, 156604 (2009).
32. J. C. Scott and L. D. Bozano, Nonvolatile memory elements based on organic materials, *Advanced Materials* 19, 1452 (2007).

33. Q. D. Ling, D. J. Liaw, C. X. Zhu, et al.,Polymer electronic memories: Materials, devices and mechanisms, *Progress in Polymer Science* 33, 917 (2008).
34. H. E. Katz, X. M. Hong, A. Dodabalapur, et al., Organic field-effect transistors with polarizable gate insulators, *Journal of Applied Physics* 91, 1572 (2002).
35. N. Setter, D. Damjanovic, L. Eng, et al., Ferroelectric thin films: Review of materials, properties, and applications, *Journal of Applied Physics* 100, 051606 (2006).
36. Y. Arimoto and H. Ishiwara, Current status of ferroelectric random-access memory, *MRS Bulletin* 29, 823 (2004).
37. W. Tang, H. Z. Shi, G. Xu, et al., Memory effect and negative differential resistance by electrode-induced two-dimensional single-electron tunneling in molecular and organic electronic devices, *Advanced Materials* 17, 2307 (2005).
38. M. Lauters, B. McCarthy, D. Sarid, et al., Multilevel conductance switching in polymer films, *Applied Physics Letters* 89, 013507 (2006).
39. T. Kever, U. Bottger, C. Schindler, et al., On the origin of bistable resistive switching in metal organic charge transfer complex memory cells, *Applied Physics Letters* 91, 083506 (2007).
40. Q. X. Lai, Z. H. Zhu, Y. Chen, et al., Organic nonvolatile memory by dopant-configurable polymer, *Applied Physics Letters* 88, 133515 (2006).
41. L. D. Bozano, B. W. Kean, M. Beinhoff, et al., Organic materials and thin-film structures for cross-point memory cells based on trapping in metallic nanoparticles, *Advanced Functional Materials* 15, 1933 (2005).
42. C. W. Tang and S. A. Van Slyke, Organic electroluminescent diodes, *Applied Physics Letters* 51, 913 (1987).
43. S. F. Alvarado, L. Rossi, P. Muller, et al., STM-excited electroluminescence and spectroscopy on organic materials for display applications, *IBM Journal of Research and Development* 45, 89 (2001); A. Kahn, N. Koch, and W. Y. Gao, Electronic structure and electrical properties of interfaces between metals and pi-conjugated molecular films, *Journal of Polymer Science Part B* 41, 2529 (2003).
44. P. T. Lee, T. Y. Chang, and S. Y. Chen, Tuning of the electrical characteristics of organic bistable devices by varying the deposition rate of Alq(3) thin film, *Organic Electronics* 9, 916 (2008).
45. A. Dima, F. G. Della Corte, C. J. Williams, et al., Silicon nano-particles in SiO2 sol-gel film for nano-crystal memory device applications, *Microelectronics Journal* 39, 768 (2008).
46. Y. Q. Zhan, X. J. Liu, E. Carlegrim, et al., The role of aluminum oxide buffer layer in organic spin-valves performance, *Applied Physics Letters* 94, 053301 (2009).
47. I. Bergenti, A. Riminucci, E. Arisi, et al., Magnetic properties of cobalt thin films deposited on soft organic layers, *Journal of Magnetism and Magnetic Materials* 316, E987 (2007).
48. I. V. Ovchinnikov and K. L. Wang, Voltage sensitivity of Curie temperature in ultrathin metallic films, *Physical Review B* 80, 012405 (2009).
49. J. G. Simmons and R. R. Verderber, New conduction and reversible memory phenomena in thin insulating films, *Proceedings of the Royal Society of London, Series A* 301, 77 (1967).
50. S. X. Wang and G. Li, Advances in giant magnetoresistance biosensors with magnetic nanoparticle tags: Review and outlook, *IEEE Transactions on Magnetics* 44, 1687 (2008).
51. J. Z. Sun and D. C. Ralph, Magnetoresistance and spin-transfer torque in magnetic tunnel junctions, *Journal of Magnetism and Magnetic Materials* 320, 1227 (2008).
52. G. Veeraraghavan, T. D. Nguyen, Y. G. Sheng, et al., An 8 × 8 pixel array pen-input OLED screen based on organic magnetoresistance, *IEEE Transactions Electron Devices* 54, 1571 (2007).

53. Z. G. Yu, M. A. Berding, and S. Krishnamurthy, Organic magnetic-field-effect transistors and ultrasensitive magnetometers, *Journal of Applied Physics* 97, 024510 (2005).
54. M. S. Meruvia, M. L. Munford, I. A. Hummelgen, et al., Magnetic metal-base transistor with organic emitter, *Journal of Applied Physics* 97, 026102 (2005).
55. S. Parkin, X. Jiang, C. Kaiser, et al., Magnetically engineered spintronic sensors and memory, *Proceedings of the IEEE* 91, 661 (2003).
56. F. N. Hooge and A. M. H. Hoppenbrouwers, Amplitude distribution of 1/f noise, *Physica* 42, 9 (1969); F. N. Hooge and A. M. H. Hoppenbrouwers, 1/f noise in continuous thin gold films, *Physica* 45, 425, 331 (1969); F. N. Hooge, 1-F NOISE, *Physica B & C* 83, 14 (1976).
57. J. M. Almeida, P. Wisniowski, and R. P. Freitas, Low-frequency noise in MgO magnetic tunnel junctions hooge's parameter dependence on bias voltage, *IEEE Transactions on Magnetics* 44, 2569 (2008); A. Gokce, E. R. Nowak, S. H. Yang, et al., 1/f noise in magnetic tunnel junctions with MgO tunnel barriers, *Journal of Applied Physics* 99, 08A906 (2006).
58. M. Pannetier-Lecoeur, C. Fermon, A. de Vismes, et al., Low noise magnetoresistive sensors for current measurement and compasses, *Journal of Magnetism and Magnetic Materials* 316, E246 (2007).
59. M. Pannetier, C. Fermon, G. Le Goff, et al., Noise in small magnetic systems—Applications to very sensitive magnetoresistive sensors, *Journal of Magnetism and Magnetic Materials.* 290, 1158–60 (2005).
60. H. Nyquist, Thermal agitation of electric charge in conductors, *Physical Review* 32, 110 (1928).
61. P. Horowitz and W. Hill, *The Art of Electronics*, 2nd ed. (Cambridge University Press, Cambridge, 1989), p. 1041.
62. T. A. Moore and J. A. C. Bland, Mesofrequency dynamic hysteresis in thin ferromagnetic films, *Journal of Physics—Condensed Matter* 16, R1369 (2004).
63. G. G. Malliaras, Y. L. Shen, D. H. Dunlap, et al., Nondispersive electron transport in Alq(3), *Applied Physics Letters* 79, 2582 (2001); B. J. Chen, W. Y. Lai, Z. Q. Gao, et al., Electron drift mobility and electroluminescent efficiency of tris(8-hydroxyquinolinolato) aluminum, *Applied Physics Letters* 75, 4010 (1999).
64. S. Pramanik, C. G. Stefanita, S. Patibandla, et al., Observation of extremely long spin relaxation times in an organic nanowire spin valve, *Nature Nanotechnology* 2, 216 (2007).
65. F. A. Cardoso, J. Germano, R. Ferreira, et al., Detection of 130 nm magnetic particles by a portable electronic platform using spin valve and magnetic tunnel junction sensors, *Journal of Applied Physics* 103, 07A310 (2008).
66. W. Shen, B. D. Schrag, M. J. Carter, et al., Quantitative detection of DNA labeled with magnetic nanoparticles using arrays of MgO-based magnetic tunnel junction sensors, *Applied Physics Letters* 93, 033903 (2008).
67. R. Guerrero, M. Pannetier-Lecoeur, C. Fermon, et al., Low frequency noise in arrays of magnetic tunnel junctions connected in series and parallel, *Journal of Applied Physics* 105, 113922 (2009).

Index

A

Acoustical phonons, 83
Ahrrenius law, 171, 172
Almeida-Thouless (AT) instability line, 183
AMR, *see* Anisotropic magnetoresistance
Anisotropic magnetoresistance (AMR), 143
Applications, *see* Organic spintronics, toward
    sensor and memory applications;
    Spintronic applications of organic
    materials
AT instability line, *see* Almeida-Thouless
    instability line

B

Band transport, 64
BAP spin relaxation mechanism, *see* Bir-Aronov-
    Pikus spin relaxation mechanism
BDR tunneling model, *see* Brinkman, Dynes,
    and Rowell tunneling model
Bias voltage, 58, 223
Bipolaron
    branching ratio, 82, 86
    formation, 84, 128, 240
    model, 81, 82, 93, 235, 240
        branching ratio, 82
        charge carrier mobility, 103
        density of states, 82
        extended Monte Carlo implementation, 88
        hyperfine fields, 82
        line width of OMAR, 110
        magnetic field, 82, 92
        Monte Carlo implementation of, 88
        organic magnetoresistance, 92
Bir-Aronov-Pikus (BAP) spin relaxation
    mechanism, 10
Bloch law, 181
Bohr magneton, 139, 267
Boltzmann transport equation, 34
Brillouin zone center, 33, 35
Brinkman, Dynes, and Rowell (BDR) tunneling
    model, 22
Bubble nucleation model, 186

C

Carrier transit time, 56
CGS phase, *see* Correlated spin glass phase
Chain model, 118, 119

Charge transfer (CT), 237
Charge transport, 16
    band-like transport, 24
    bipolarons, 204
    current continuity equation, 40
    disordered organic materials, 82, 115
    hopping among [TCNE]s, 170
    measurement, 3–4
    models, 10
    organic semiconductors, 190, 191
    pentacene, 14
    recombination limited regime, 202
    sign change, 108
    spin-polarized, 167
    thermal evaporation and, 23
    triplet exciton-charge reaction, 104
    unipolar, 192
Chemically induced dynamic electron
    polarization (CIDEP), 88
Chemical vapor deposition (CVD), 160, 164, 180
CIDEP, *see* Chemically induced dynamic
    electron polarization
Conductivity mismatch problem, 114, 142
Correlated spin glass (CGS) phase, 167
Coulomb's law, 83
CP, *see* Current polarization
CT, *see* Charge transfer
Current polarization (CP), 60, 62
CVD, *see* Chemical vapor deposition

D

Density operator, 270
Density of states (DOS), 82, 88, 115, 120
Diffusive transport process, 54
Disordered organic materials, 82, 115
DOS, *see* Density of states
DP mechanism, *see* D'yakonov-Perel'
    mechanism
Drift diffusion equations, 32
Drude-like transport, 81
D'yakonov-Perel' (DP) mechanism, 10, 115, 116

E

EL, *see* Electroluminescence
Electroluminescence (EL), 4
    current efficiency, 107
    efficiency, 4

A new face for organics

In 2004, after two decades' worth of experience investigating the photophysical properties of conducting polymers, Z. Valy Vardeny demonstrated a spin valve with an organic active layer. *Nature Materials* asked him about his views on the achievements in organic spintronics and the future of the field.

■ What inspired you to study physics?
At the time I had to go to university, the Technion in Haifa was possibly the most famous engineering school in Israel, and nuclear physics was very fashionable. So initially that's what I wanted to do. However, once I began to study, I realized that condensed matter was much more complex and intriguing. I had to join the military service temporarily, and when I came back to the Technion I had made up my mind that I would become a condensed-matter physicist, rather than a nuclear one.

■ Were you already interested in organic compounds?
Not really. My general training was in photophysical phenomena, but at the beginning of my career my research was focused on amorphous silicon. In particular, after my PhD on Raman scattering spectroscopy of copper halide crystals, I took a postdoctoral position at Brown University, in Providence, Rhode Island, in the group of Jan Tauc, where I developed a set-up for ultrafast and continuous-wave spectroscopy on hydrogenated amorphous silicon. Incidentally, there are plenty of similarities between this material and organic semiconductors. But my interest in the organics arose later, after attending a seminar by Alan Heeger in 1981. He was talking about transport in conducting polymers, namely *trans*-polyacetylene (on which he was awarded the Nobel Prize in Chemistry in 2000, together with Alan McDiarmid and Hideki Shirakawa), and I suddenly realized that there was an opportunity for me to study photophysics with these materials. I proposed that to Heeger who then gave me some samples, and we started a fruitful collaboration that lasted for several years afterwards.

■ So that became your main interest?
Indeed, especially because my supervisor went on sabbatical in 1982, and not only did he ask me to take care of his research group at Brown University, but also gave me the freedom to choose my own future research direction. Later, when I took my position at the University of Utah, there was no research programme on organic semiconductors, so I

© HEIDI FRANK, PHYSICS DEPARTMENT, UNIV. OF UTAH

started one. And I have kept on working on these materials for so long because they have not allowed me to become bored. Through the years, the organic semiconductors have changed face several times, and I had to learn new aspects of their properties. Initially they were studied mainly for batteries and electromagnetic shielding on doping. In the 1980s the focus moved to nonlinear optics. Then in the early 1990s, when Richard Friend demonstrated organic light-emitting diodes (OLEDs) with a π-conjugated polymer as the active material, I could see that this was my natural field. After all, once you inject an electron and a hole in the polymer, they form a polaron pair and subsequently an exciton, and the rest is photophysics.

■ And now even spintronics. How did the idea of an organic spin valve come about?
Well, in 2002 I saw the paper from the group in Bologna claiming spin injection from a ferromagnetic (FM) electrode into an organic semiconductor[1]. At that time, Jing Shi had just joined our department at the University of Utah. Coming from a company such as Motorola meant he had vast experience in inorganic spintronics. So we combined our different expertises, and we realized a vertical spin valve with an organic active layer[2].

■ Do you think our understanding of spin-related phenomena in organic semiconductors has increased considerably since these first results?
Definitely. For example, when magneto-resistance was observed in organic spin-valve

devices a few years ago, no one could be sure whether this was due to the fact that spin-polarized carriers were injected in the organic semiconductor or that they were tunnelling through it. The problem is that there is no easy way to check. Let's compare organic semiconductors with inorganic III–V semiconductors, such as GaAs, which is a classical direct-gap semiconductor. By looking at the circular polarization of the emitted light in a two-terminal device, it is possible to infer the spin polarization of carriers injected in the material from the FM electrodes. By contrast, in organic semiconductors the spin–orbit interaction is so weak that the polarization of the emitted light is lost when excitons recombine. So we need other, more complex techniques. And recently, low-energy muon spin rotation and two-photon photoemission have confirmed that spin-polarized carriers are actually injected in the organic semiconductors from FM electrodes[3,4].

■ What about fundamental issues that need solving and are currently under intense investigation?
One important question is the role of the hyperfine interaction (HFI) in the spin-polarized carrier transport in organic spin devices. We know that the spin–orbit interaction is weak in organic semiconductors, leading to a spin-lattice relaxation rate in these materials that is much smaller than in inorganic semiconductors. However, the role of the HFI in determining the spin-lattice relaxation rate at the magnetic field strength used in spintronics devices is not clear. This is an unanswered question that is the focus of intense theoretical[5] and experimental endeavours (T. D. Tho *et al.* manuscript in preparation) at the present time. Another unanswered question is the understanding of the relatively large magnetic field effect (MFE) in electroluminescence and conductivity that has been obtained in regular OLEDs with no FM electrodes[6]. The physics of the MFE seems to be coupled with the interaction of same-charge and opposite-charge polaron pairs; and this is an opportunity to thoroughly investigate these species in the organic semiconductors, which was not available to us before. And we should

keep in mind that HFI could also affect the organic spin-valve response[5], which would lead to a fascinating connection between these seemingly unrelated effects, that is, magnetotransport and MFE.

But what is the advantage of working with organic spintronics when one could study more stable and reliable inorganic materials?

This is an old question, which precedes organic spintronics and relates to organic electronics in general. First of all, organic structures are much cheaper to fabricate, which has obvious advantages for both research and applications. Scientists can enter the field even without huge amounts of funding, and we can foresee devices that can be replaced after a short period of time as the cost is low. In addition, the active material in spintronics devices is very versatile. Chemists can really adapt a compound quite easily to the functionality that they have in mind.

Apart from spin valves, are there other types of devices for which organic compounds are suitable?

Surely. We need to remember, first of all, that when we talk about organic spintronics we do not necessarily refer to spin-transport phenomena, but more generally to magnetic field effects. We can imagine devices of different types, with two FM electrodes — such as spin valves or spin transistors — but also with one FM electrode. For example, an organic diode with one ferromagnet can actually act as a detector of the polarization of the ferromagnet electrode itself[7]. And finally, OLEDs without any FM electrodes can take advantage of the huge electroluminescence and conductance response to a magnetic field. For example, a magnetic field of only about 100 gauss can change the electroluminescence intensity by 20 or 30% (ref. 8). All these aspects were well-represented in the latest SPINOS conference that took place in Salt Lake City, in February 2009.

Are there spintronics applications that are unique to organic semiconductors?

Yes, and they stem from the fact that organic semiconductors can emit both fluorescence (from the recombination of singlet excitons) and phosphorescence (from triplet excitons), which have different colours. So you can imagine having an OLED with two FM electrodes, and by applying even a small magnetic field you could change both the intensity and colour of the emitted light. That would be amazing, and unique for organics.

What about using organic magnets, similar to the case of dilute magnetic semiconductors in inorganic spintronics?

The problem, as far as I can tell, is that the spin polarization in organic magnets has not been measured as yet; thus it may not be really high enough to inject substantial spin-polarized carriers into an organic interlayer. The other issue is that in inorganic spintronics there is a well-known hurdle concerning the conductivity mismatch between metallic FM electrodes and the active semiconductor, which is a problem for spin injection[9–11]. In organics this is not really an important problem because carrier injection is mainly due to tunnelling; and the tunnel barrier may be strongly magnetic-field dependent for an injecting FM electrode. Consequently, it may act as a spin-dependent filter for carrier injection[12].

Do you think the interest in organic spintronics from both academia and industry has risen since the Bologna paper and yours?

I would say it has. Of course I — like anyone else in the field — might be biased, but just take a look at the numbers. In 2007, Carlo Taliani organized the first SPINOS 2007 conference in Bologna, and there were about 50 participants. At the SPINOS 2009, there were already about 100 delegates. In addition, I know that there are international collaborations sponsored by the European Union on the topic. So at least from an academic perspective, the interest has certainly risen. For industry the situation is different. The field is still in its infancy, and particularly in the current economic situation it is understandable that there would not be investments in such a young field. We need to wait until considerable magnetoresistance is observed at room temperature, and then I am sure things will change.

Do you think that the different languages of scientists with different backgrounds is a serious obstacle to the development of the field?

Not more than in other fields. Taking my experience as an example; when I started working on conducting polymers I came from a background of physics of amorphous materials. Others, like Heeger and McDiarmid, were chemists. It took us some time to simply understand each other's language, namely definitions and concepts. The chemists used radicals, the physicists used spin-½ excitations. We physicists talked about valence and conduction bands, the chemists said highest occupied molecular orbital (HOMO) and lowest unoccupied molecular orbital (LUMO). The same happened with OLEDs. Richard Friend came from a semiconductor-physics background, but it took some time to understand what

electron injection into organics meant. There are always language barriers initially, but they shrink with time and eventually we contribute together to the development of a field using different points of view. I am sure that scientists with backgrounds in spintronics think organic-semiconductor researchers often do not know what they are talking about, and vice versa. But eventually we will understand each other. Collaboration is probably the best solution. For example, I believe that the research group at THALES has started collaborating with that in Bologna; and I think this can produce important results.

What are the advances in the field that you foresee in the near future?

I think they will be mainly in two areas. First, on the spin-injector front. At present, $La_{0.7}Sr_{0.3}MnO_3$ (LSMO) is the most widely used material because it is a half-metal (namely, at the Fermi level the carriers are almost fully spin polarized), but I do not think this is necessarily the best possibility, because its spin-aligned carrier-injection capability drastically decreases at room temperature. The magnetic properties of the interface between the FM electrode and the organic semiconductor need better understanding. Also, chemists will be able to synthesize compounds with much higher fluorescence and phosphorescence emission efficiencies, which will be fundamental in realizing the devices I mentioned before. What I think could revolutionize the field is if theorists achieve better understanding of spin injection into organics, in order to actually predict the spin-injection capability of given electrode–semiconductor pairs, which experimentalists could then verify, for example, by measuring spin-dependent photoemission[4].

References

1. Dediu, V., Murgia, M., Matacotta, F. C., Taliani, C. & Barbanera, S. Solid State Commun. 122, 181–184 (2002).
2. Xiong, Z. H., Wu, D., Vardeny, Z. V. & Shi, J. Nature 427, 821–824 (2004).
3. Drew, A. J. et al. Nature Mater. 8, 109–114 (2009).
4. Cinchetti, M. et al. Nature Mater. 8, 115–119 (2009).
5. Bobbert, P. A., Wagemans, W., van Oost, F. W. A., Koopmans, B. & Wohlgenannt, M. Phys. Rev. Lett. 102, 156604 (2009).
6. Francis, T. L., Mermer, O., Veeraraghavan, G. & Wohlgenannt, M. New J. Phys. 6, 185 (2004).
7. Wu, D., Xiong, Z. H., Li, X. G., Vardeny, Z. V. & Shi, J. Phys. Rev. Lett. 95, 016802 (2005).
8. Nguyen, T. D., Sheng, Y., Rybicki, J. & Wohlgenannt, M. Phys. Rev. B 77, 235209 (2008).
9. Schmidt, G., Ferrand, D., Molenkamp, L. W., Filip, A. T. & van Wees, B. J. Phys. Rev. B 62, R4790 (2000).
10. Rashba, E. I. Phys. Rev. B 62, R16267 (2000).
11. Smith, D. L. & Silver, R. N. Phys. Rev. B 64, 045323 (2001).
12. Yunus, M., Ruden, P. P. & Smith, D. L. J. Appl. Phys. 103, 103714 (2008).

INTERVIEW BY FABIO PULIZZI